Network Security

A Decision and Game-Theoretic Approach

Covering attack detection, malware response, algorithm and mechanism design, privacy, and risk-management, this comprehensive work utilizes unique quantitative models derived from decision, control, and game theories to address diverse network security problems. It provides the reader with a system-level theoretical understanding of network security, and is essential reading for researchers interested in a quantitative approach to key incentive and resource allocation issues in the field. It also provides practitioners with an analytical foundation that is useful for formalizing decision-making processes in network security.

Tansu Alpcan is an Assistant Professor at the Technical University of Berlin, and is concurrently affiliated with Deutsche Telekom Laboratories. His research involves applications of distributed decision-making, game theory, and control to various security and resource allocation problems in complex and networked systems. Dr. Alpcan, who has numerous publications in security, networking, control, and game theory, is the recipient of multiple best paper awards from IEEE and research achievement awards from the University of Illinois. He has chaired and played an active role in the organization of various conferences, including GameSec, GameComm, and GameNets, and is currently chairing the Interest Group on Security in Media Processing and Communications within the IEEE Technical Committee on Multimedia Communications.

Tamer Başar holds several academic positions at the University of Illinois at Urbana-Champaign, including the titles of Swanlund Endowed Chair and Center for Advanced Study Professor of Electrical and Computer Engineeering. He is currently the Editor-in-Chief of Automatica, Series Editor for Systems and Control: Foundations and Applications, and Managing Editor of the Annals of the International Society of Dynamic Games (ISDG). He is a member of the US National Academy of Engineering, Fellow of the IEEE and IFAC, Founding President of the ISDG and Current President of the AACC. Dr. Başar has won a number of awards, including the Isaacs Award of ISDG, Bellman Control Heritage Award of the AACC, the Bode Lecture Prize of the IEEE CSS and the Quazza Medal and Outstanding Service Award of IFAC.

Network Security
A Decision and Game-Theoretic Approach

Tansu Alpcan
Deutsche Telekom Laboratories,
Technical University of Berlin, Germany

and

Tamer Başar
University of Illinois at Urbana-Champaign, USA

CAMBRIDGE
UNIVERSITY PRESS

CAMBRIDGE UNIVERSITY PRESS
Cambridge, New York, Melbourne, Madrid, Cape Town, Singapore,
São Paulo, Delhi, Dubai, Tokyo, Mexico City

Cambridge University Press
The Edinburgh Building, Cambridge CB2 8RU, UK

Published in the United States of America by Cambridge University Press, New York

www.cambridge.org
Information on this title: www.cambridge.org/9780521119320

First published 2011

Printed in the United Kingdom at the University Press, Cambridge

A catalog record for this publication is available from the British Library

Library of Congress Cataloging in Publication data
Alpcan, Tansu, 1975–
Network security : a decision and game-theoretic approach / Tansu Alpcan, Tamer Basar.
    p.   cm.
Includes bibliographical references.
ISBN 978-0-521-11932-0 (hardback)
1. Computer networks–Security measures.   2. Game theory.
I. Basar, Tamer.   II. Title.
TK5105.59.A45   2010
005.8–dc22

                                                    2010027364

ISBN 978-0-521-11932-0 Hardback

To

Alper, Altay, and Özlem (T.A.)

and

Tangül, Gözen, Elif, and Altan (T.B.)

Contents

Preface

We are a lucky generation for witnessing the microprocessor and Internet revolutions, the type of technological marvels that mark the start of a new era: *the information age*. Just like electricity, railroads, and automobiles, the information technologies have a profound effect on our way of life and will stay with us for decades and centuries to come. Thanks to these advances, we have been building complex communication and computing networks on a global scale. However, it is still difficult today to predict how this information age will progress in the future or to fully grasp its consequences. We can hope for a complete understanding perhaps in decades to come, as past history tells us.

Although we have engineered and built the *Internet*, the prime example of the information revolution, our (mathematical) understanding of its underlying systems is cursory at best, since their complexity is orders of magnitude greater than that of their predecessors, e.g. the plain telephone network. Each disruptive technology brings its own set of problems along with enormous opportunities. Just as we are still trying to solve various issues associated with automobiles, the challenges put forward by the information and communication networks will be there not only for us but also for the next generations to address.

An important challenge today is *security* of complex computing and communication networks. Our limited understanding of these systems has a very unexpected side-effect: partial loss of "observability" and "control" of the very systems we build. Who can claim today full knowledge and control of all the running computing and communication processes on their laptop, corporate network, or country at all times? The science-fiction literature has always focused on fears of losing control of "intelligent machines." It is ironic that very few people imagined losing control of our dumb but complex and valuable systems to our malicious yet very own fellow human beings.

Security is a challenge stemming not only from the complexity of the systems surrounding us but also from the users' relative lack of experience with them. Unlike other complex systems, such as vehicle traffic, ordinary users receive very little training before obtaining access to extremely powerful technologies. Despite this (or maybe because of it), users' expectations of their capabilities are very high. They often expect everything to function as simply and reliably as, for example, the old telephone network. However, even for advanced users, bringing all aspects of a connected computer under control is a very time-consuming and costly process. The most diligent efforts can unfortunately be insufficient when faced with a determined and intelligent attacker.

These facts when combined with unrealistic expectations often result in significant disappointment in the general public regarding security.

In spite of its difficulty, securing networked systems is indisputably important as we enter the information age. The positive productivity benefits of networks clearly overcome the costs of any potential security problems. Therefore, there is simply no turning back to old ways. We have to live with and manage the security risks associated with the new virtual worlds which reflect our old selves in novel ways.

The emerging security challenges are multifaceted ranging from complexity of underlying hardware, software, and network interdependencies to human and social factors. While individuals and organizations are often very good at assessing security risks in real life, they are quite inexperienced with the ones they encounter on networked systems, which are very different in complexity and timescale. Although many lessons from real-world security can be transferred to the network security domain, there is a clear need for novel and systematic approaches to address the unique issues the latter brings about. It is widely agreed by now that security of networked systems is not a pure engineering problem that can be solved by designing better protocols, languages, or algorithms. It will require educating users and organizations, changing their perspectives, and equipping them with better tools for assessing and addressing network security problems.

Although many aspects of the network security problem are new, it also exhibits constraints familiar to us, which we often encounter in real life. Many resources available to malicious attackers and defending administrators of networks are limited. They vary from classical resources, such as bandwidth, computing speed and capability, energy, and manpower, to novel ones such as time, attention span, and mental load. Network security involves decision making by both attackers and defenders in multiple levels and timescales using the limited resources available to them. Currently, most of these decisions are made intuitively and in an ad-hoc manner.

This book, which is the first of its kind, aims to present *a theoretical foundation for making resource allocation decisions that balance available capabilities and perceived security risks in a principled manner*. We focus on analytical models based on game, information, communication, optimization, decision, and control theories that are applied to diverse security topics. At the same time, connections between theoretical models and real-world security problems are highlighted so as to establish the important feedback loop between theory and practice. Hence, this book should not be viewed as an authoritative last word on a well-established field but rather as an attempt to open novel and interesting research directions, hopefully to be adopted and pursued by a broader community.

Scope and usage

This book is aimed mainly at researchers and graduate students in the field of network security. While the emphasis is on theoretical approaches and research for decision-making in security, we believe that it would also be beneficial to practitioners, such as system administrators or security officers in industry, who are interested in the latest theoretical research results and quantitative network security models that build on control, optimization, decision, and game-theoretic foundations. An additional objective is the introduction of the network security paradigm as an application area to researchers well versed in control and game theory.

The book can be adopted as a reference for graduate-level network security courses that focus on network security in diverse fields such as electrical engineering, computer science and engineering, and management science. A basic overview of the mathematical background needed to follow the underlying concepts is provided in the Appendix.

Part I of the book is a very basic introduction to relevant network security concepts. It also discusses the underlying motivation and the approach adopted, along with three example scenarios. It is accessible to a general audience.

Part 2 presents security games and illustrates the usage of various game-theoretic models as a way to quantify the interaction between malicious attackers and defenders of networked systems. Deterministic, stochastic, and limited-information security games are discussed in order of increasing complexity.

Part 3 focuses on decision making for security and provides example applications of quantitative models from optimization and control theories to various security problems. Among the topics presented are "security risk-management," "optimal allocation of resources for security," and social side of security: "usability, trust, and privacy." Chapters in this part are not dependent on each other and can be read independently.

Part 4 studies distributed schemes for decentralized malware and attack detection. First, a distributed machine learning scheme is presented as a nonparametric method. Subsequently, centralized and decentralized detection schemes are discussed, which provide a parametric treatment of decentralized malware detection. Hence, this part builds a bridge between security and statistical (machine) learning.

Acknowledgments

We would like to thank several individuals, particularly Kien Nguyen, Michael Bloem, Jeff Mounzer, Yalin Sagduyu, Stephan Schmidt, Jean-Pierre Hubaux, Nick Bambos, Christian Bauckhage, Sonja Buchegger, Walid Saad, M. Hossein Manshaei, Ann-Miura Ko, Maxim Raya, Albert Levi, and Erkay Savaş, whose research has provided a basis for various sections of this book. They have also kindly supported us in the writing process and provided feedback on relevant parts. We also thank Florin Ciucu for his careful reading of some of the chapters and for providing feedback.

Tansu Alpcan wishes to thank Deutsche Telekom Laboratories and its managing director Peter Möckel for their kind support in the writing of this book.

Finally, we would like to thank our editor Julie Lancashire and the team at Cambridge University Press, particularly Sarah Finlay and Sabine Koch, for their continuous and kind support during the preparation of this manuscript.

Tansu Alpcan   Tamer Başar

Berlin, Germany   Urbana, Illinois, USA

March, 2010   March, 2010

Artwork

The cover image is created by Tansu Alpcan using *Google SketchUp 7*[1] software for 3D design and *Kerkythea*[2] open source software for 3D rendering.

Almost all of the graphics in the book, except for scientific graphs and those listed below, are created originally by Tansu Alpcan using the *Inkscape*[3] open source software and (in some cases) public domain graphics from the *Open Clip Art Library*.[4]

The scientific graphs in the book (e.g. in Figures 3.3, 3.6 to 3.9, 4.2, 10.5, etc.) are generated using the *MATLAB* software by Mathworks Inc.

The image in Figure 1.1 is courtesy of the Intel Corporation and included with permission. The image in Figure 1.2 is a partial representation of the Linux 2.6.22 kernel and is generated by Tansu Alpcan using the scripts provided by *Linux Kernel Graphing Project*.[5]

Figure 6.1 is after an idea by Nick Bambos and Figure 6.6 is based on an earlier drawing by Jeff Mounzer. Figures 7.1 and 8.5 are inspired from earlier images by Michael Bloem. The figures in Chapter 10 are made in collaboration with Kien Nguyen.

---

[1] http://sketchup.google.com
[2] http://www.kerkythea.net
[3] http://www.inkscape.org
[4] http://www.openclipart.org
[5] http://fcgp.sourceforge.net/lgp

Notation

Some of the notational conventions adopted in this book are listed below.

Symbol	Description
x	vector or scalar as a special case
x_i	i-th element of vector x
$y = [y_1, \ldots, y_n]$	row vector y
M	matrix
$M_{i,j}$	entry at the i-th row and j-th column of matrix M
x^T, M^T	transpose of a vector x or matrix M
$\mathcal{S}$	set
$\mathcal{P}^A$	(set of) attacker player(s) in security games
$\mathcal{P}^D$	(set of) defender player(s) in security games
$\mathbb{R}$	set of real numbers
$\{0,1\}$	set with two elements: 0 and 1
$[0,1)$	right-open line segment $\{x \in \mathbb{R} : 0 \leq x < 1\}$
$[0,1]$	closed unit interval $\{x \in \mathbb{R} : 0 \leq x \leq 1\}$
$f(x)$, $V(x)$	real valued functions or functionals with argument x (vector or scalar)
$\mathrm{diag}(x)$	diagonal matrix with diagonal entries x
$\approx$	approximately equal
$:=$	definition; term on the left defines the expression on the right
I	identity matrix, $I := \mathrm{diag}([1, \ldots, 1])$
$\|x\|$, $\|M\|$	norm of a vector x or of a matrix M
min, max	minimum and maximum operations
inf, sup	infimum and supremum operations
NE	Nash equilibrium (or Nash equilibria)
FP	fictitious play

Please see Appendix A for further information and definitions.

Part I

Introduction

1 Introduction

Chapter overview

1. Network security
   - importance and relevance of network security
   - challenges of hardware, software, and networking complexity
   - multifaceted nature of network security
2. Approaches
   - why and how are decision and game theories relevant
   - interplay between theory and practice
   - detection, decision, and response
3. Motivating examples
   - security games
   - security risk-management
   - optimal malware epidemic response

Chapter summary

Network security is an important, challenging, and multi-dimensional research field. In its study, theory and practice should function together as parts of a feedback loop. Game, optimization, and control theories, among others, provide a mathematical foundation to formalize the multitude of decision-making processes in network security. The analytical approaches and quantitative frameworks presented lead to better allocation of limited resources and result in more informed responses to security problems in complex networked systems and organizations.

1.1   Network security

Networked computing and communication systems are of **vital importance** to the modern society simply because the civilization we know today would cease to exist without them. A good illustration of this fact is provided by the Internet, the epitome of networks that has evolved to a global virtual environment and become an indispensable part of our lives. Nowadays our communication, commerce, and entertainment are all based on networked systems in one way or another. Once they are disrupted its cost to society is hard to measure, but enormous, for sure. As an example, the Code Red worm, which infected about 360,000 servers in 2001, has cost – according to estimates – hundreds of millions of dollars globally in lost productivity and clean-up of systems afterwards [115].

The security of computers and networks has become an increasingly important concern as they have grown out of research environments where they fulfilled only specific duties at the hands of well-trained specialists. Security problems emerged once such systems entered general public life and started to be used for a multitude of different purposes in business, entertainment, and communication. Today, an overwhelming majority of users are no longer trained professionals who would know the nature and limitations of these systems. To complicate matters further, there is no shortage of malicious individuals and groups to exploit weaknesses of networked systems and their users for financial gain or other purposes.

Despite its vital importance and the ongoing research efforts, network security remains an **open problem**. This is partly because networked systems are difficult to *observe* and *control* even by their legitimate users and owners due to their *complexity* and *interconnected nature*. A regular user has only limited observational capabilities, for the user interface is only the tip of the iceberg. A significant number of automated system processes run hidden in the background, since it is simply infeasible to expose users to all of them. Furthermore, many of these processes involve communication with multiple other computing systems across networks, creating tightly coupled supersystems. As a simple example, the system clock of a computer is usually controlled by a specific program that runs in the background and corrects it by connecting to a time server possibly on the other side of the world, all of which is unknown to most regular users.

The first factor that is responsible for loss of observability and control of networked systems is **complexity**. Complexity is due to both hardware and software. The complexity of system and application software has increased significantly as a result of enormous advances in hardware in the last few decades. The microprocessor revolution is probably the best example of these advances (Figure 1.1). Thanks to the progress in microprocessors, the personal computers today are as powerful as the supercomputers of two decades ago. As a natural consequence, the software running on computers has become more layered and complex. The widely used Microsoft Windows and Linux operating systems on personal computers consist of tens of millions of lines of code (Figure 1.2).

A related issue contributing to the complexity is the unintentional flaws (bugs) in software. In current software architectures, there is always a mismatch between the

Figure 1.1 Intel processors with Nehalem architecture have multicores and between 500 and 1000 million transistors. Many of them are sold for home and small office usage. (Image courtesy of Intel Corporation.)

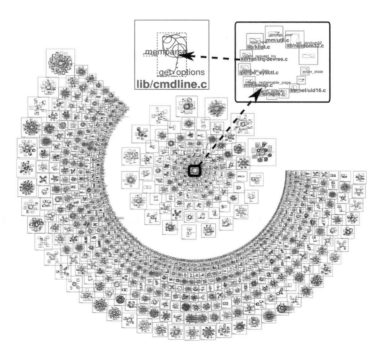

Figure 1.2 Linux 2.6 operating system kernel, visually depicted here, is widely used on a variety of devices and has more than 10 million source lines of code.

intentions of the developer and the actual behavior of the program, which exhibits itself as **software bugs**. Unfortunately, this is another permanently open problem due to the fundamental miscommunication between humans, who are by nature imprecise, and computers, which are based on rigid mathematical principles.

The second factor that contributes toward making networked systems difficult to observe and control is their **interconnected** nature. The distributed architecture of contemporary networks along with the complexity of the underlying computing and communication environments prevent systems administrators and organizations from having absolute control over their networks. Network boundaries are often vague and administrators cannot exercise control outside their local domain, which leaves networked systems vulnerable to distant security attacks due to global connectivity. This issue is often half-jokingly captured by the phrase "the most secure computer is the one unplugged from the network" (Figure 1.3).

The difficulty of observing and exercising control on networked systems can be illustrated by the following everyday **example**. Consider a laptop computer connected to the Internet. Although the connection makes the laptop a part of the most complex systems ever built, it still shares a simple property of the simplest man-made tools: it can be used for "good" purposes as well as exploited for "bad" ones. A malicious *attacker* can potentially run a malicious program (*malware*) on this laptop without permission of its owner by exploiting various *vulnerabilities*. While the laptop is physically next to its rightful owner, it can thus be partially controlled by the attacker who may be on the other side of the world. Furthermore, the owner may not even know or *observe* the security problem, allowing the attacker to maintain partial control over an extended time period.

In addition to limited observability and control of the underlying complex systems, another defining aspect of network security is its social dimension or so-called human factors. Network security is not a problem that can be solved once and for all by engineering a solution. It should be seen as a **problem** that needs **to be managed**, similarly to the security of a city. In other words, there is no such thing as a fully secured network, just as there is no such thing as a city without crime. Networks are man-made systems.

Figure 1.3 *"The most secure computer is the one unplugged from the network."* The US Department of Defense C2 rating of Windows NT 3.5 only applied to a computer unplugged from the network!

Since the system engineers are human beings who make mistakes, future networks will have vulnerabilities in some form, no matter how carefully they are designed. As long as there are people who would benefit from exploiting vulnerabilities for selfish reasons, there will always be security threats and attacks.

Given the **dynamic nature** of network security, attackers cannot be stopped by purely static measures such as classical firewalls. When targeting the vulnerabilities of networks, attackers update their strategies from day to day. Hence, it is crucial for the defense side to also take dynamic measures and address security both in the design phase and afterwards. Many existing defense mechanisms already adopt such a dynamic approach and offer various *security services*. Automatic (patching) updates of antivirus programs, browsers, and operating systems are well-known examples. In a sense, network security is like a "game" played between attackers and defenders on the board of complex networks using attacks and defensive measures as pieces (Figure 1.4).

The complexity, multi-dimensionality, and importance of network security makes it hard to describe with a single definition. The **multifaceted nature** of network security puts one in mind of an ancient story about *blind men and an elephant*. According to the tale, a group of blind men, who have never heard of elephants, want to learn about them by observing an elephant that comes to their village for the first time. Once they approach the elephant, each blind man touches a different part of the animal and tells others his opinion: the first man touches the body and says "elephant is like a wall," the second one touching the tusk says "elephant is like a spear," the third one holds the trunk and says "elephant is like a snake," and so on …

The field of network security can be compared to the elephant in the story (Figure 1.5) and security researchers to the blind men trying to understand it. For researchers in the field of cryptography, security is all about cryptographic algorithms and hash functions.

Figure 1.4 Network security is like a "game" played between attackers and defenders.

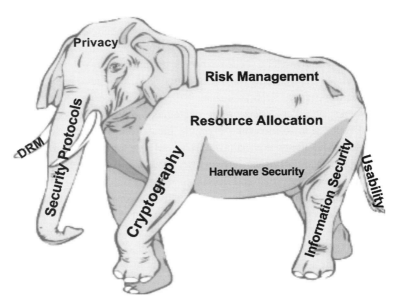

Figure 1.5 Elephant as metaphor for network security, from the ancient tale of *blind men and an elephant*.

Those who are in information security focus mainly on privacy, watermarking, and digital rights management systems. For researchers with an interest in hardware, security is about tamper-resistant architectures and trusted computing. Network security encompasses all these aspects and more. Researchers, unlike the blind men, are of course aware of this fact regardless of their specific focus. However, a wide field such as this calls for specialization and different perspectives.

This book also adopts a specific view of network security and introduces a **decision and game-theoretic approach**. The upcoming chapters study incentive mechanisms, analytical models, and resource allocation aspects of network security in terms of detection, decision making, and response. But first, the next section presents the adopted decision and game-theoretic approach, which is subsequently illustrated with examples in Section 1.3.

1.2   The approach

There is a fundamental relationship between **security and decision making**. Whether it is about buying a simple lock versus installing an expensive alarm system in a house, deploying a security suite on a personal computer, or applying a patch to a production server, decisions on allocating limited resources while balancing risks are at the center of network security. Making such decisions in a principled way instead of relying on heuristics provides numerous advantages and simply corresponds to following the celebrated *scientific method*.

It is, therefore, not surprising to observe **theoretical models** at the system level play an increasing role in network security as the field matures from its earlier qualitative and empirical nature. The increasing number of books, journal articles, and conference publications that study the problem analytically is clear evidence of the emerging interest in this approach to network security. The mathematical abstraction provided by quantitative models is useful for generalization of problems, combining the existing ad-hoc schemes under a single umbrella, and opening doors to novel solutions. Hence, the analytical approach provides a unique advantage over heuristic schemes that are problem specific. One of the main objectives of this book is to develop a deeper understanding of existing and future network security problems from a decision and game-theoretic perspective.

Securing information, controlling access to systems, developing protocols, discovering vulnerabilities, and detecting attacks are among the well-known topics of network security. In practice, all these involve decision making at multiple levels. **Security decisions** allocate limited resources, balance perceived risks, and are influenced by the underlying incentive mechanisms. Although they play an important role in everyday security, they are often overlooked in security research and are usually made in a very heuristic manner.

The human brain is undoubtedly a wonderful decision-making engine which current technology has not been able even remotely to replicate. A security expert can make very balanced decisions taking into account multiple factors and anticipating future developments. Admittedly, none of the techniques discussed in this book can come close to the performance of such an expert who relies on "intuition" which combines enormous pattern recognition capabilities with years of experience.

However, relying on **human expertise** in this manner has its own set of **shortcomings**. The first one is scale. Given the complexity and the number of networked systems around us, a security expert cannot oversee all systems all the time. A second issue is the availability of good experts. The number of experts is very limited due to the long training period required. In many cases, an organization has to work with available people of limited knowledge in less than ideal circumstances. A third problem is the timescale. Computers operate on a much faster timescale than humans and some security problems (e.g. malware epidemics) require an immediate response of the order of seconds. The human brain, despite its wonderful properties, operates on a much slower timescale.

It is hence unavoidable to have to rely on **computer assistance** in some form to address network security problems. Consider, for example, an organization that employs multiple experts to secure its networked systems. Given the scale and availability limitations, the organization has to naturally direct the attention of its experts primarily to the most important systems. This strategic resource allocation decision is often made by the management again relying on human expertise, but manifests itself electronically through scheduling systems or spreadsheets. In addition, the organization has to equip its security experts with a variety of computational tools to effectively observe and respond to security problems in a timely manner. Thus, computer assistance is essential in network security regardless of the degree of reliance on human expertise.

Both computerized support systems and security managers *implicitly* make numerous strategic decisions in terms of resource allocation, detection, or response. Decisions such as a log file viewer not showing some fields due to limited screen estate or a manager ordering a security administrator to patch a certain server have nontrivial consequences for overall security. In most of the existing security structures, such decisions are made in a heuristic manner. Therefore, the issues with human experts discussed above directly apply.

An alternative to implicit and heuristic decision making is the **analytical approach** based on mathematical models. For example, a manager can pose the problem of allocation of limited security experts within the organization as one of optimization where the available resources (e.g. in terms of man-hours) and degree of importance of specific subsystems (e.g. in terms of monetary loss when attacked or down) are quantified explicitly. Then, the problem can be solved automatically with computer assistance on a large scale. Another example is a packet-filtering system to decide on whether or not to drop a packet, based on a preset threshold. Unlike the previous case, this decision is made within milliseconds and repeated millions of times. How to (dynamically) determine the threshold value used as decision criterion can be investigated analytically and solved within an optimal and robust control framework based on given preferences.

The **quantitative approach** described has multiple **advantages** over the ad-hoc ones. First, the knowledge of the decision maker is expressed through mathematical models in a transparent and durable manner. Second, the decision making can now be made on a large scale. Third, it can be made as fast as numerical solution methods allow, in many cases of the order of milliseconds. While some security decisions, such as ones on investments or policies, are made over days or months, there are many security decisions made on much smaller timescales, as in the packet-filtering example mentioned. Finally, the decision-making process captured by the model can now be checked experimentally and improved upon, providing a way of aggregating the knowledge of multiple experts.

Developing a sound decision and game-theoretic framework for security requires building a **feedback loop** between high-quality theoretical research and real-life problems experienced by practitioners on a daily basis, as depicted in Figure 1.6. Actual problems lead to deep questions for fundamental research whereas theoretical advances provide novel algorithms and solutions. It has been repeatedly observed in many fields of science that weakening this feedback loop is detrimental to both theoretical and practical efforts. Therefore, the intention here is to bridge the gap between theory and practice as much as possible through simulations and proof-of-concept demonstrations addressing existing and emerging network security topics.

A variety of well-established **mathematical theories** and tools can be utilized to model, analyze, and address network security problems. Adopting a defensive approach, many resource allocation aspects of protecting a networked system against malicious attacks can be formulated as optimization problems. Especially convex **optimization** problems are well understood and a plethora of tools exists to solve them efficiently. When the underlying system dynamics play a significant role in security, **control theory** provides a large field of expertise for extending static optimization formulations to control of dynamic systems. Fundamental concepts such as observability and controllability

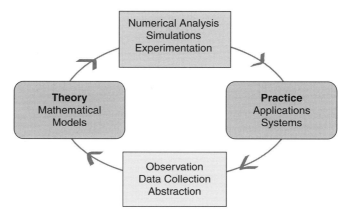

Figure 1.6 Feedback loop between theory and practice is of fundamental importance for the success of both.

relevant to security have been mathematically formalized and utilized in control theory for decades.

Beyond single-person decision making, **game theory** provides a rich set of mathematical tools and models for investigating multi-person strategic decision-making where the players (decision makers) compete for limited and shared resources. As a special case, security games study the interaction between malicious attackers and defenders. Security games and their solutions are used as a basis for formal decision-making and algorithm development as well as to predict attacker behavior. Security games vary from simple deterministic ones to more complex stochastic and limited information formulations and are applicable to security problems in a variety of areas ranging from intrusion detection to social, wireless, and vehicular networks.

Despite the broad scope and extent of available mathematical models in optimization, control, and game theories, one should not consider these fields as static. There are many open problems in each of them and they are themselves progressing. For example, how to incorporate information structures and formalize decision making under information limitations in single- and multiple-person dynamic problems are active research areas of great relevance to security applications. The models in this book should be interpreted merely as first steps toward developing realistic frameworks for decision making in security. Hence, establishing mature and relevant models is one of the important research challenges in the decision-theoretic approach to security.

Organization of the book

Based on the presented approach, the remainder of the book is organized into three parts encompassing detection, analysis, and optimized response as illustrated in Figure 1.7. Decision and game-theoretic schemes utilizing optimization, control, and machine learning are applied to a variety of network security topics. Principles for scalable, robust, effective security architectures for autonomous operation as well as computer

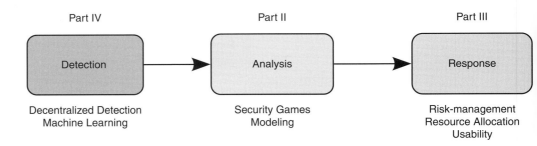

Figure 1.7 Organization of the book.

assistance systems are investigated. In addition to system- and network-level problems, organizational issues and human factors are also discussed, e.g. in the context of risk-management and usability.

Analysis, modeling, and decision making based on game theory are discussed in *Part II* **"Security games."** Security games provide a quantitative framework for modeling the interaction between attackers and defenders. They are used as a basis for formal decision-making and algorithm development as well as to predict attacker behavior. Stochastic or Markov security games model the unknown or uncontrollable parameters in security problems and capture the dynamic nature of the underlying game parameters, interdependencies, and external factors. The Bayesian game approach and fictitious play analyze the many cases where the players, attackers, and especially defenders do not have access to each others payoff functions and adjust their strategies based on estimates of opponent's type or observations of opponent actions. The security game framework is applicable to security problems in a variety of areas ranging from intrusion detection to social, wireless, and vehicular networks.

Part III **"Decision making for network security"** presents an optimization and control-theoretic approach for analysis of attack prevention and response, IT risk-management, and usability in security. The decision and control-theoretic approach quantifies implicit costs and formalizes decision-making processes for resource allocation processes in network security. The introduced approach is illustrated with three example scenarios where various optimal and robust control methods are utilized for placement of network filters, when responding to malware epidemics, and dynamic filtering of suspicious packets.

Similarly, security risk assessment and response are posed as dynamic resource allocation problems. First, a quantitative risk-management framework based on probabilistic evolution of risk and Markov decision processes and then a noncooperative game model for long-term security investments of interdependent organizations are presented. Subsequently, a cooperative game is studied to develop a better understanding of coalition formation and operation between divisions of large-scale organizations.

Social aspects of security and human factors are also captured using decision and game-theoretic models. The complex relationship between security and usability is discussed and two example schemes, one for improving usability of security alert dissemination and one for effective administrator response, are investigated. Finally, the community effects in evolution of trust to digital identities in online environments are studied using a specific noncooperative trust game.

Part IV presents a variety of approaches ranging from distributed machine learning to decentralized detection to address the problem of **"Security attack and intrusion detection."** Detecting the presence of security threats and compromises is a challenging task, since the underlying paradigms often do not fit into existing parametric detection models used in signal processing. Furthermore, normal usage data as well as attack information for the purpose of training detection algorithms are often limited or not available. The threat analysis as well as monitoring have to be done in near real time and should not bring excessive overheads in terms of resource usage. To counter these challenges, decentralized and distributed machine-learning schemes are studied. Recent advances in networking, multiprocessor systems, and multicore processors make parallel and distributed approaches particularly suitable for network security owing to their efficiency, scalability, and robustness. Moreover, monitoring and data collection processes are inherently distributed in networked systems. Hence, distributed machine learning and detection schemes are especially relevant in the network security context. In decentralized network structures, where a number of sensors report directly to a fusion center, the sensors receive measurements related to a security event and then send summaries of their observations to the fusion center. Optimization of the quantization rules at the sensors and the fusion rule at the fusion center are also studied.

1.3 Motivating examples

The decision and game-theoretic approach to security presented in this book is illustrated with three specific examples. Each of these motivating examples summarizes a specific model and application area which will be further analyzed in depth later in the book in its respective chapters.

1.3.1 Security games

Security games provide an analytical framework for modeling the interaction between malicious attackers, who aim to compromise networks, and owners or administrators defending them. The "game" is played on complex and interconnected systems, where attacks exploiting vulnerabilities as well as defensive countermeasures constitute its moves. The strategic struggle over the control of the network and the associated interaction between attackers and defenders is formalized using the rich mathematical basis provided by the field of game theory.

The underlying idea behind the game-theoretic models in security is the *allocation of limited available resources* from both players' perspectives. If the attacker and defenders had access to unlimited resources (e.g. time, computing power, bandwidth), then the solutions to these games would be trivial and the contribution of such a formalization would be limited. In reality, however, both attackers and defenders have to act strategically and make numerous decisions when allocating their respective resources. Security games allow the players to develop their own strategies in a principled manner based on formal quantitative frameworks.

A security game can be defined as having four components: *the players*, the set of possible actions for each player, the outcome of each player interaction (action–reaction), and information structures in the game. In a *two-player* noncooperative security game, one player is the *attacker* who abstracts one or multiple people with malicious intent to compromise the defended system. Although different malicious individuals may have varying objectives, defining a single metaphorical "attacker" player simplifies the analysis at the first iteration. The other player represents system administrators and security officers protecting a network who usually share a common goal.

Consider the following example security game, which provides basic insights to security decision making and the interaction between attackers and defenders in the context of intrusion detection. A network is monitored by an intrusion detection system and targeted by malicious attackers. In the simplest case, the action set of the attacker consists of "attack" and "not attacking." Similarly, the action set of the defender includes "intensified monitoring" and "default response," i.e. no major defensive action (Figure 1.8). It is reasonable to assume no order of play between players, yet if there is one, it can be easily captured within the described model.

For each action and corresponding reaction, there is an outcome in terms of costs and benefits for the players. If the players have an estimate on these outcomes, they can adjust their strategies accordingly and compute best-responses. At the intersection of the best-responses lies the Nash equilibrium (NE) solution, where no player has any incentive to deviate from the solution as it would result in a worse outcome. Hence, NE provides a powerful solution concept for security games as is the case in other applications of game theory. If the defender adopts an NE strategy, it does not matter whether the attacker is rational, since any deviation from the NE solution will decrease the cost of the defender and benefit of the attacker. Therefore, when properly implemented by the members of the organization or by security systems, NE solutions represent worst-case defense strategies against competent attackers.

The outcome of the defined game depends on factors such as the defender's gain for detecting the attack, the costs of a false-alarm and missing an attack, and the detection

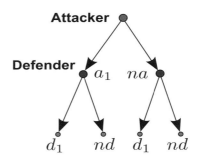

Figure 1.8 Simple security game.

penalty for the attacker. Missing an attack is associated with a cost for the defense, while false-alarms cost nothing to the attacker. Naturally, these cost factors directly affect the NE solution of the game. The probability of the attacker targeting the network at the NE decreases proportionally with the false-alarm cost, while the defender has extensive resources. This naturally discourages attacks. The probability of the intensified monitoring is affected by the attacker's gain from a successful intrusion. If the potential benefit of the attack significantly outweighs the detection penalty, then the defender is inclined to monitor with increased frequency.

The players of security games often do not have access to each other's payoff functions. Then, they adjust their strategies based on estimates of the opponent's type or observations of opponent actions using various learning schemes. These observations may not be accurate due to imperfect "communication channels" that connect the players, e.g. sensors with detection errors in the case of intrusion detection games. Security games such as the one discussed above as well as stochastic and limited information security games are presented in Part 2, specifically Chapters 3, 4, and 5.

1.3.2 Security risk-management

Today, networked systems have become an integral and indispensable part of daily business. Hence, system failures and security compromises have direct consequences in multiple dimensions for organizations. In some cases, network downtime translates to millions of dollars lost per second, in others stolen customer data may turn into a public-relation nightmare. Organizations and enterprises are becoming increasingly aware of emerging IT and security risks and are learning how to manage them with the help of security risk-management, a young and vibrant field with many research challenges and opportunities.

Early IT and security risk-management research has been mostly empirical and qualitative in nature. The situation is rapidly changing as the field is enriched by quantitative models and approaches. Analytical frameworks formalize risk-management processes and provide a foundation for computer-assisted decision-making capabilities. Hence, they not only improve scalability and efficiency of solutions but also increase transparency and manageability.

As an example analytical risk model, the probabilistic risk-management based on graph-theoretic concepts provides a unified quantitative framework for investigation of the interdependence between various *business units*, the potential impact of various *vulnerabilities* or threats, and the risk implications of relationships between *people*. These entities are represented by the nodes of a graph that also quantifies their interdependencies with edges between them. Thus, the probabilistic risk framework is used to model how risk cascades and gradually spreads (diffuses) in an organization. In order to explore how business units, security vulnerabilities, and people affect and relate to each other with respect to risks, an approach based on "diffusion processes" over a graph is utilized.

In this graph-theoretic model, the risks are represented in relative terms through relative risk probability vectors which evolve according to the linear (Markovian) diffusion process over the graph. By adjusting the diffusion parameters of the model, it is possible to explore the entire spectrum between initial relative risk probabilities and the final ones after the diffusion process stabilizes. The relative risk probability vector can then be used to rank the nodes with respect to their immediate as well as cascaded risks.

The model is applicable to a variety of scenarios with diverse decision makers, factors, and timescales. For example, if an employee in an enterprise fails to perform a server maintenance function such as applying a security patch, this may result in a compromise of the server by a malicious attacker. The attacker may then disable some check functions which may lead to potential failures of other employees to perform further maintenance functions. Thus, the risks cascade to other business units and processes.

Subsequent to the risk assessment phase, the probabilistic risk model can be used for taking appropriate action to perform *risk mitigation and control*. More specifically, the question of how to control the risk-diffusion process over time is addressed in order to achieve a more favorable risk distribution across the assets of an organization. The control actions available to the decision maker (e.g. risk manager) range from policies and rules to allocation of security resources or updating system configuration. These actions change the dependencies in the organization (between the nodes in the graph), and hence directly affect the evolution of the relative risk probability vectors.

Continuing with the previous example, patching a server decreases the weight between a virus threat node and the server node in the graph model. This in turn results in a different risk probability vector and risk ranking after the risk diffuses through the organization graph. If a strategic server is patched, then it may prevent service failures and decrease potential losses. The analytical model helps by identifying the right server based on the collected information on organizational and systematic interdependencies.

Stopping potential risks at their source is clearly a good strategy and one may ask, why not take such actions for any perceived risk? The answer lies in the hidden and open costs of each risk control action as well as limited aggregate resources. Risk managers have to "manage risks" by balancing available resources with potential costs all the time. Analytical frameworks such as the one discussed can facilitate risk-management by providing principled, scalable, and transparent methods.

The graph theoretic model mentioned above is only one specific analytical framework for balancing risks and limited available resources. A detailed analysis along with numerical examples as well as additional aspects such as security investment strategies, cooperation between organizational divisions, and incentive mechanisms are presented in Chapter 6.

1.3.3   Optimal malware epidemic response

The malware attacks to computer networks are growing more sophisticated and coordinated as indicated by the increasing number of botnets. Self-spreading attacks such as worm epidemics or botnets are costly not only because of the damage they cause but also because of the challenge of preventing and removing them. Malware attacks often exploit the inherent difficulty of differentiating legitimate from illegitimate network use, network security resource constraints, and other vulnerabilities. When such a malware epidemic infects a network, a timely and efficient response is of crucial importance.

Administrators responding to a malware epidemic usually have limited resources in terms of manpower and time. While the infected hosts (servers or computers) are costly for a network, removal rates are constrained and system update itself has a nonzero cost. For example, productions systems must be tested extensively to ensure that they will function well after a patch has been applied. Thus, the administrators have to make multiple resource allocation decisions on priority and schedule.

Classical epidemic models have been successfully applied to model the spread of computer malware epidemics. In these models, differential equations describe the rate of change in the number of infected hosts, which is proportional to the multiplication of currently infected and not yet infected host numbers. Such dynamic epidemic models provide a basis for optimization of patching response strategies to a worm epidemic within a quantitative cost–benefit framework.

A general optimal and robust control framework based on feedback control methods for dynamic malware removal is presented. Optimal control theory allows the costs of infected hosts and the effort required to patch them to be explicitly specified. The resulting quadratic cost function is used in conjunction with the dynamic epidemic model differential equations to derive the optimal malware removal strategies.

In addition, H^∞-optimal control theory can be utilized owing to the challenging nature of detecting malware, i.e. expected inaccuracies in detection. This justifies the need for a robust response solution capable of capturing model inaccuracies and noisy measurements that have a non-negligible impact on performance. Hence, H^∞-optimal control with its focus on worst-case performance is directly applicable to the problem domain.

Consider a network with multiple partially infected subnetworks. Depending on the value of each subnetwork and the cost of removal actions, a feedback controller based on the current number of infected hosts can be derived using optimal control theory. Even in its approximate linear form, where individual patching rates are proportional to infected hosts, such a controller naturally outperforms an arbitrary one. The H^∞-optimal control theory likewise results in a linear feedback controller, albeit with a different parameter that is computed by taking into account modeling errors and observation noise.

The practical implementation of both dynamic controllers is rather straightforward. Once the model parameters are identified through domain knowledge or observations, both the optimal or H^∞-optimal feedback controllers' respective feedback parameters can be computed offline. Then, the malware removal rates are obtained by multiplying parameters with the number of infected hosts. If needed, the parameters can be updated periodically in certain time intervals.

In addition to the malware removal problem above, Chapter 7 presents optimal and robust control frameworks for similar dynamic resource allocation problems in security such as the placement of malware filters and dynamic filtering of packets. In each case, the analytical frameworks provide the basis for scalable and quantitative security algorithms.

1.4   Discussion and further reading

Due to its multifaceted and interdisciplinary nature, it is difficult to cite a single source summarizing all of the issues in network security. There fortunately exist multiple books by various distinguished researchers providing a stimulating introduction to the subject that may serve as a starting point [23, 50, 161]. The entire text of reference [50] as well as essays in the book [161] are also available online (see reference [160] for more on the latter).

The challenging nature of network security stems paradoxically from the three big achievements of the last century: the Internet (networking and communication revolution), the integrated circuits (microprocessor revolution), and the software revolution. The enormous advances in these fields have given birth to very complex and interdependent systems that are not only difficult to control but also impossible to fully observe or monitor. The problem of network security is difficult to appreciate without at least a cursory knowledge of these three revolutions that have changed the (technology) world.

The approach adopted here is based on optimization, control, and game theories, which have a long and successful history. Due to their fundamental nature, they have been applied over decades to a vast variety of scientific areas ranging from economics and engineering to social and life sciences. Utilizing many results from these three fields, this book comfortably stands on a deep and solid mathematical foundation. The books [35, 38] in (distributed) optimization, [134] in control, and [31] in game theory, respectively, are well known in engineering and may serve as a starting point for the interested reader.

Although decision, control, and game theories have been applied to networking problems for more than two decades, the introduction of these frameworks and methods to the area of network security is very recent and still in progress. A useful collection of articles specifically on applications of game theory to networking has been published as part of the GameNets conferences and GameComm workshops (e.g. [20, 28]) as well as in various journals including reference [104]. Additional venues publishing relevant studies include the Workshop on the Economics of Information Security (WEIS) and the recently started Conference on Decision and Game Theory for Security (GameSec).

Given this background, the main premise of this book can be better appreciated. The previous research efforts, which have resulted in successful formalization of various decision-making processes in networking, also open doors to a better understanding of network security. Hence, along with reference [50], this book is the first to present a foundation for future network security research that adopts a decision, control, and game-theoretic approach.

The ancient story "Blind men and an elephant" mentioned in this chapter originates from India but has been adopted by many cultures in Asia and Middle East. The story is used in general – as here – to indicate that reality may be viewed differently depending upon one's perspective and that there may be some truth in what someone says. It is certainly a good lesson that has to be kept in mind given the narrow specialization in many fields of science and the need for more interdisciplinary research.

2 Network security concepts

Chapter overview

1. Networks and security threats
   - a brief overview of networks and the World Wide Web
   - security vulnerabilities and the ways to exploit them
2. Attackers, defenders, and their motives
   - who are the attackers and what are their objectives
   - who are the defenders and how are they motivated
3. Defense mechanisms
   - firewalls, intrusion detection, antimalware methods
   - survivability, cryptography
4. Security tradeoffs and risk-management
   - tradeoffs: usability, accessibility, overhead, and economics
   - security risk-management

Chapter summary

Network security is a wide field with multiple dimensions. This chapter provides a brief overview of basic network security concepts, security tradeoffs, and risk-management from a decision and game-theoretic perspective.

2.1 Networks and security threats

2.1.1 Networks and the World Wide Web

A (computer) network is simply defined as a group of interconnected computing devices which agree on a set of common communication protocols. Along with the microprocessor revolution, the creation of networks permeating all segments of modern life is arguably among the biggest accomplishments of engineering. There exists already a great variety of networks in terms of scale, connection methods, and functions. However, networks continue to evolve and diversify with unabating speed. The devices, or nodes, on the network can also be heterogeneous and may vary from small sensor motes to smartphones, laptops, desktops, and servers.

The Internet, as a network of networks that is unique in its scale and global reach, illustrates the above-mentioned diversity and heterogeneity. The set of protocols connecting the devices to each other on the Internet is called Internet protocols, and includes IP (Internet protocol), TCP (transfer control protocol), UDP (user datagram protocol), and HTTP (hypertext transfer protocol). Compatible versions of these protocols have to be installed on each device that connects to the Internet in order to establish communication with others.

Networks can be classified based on their **features** such as scale, connection method, functional relationship of nodes, and access control. A brief overview of networks based on these features is provided in Table 2.1. A detailed discussion of networks, and the underlying technologies and protocols, would easily fill an entire book. There are already many good books which present the topic at various levels and depth. We refer to [171] as a classical one.

The **applications** of and on networks are even more diverse than their types. The main goal of networks can be succinctly stated as "moving bits from here to there." However, in a digital world where these bits represent all kinds of information and data (documents, multimedia, software, etc.), the various uses of networked systems are exponentially growing. Consequently, an increasing number of software applications are connected to a network through a variety of protocols including the widely adopted

Table 2.1 Types of network

Feature	Examples
Scale (small to large)	Personal area network (PAN)
	Local area network (LAN)
	Wide area network (WAN), the Internet
Connection method	Wired (optical networks and Ethernet), wireless (GSM, 802.xx)
Functional relationship	Client–server, peer-to-peer, hybrid
Access control	Intranet, virtual private network (VPN)

Internet protocols. This leads to a very complex coupled system that exhibits security vulnerabilities due to flaws in both software applications and the implementation of protocols.

Networks are today **ubiquitous** and already constitute an indispensable part of modern life. Thanks to recent developments in wireless technologies, especially around the set of protocols IEEE 802.11x also known as Wi-Fi, many people have wireless networks deployed in their homes. Businesses rely on Ethernet-based wired networks for their intranets in addition to wireless LANs. Mobile (cellular) phones utilizing the Global System for Mobile communications (GSM) and other wireless technologies have reached and are actively used by more than half the world's population. Although currently used mostly for voice communications, smartphones and their evolution indicate the coming age of wireless networked personal mobile computing devices. On a smaller scale, sensor networks, near field communication (NFC), and radio-frequency identification (RFID) tags make objects other than computing devices parts of networks and open the door to *ubiquitous (pervasive) computing* and possibly an *Internet of things*.

As one of the most popular applications on the Internet, the **World Wide Web (www)**, also known simply as the Web, started as a system of interlinked static documents. Today, thanks to the reach of the underlying Internet and the emergence of the **Web 2.0** paradigm, it has become an interconnected computing platform enabling communication, business, and entertainment on a global scale. Using the Web browser as a standard thin client for server applications constitutes the main idea behind Web 2.0. This model has various advantages in terms of software maintenance and updates. It changes, at the same time, the nature of the Web and turns it into a much more interactive and social platform. Consequently, novel Web 2.0 concepts such as social networks, publishing schemes (online media, blogs), and communication schemes (twitter) have emerged and enjoy widespread popularity.

The Web 2.0 applications bring their own vulnerabilities, which naturally have implications for network security. The underlying asynchronous JavaScript and XML (AJAX) technology powering these applications follow (approximately) a classical client–server programming and communication approach. The client program is hosted and run on the client-side usually in a very accessible format. The server-side application interface accessed by the client is automatically exposed to the entire network. Both of these issues necessitate a very careful design of both the client and server interfaces with security concerns taken into account. In practice, however, these **AJAX security** issues with significant potential implications are ignored due to either economic reasons or lack of expertise.

Another application related to Web 2.0 is **cloud computing**, which is a vague concept referring to Internet-based information technology (IT) services and resources. It can be described as a way to increase IT capacity or to add IT-related capabilities on the fly without investing in new infrastructure, personnel, or software. The term is also used in connection with any subscription-based or pay-per-use service that extends IT's existing capabilities in real time over the Internet. The recently emerging paradigm is built upon the rapidly advancing networking technologies and commonly agreed protocols as well as the interactive platform aspect of Web 2.0. In addition to advantages such as building whole businesses on "virtual infrastructure" on the cloud, outsourcing and exposing computing and networking infrastructure in this way also brings a new set of security risks for organizations.

2.1.2   Security threats

Attack types

Attacks in the context of network security can be defined as attempts to compromise the confidentiality, integrity, and availability, or to obtain (partial) control of a computer or communication network. Security threats, which can be defined similarly, are as diverse as the uses of networks. This book adopts the convention where the term "attack" refers to a realization of a "threat" by a malicious party.

Security threats and attacks can be classified equally well in a variety of ways. The following **classification** scheme summarized in Table 2.2 is adopted in this book without loss of generality [26]. Each class of attacks will be discussed next individually along with examples.

The attacks compromising **confidentiality** of a system aim for unauthorized access to confidential information of a person or organization. The simplest form of this is information collection such as learning the IP address and operating system of a server, or the architecture of an intranet. Preliminary information acquired on the system is often used by attackers to launch more significant attacks and access valuable data such as corporate secrets or customer data. Such losses can be very expensive for businesses both monetarily and in terms of reputation. Attackers may use the information collected about individuals for **identity theft**, namely impersonating people without their knowledge to commit fraud. A common example is theft by stealing credit card and address information on the Internet.

Attacks threatening the **integrity** of databases and systems are among the most common. They could be as simple as defacing a website or as harmful as implanting backdoors to a system for later use. For example, attackers may compromise multiple nodes of a network without the knowledge of their owners through malicious software

Table 2.2 Types of Security Threat and Attack

Security service	Sample threats and attacks
Confidentiality	Information theft, unauthorized access, identity theft, stealing corporate secrets
Integrity	Altering websites, compromising data, implanting backdoors and Trojans
Availability	Denial-of-service (DoS) attacks
Control	(Partial) control of system, e.g. account and root access

or **malware** such as *worms*, *viruses*, and *Trojan*s. Then, they use the installed malware on these computers as a botnet for sending spam (unwanted email messages) or to steal information.

Other attacks compromise **availability** and **control** of a network. Botnets created through malware are often used to launch distributed *denial-of-service* (DoS) attacks in coordination, which affect the availability of online services. DoS attacks can be motivated, for example, by commercial purposes, e.g. secretly sponsored by a company who profits from service disruptions of the other competitor. Malware is also used to gain partial control of larger systems such as servers. Attackers then use existing or secret accounts on those systems to control system behavior, which constitutes one of the most dangerous security threats.

Vulnerabilities and attack methods

Security attacks use various methods to exploit the vulnerabilities of networked systems and their users. Since they are closely related to each other, attack methods can be classified by the underlying vulnerabilities: software, physical (hardware), communication, and psychological/social.

Physical and hardware-based methods: It is ironic how millions' worth of network security measures can be circumvented simply by attackers impersonating computer technicians and carrying away sensitive hardware, or by an employee of a government organization forgetting disks with valuable information in a train on the way home, or by a disgruntled employee stealing the entire customer database of a multinational company stored on a simple CD. Unsurprisingly, the encryption of sensitive information on hard disks and other storage media is strongly recommended uniformly by all security experts.

Another interesting class of hardware-based attacks exploit physical access to read or patch system memory. It has been shown that even secret keys of complex encryption schemes can be estimated using such methods. As documented in the security literature, they can even be used to retrieve the keys of encrypted hard disks in some cases. Therefore, the physical security of a device or network provides the foundation for the rest of the security measures.

Software-based methods: Software-based attack methods exploiting vulnerabilities in software are the most diverse and common of all. This is partly because the interconnected nature of networks, especially the Internet, increases the number of potential attackers by orders of magnitude when compared to physical attacks which require the physical presence of the attacker. Another important reason is the presence of software flaws or *bugs*.

Software flaws leading to security vulnerabilities are caused by two factors: one is the sheer scale and complexity of today's software applications, made possible by Moore's law.[1] At any given moment, the number of lines of software running or installed on a

---

[1] Moore's law by Gordon E. Moore, co-founder of Intel, states roughly that the number of transistors, i.e. computing power, that can be placed inexpensively on an integrated circuit doubles approximately every two years.

1970s
supercomputer

2000s
personal laptop

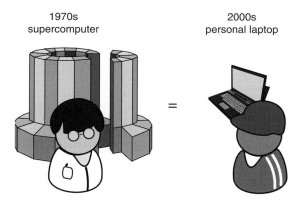

Figure 2.1 An ordinary personal laptop today is as powerful and complex as a supercomputer from the 1970s. However, it is not in the hands of specialists but members of the general public with limited knowledge of computers.

Figure 2.2 There is a fundamental miscommunication between human beings, who are naturally imprecise, and mathematically precise computing systems.

networked device is of the order of tens to hundreds of millions (Figure 1.2). Moreover, an ordinary personal laptop today is at least as powerful as a supercomputer from the 1970s (Figure 2.1). This, combined with the network effects, results in enormous complexity. The other factor leading to software-based security vulnerabilities is the miscommunication of intended application behavior between human developers, who are naturally vague, and computing systems, which operate under strict mathematical precision (Figure 2.2). Given the software development practices of today and economic pressures, the number of software flaws on networks that can be exploited by attackers can be frighteningly high.

Until this point, we have used the terms virus, worm, and Trojan without distinction under the umbrella of malware (malicious software). Although these are dynamic concepts evolving rapidly, there are accepted differences between these terms. A *computer virus* is a malicious software that hides in an existing program on the system and copies itself without permission or knowledge of the owner. *Worms* are similar to

Table 2.3 A phishing email

REQUEST FOR URGENT BUSINESS RELATIONSHIP

I am making this contact with you on behalf of my colleagues after a satisfactory information we gathered from an international business directory.

My colleagues and I are members of the Contractor Review Committee of the Nambutu National Petroleum Corporation (NNPC). I have been mandated by my colleagues to look for a trustworthy company/individual into whose account some funds is to be transferred. The funds in question is $25.5M (twenty-five million, five hundred thousand US dollars) now in a dedicated account with the Central Bank of Nambutu (CBN). The above funds arose from the over-invoicing of some supplies and oil drilling works which have been executed and concluded. The fund is therefore free to be transferred overseas.

The underlisted shall be required from you immediately by fax:- **the beneficiary's name and confidential telephone and fax numbers, the full name and address of company/beneficiary**. All necessary particulars of the bank account where you wish the contract sum to be transferred (**account number, bank address, the telephone, fax and telex numbers of the bank**).

Immediately we receive the requested information from you, we shall put up an application for fund & transfer to the appropriate ministries and departments in favor of the beneficiary (you or your company).

Please, we implore you to treat this deal with utmost confidentiality. As civil servants, we would not want any exposure. Do not go through the international telephone operator when lines are busy. Always dial direct.

Thanks for your anticipated co-operation.
Best regards,
XXX

viruses with the distinction being that they are stand-alone programs and can spread using the network. On the other hand *Trojans* are not self-spreading. They are defined as innocent-looking programs which secretly execute malicious functions. Other less dangerous but annoying software categories include *spyware*, which collects personal information without consent, and *adware*, which automatically displays advertisements.

Communication-based methods: The *man-in-the-middle* attack is one of the best examples of this category. This attack is an active eavesdropping scheme where the attacker intercepts and manipulates the communication between two or more parties without their knowledge. The victims believe that they are talking to each other in a secure way while the entire communication is controlled by the attacker. Since most of the real-life methods used in assessing identities are not available on the Internet, such attacks can be launched easily and with great success. Against this class of attacks, cryptographic measures such as public key infrastructure and mutual authentication can be used as defensive measures.

Psychological and social engineering methods: Social engineering methods rely on, so to say, the oldest tricks in the criminal's book: manipulation through deception. Arguably the most famous of such methods is *phishing*, which is defined as the fraudulent process of acquiring sensitive (personal) information through various means of deception. The deception is done using carefully crafted emails (see Table 2.3 for an example), fake websites, or combinations of both. In all cases, the target users are deceived into thinking that the email or website is authentic. Then, they are manipulated to reveal sensitive information about themselves, e.g. by filling out a fake form with bank account details. Alternatively, such email messages may claim to have images of popular people but contain malware attachments. A classical example from the year 2001 was the "Kournikova" email, which installed a worm instead of showing a picture of the famous tennis player "Anna Kournikova" when the user clicked the attachment.

Phishing and social engineering attacks are very common and are expected to remain so as they exploit fundamental human weaknesses. The difficulty of assessing identities on the Web helps attackers to craft elaborate schemes while anonymity on the Internet often protects them from prosecution. Furthermore, the general public knows little about network security and is not capable of assessing risks on the Internet. All these factors bring to light the need for more user education and usability as part of network security.

2.2   Attackers, defenders, and their motives

2.2.1   Attackers

The term **attacker** is used in this book to personalize and abstract people with malicious intent who try to compromise confidentiality, integrity, availability, or control of a computer network without its owner's consent or knowledge. People with very different backgrounds and motivations fall under this umbrella. One possible classification [165] is presented in Table 2.4.

In the early days of networks and the Internet, the attacker scene was dominated by "gray hat" **hackers**, who are skilled and often asocial individuals motivated by curiosity and fame. The so-called *script kiddies*, semiskilled juveniles imitating hackers for fame using self-developed tools, and *crackers*, who remove copy protection mostly

Table 2.4 Attacker types

Actor	Description
Script kiddie	Often young, no sophisticated skills, motivated by fame
"Black hat" hacker	Semi-professional, criminal intent, sophisticated attack tools and programs
Cracker	Modifies software to remove protection
Malicious user	Inside organization, criminal intent
Malicious sysadmin	Control of network, criminal intent, potentially significant damage

from games, have also been prevalent. As the Internet evolved to a more commercial medium, however, the attacker profiles shifted and "black hat" hackers with criminal motivations have started to dominate. Today, the once romantic image of the hacker has been replaced by one that has ties to organized crime, blackmails individuals and financial organizations, and steals money from bank accounts.

Another important class of attacker are **malicious users and system admins**[2] inside an organization. Since users often have unrestricted access to networks and systems, they can potentially create great damage, e.g. by stealing corporate secrets or customer data. A malicious system admin can even hold a whole city hostage. A disgruntled computer engineer in 2008 locked out everyone but himself from San Francisco's new multimillion-dollar computer network, which stores city government records such as officials' emails, city payroll files, and confidential law enforcement documents.

While attackers to network security share some common points with petty criminals, there are also fundamental differences. First, the majority of attacks require a certain level of sophistication and skill, which can be acquired only through (formal) education. This makes network security attacks more a "white collar" crime with almost no violence. Second, the environment in which these attackers operate, i.e. networks and the Internet, can be described as a "virtual world" rather than the real one. This virtual environment provides unique advantages to the attackers such as anonymity of disguise due to the difficulty of assessing identities (on the Internet) and a global reach.

2.2.2   Defenders

Ideally, any individual using a networked system should be aware of security issues and be able to take defensive actions. In reality, however, the task of securing networks and systems falls squarely on the shoulders of **system administrators**. Considering that they are usually overloaded and underappreciated, it is almost a wonder that today's networks are as secure as they are. System administrators are often not only responsible for configuring and monitoring the networks against attacks but also take an active role in enforcing formal and informal security policies and educating the users on possible vulnerabilities.

One of the main problems with network security from the defense perspective is lack of **motivation**, which partly stems from the difficulty of *quantifying the value added by network security*. Large organizations have started to realize the importance of network security only very recently after the networks have become an integral part of their core business. However, there is still a lot of confusion about how to assess and quantify network security. This lack of quantification naturally affects the decision-making process regarding security investments. Hence, attitudes towards security seem to go back and

---

[2] "System admin" or "sysadmin" are often used abbreviations for system administrator.

forth between "we are doomed if we don't invest big time in security" and "no need to worry too much, it all seems fine," depending on the economic situation.

Another reason behind the lack of motivation is **misalignment of incentives**. People affected by security breaches, people responsible for management, people selling security services, security researchers, and people actually implementing security measures could all be different with misaligned incentives. The management usually has an incentive to cover up security breaches (as a short-term approach) while security companies and many researchers exaggerate the risks for their own benefit such as increasing security product sales or receiving research funding. There is an even greater disconnect between people suffering security breaches, e.g. customers of a bank, and system administrators, who are at the front of defenses but at a quite low level in organizational hierarchy: they simply do not know or care about each other. This makes *network security a complex social and management problem in addition to being a scientific one*.

The lack of motivation may lead to a "**security theater**" that is easier and cheaper than actually securing a system. Unless used in the right dosage for positive psychological effect (see "In praise of security theater" in reference [160]), security theater results in a waste of resources and leads to a false sense of security. In many cases, it is used as a tool to cover fraudulent behavior or to escape security related responsibilities. For example, a company management declaring a new set of security measures without backing them up with real investments demoralizes system administrators by overloading them and risks a big disappointment if a real attack creates publicly known damage.

Despite existing challenges, there are some recent **positive developments** for network security defense. First, there is increased awareness of network security in governments, businesses, development community, researchers, and the general public. This awareness improves security investments and educational efforts. Second, security is ever-increasingly perceived as an important feature, which creates demand for secure networks and systems. Consequently, a number of security companies have emerged providing *security services* to both individuals and organizations. Third, emerging security services, by their nature, support dynamic prevention and response improving defenses. Hence, automatic patching and updating of operating systems, antivirus programs, and firewalls have become commonplace. Finally, security research is expanding beyond its original community. Security related decision-making processes are increasingly studied and formalized using interdisciplinary models and approaches.

2.3   Defense mechanisms

Given the importance of network security and the existing vulnerabilities, a variety of defense mechanisms have been developed to secure networks. Major categories encompassing widely deployed solutions include firewalls, antivirus software, and intrusion detection systems.

Firewalls inspect network traffic passing through them and filter suspicious packets, usually based on a rule-set analyzing their properties. By regulating the incoming and outgoing traffic, firewalls maintain separated trust zones in and between networks. While simple firewall variants use basic packet properties such as source/destination address and port number, more sophisticated ones investigate packet types and even content for filtering decisions. Firewalls can be implemented in software or on dedicated hardware.

Intrusion detection and prevention systems (IDPSs) aim to detect and prevent attacks directed at networked systems. They consist of sensors which observe security events, a console to communicate with system admins, and a decision engine to generate alerts or take responsive actions. IDPSs can be network based, monitoring network traffic, or host based, to monitor operating system behavior, or application based, to monitor individual applications for signs of intrusions. While passive IDPS variants, also known as intrusion detection systems (IDSs), only generate alerts the active versions react to detected attacks by, for example, reconfiguring the system or filtering malicious packets and malware. An IDPS can be implemented in software or dedicated hardware and deployed in combination with firewalls. A detailed discussion on IDPSs is provided in Section 9.1.1.

Antivirus software is a widely adopted security solution that identifies and removes all kinds of malware including viruses on a computer. Such software scans storage medium and memory for signs of malware in regular intervals and removes them. Since the detection process involves recognition of malware characteristics (signatures) and these change with each version (generation) of malware, the signature database the antivirus software often relies on needs to be updated regularly. This makes antivirus software one of the earliest examples of *security services*.

All of the discussed defense systems incorporate some type of **detection mechanism**, which can be classified into two groups: signature based and anomaly based. **Signature-based** detection is a widely used method to check monitored events (e.g. packets on the network) against a known list of security attack signatures. This approach has the advantage of enjoying a relatively small false-alarm rate and ease of implementation. The disadvantages are the need to maintain and update the attack signature database, and the restriction to detection of only the known attacks documented in the database. The second is the **anomaly detection**, where changes in the patterns of nominal usage or behavior of the system are detected as an indicator for attacks. Although this approach increases the probability of detecting undocumented new attacks, it is difficult to implement, and often has a higher false-alarm rate. Both of these methods are further discussed in Section 9.1.1.

As defense mechanisms evolve and become more complex, they increasingly rely on theoretical models. The fields of *machine learning* and *pattern recognition* play an important role in providing the underlying scientific basis for various attack detection mechanisms used in all defense systems. Advanced classification and clustering techniques are utilized to differentiate increasingly sophisticated attacks from normal behavior. Similarly, *game theory* and *control theory* are emerging as candidates to provide a basis for analyzing and developing formal decision-making schemes in network security. They address issues such as where to allocate limited defense resources or how to configure filtering schemes in an optimal way. In addition, some of the well-known notions from control theory such as *"observability"* of a system are directly applicable in the monitoring and detection context.

One branch of science that has a special place in network security is **cryptography**. Cryptography can be defined as the practice and study of hiding information. It provides a basis for crucial security services between the sender and receiver of a message such as

- **confidentiality**: ensuring that the message content is kept secret except to the intended receiver;
- **integrity**: assuring the receiver that the received message has not been altered;
- **authentication**: the process of proving one's identity;
- **nonrepudiation**: ensuring that the sender has really sent the message.

Cryptographic algorithms can be classified into

- **secret key cryptography** (classical cryptography), where a single secret key is used for both encryption and decryption;
- **public key cryptography**, where one key is used for encryption and another for decryption;
- **hash functions**, which compute a fixed-size bit string, *the hash value*, from given data such that a change in the data almost certainly changes the hash value. Hence, they ensure integrity.

While cryptography provides a foundation for network security, it can also be a *double-edged sword*. Attackers can use it for their own benefit and encrypt malware as well as attack traffic. Since current firewalls and IDSs cannot detect malware in encrypted traffic, use of cryptographic methods by attackers may render many existing defense mechanisms useless.

Cryptography can also be characterized as a topic that has many inherent asymmetries. Encryption usually takes more time than decryption. The basic cryptographic primitives rely on one-directional functions such as modular arithmetic or factorization of primes which are easy to compute in one direction and very difficult in the other. Creating an encryption scheme is much easier than checking whether it is indeed secure or not.

A perspective different from those discussed above is provided by the concept of **survivability**, defined as "the capability of a system to fulfill its mission, in a timely manner, in the presence of attacks, failures, or accidents" [61, 62, 89]. Hence, instead of focusing on the prevention and detection of attacks, survivability aims to increase resilience or robustness of the system despite the existence of attacks. Since in many cases full control of a network is not possible, this realistic approach is very valuable and acts like an insurance when other defenses fail. Simple methods such as mirroring services in off-location servers and regular backups may prove to be invaluable to network owners when faced with a surprise attack or failure.

2.4   Security tradeoffs and risk-management

2.4.1   Security tradeoffs

There are several fundamental **tradeoffs** in designing a network security system as summarized in Table 2.5. Whether it is a simple mechanical door-lock protecting a house or a firewall, a basic tradeoff is the one between *security risk* and *ease of use*. In many cases, additional security mechanisms bring a usability overhead, hence making the system less convenient for its users. A real-life example is keeping a frequently used door locked against intruders versus leaving it unlocked as a security hole. Similarly, users of computer networks want to access data easily and instantly while expecting the sensitive data to be protected from unauthorized access. Achieving security requires sacrificing some convenience, since it requires deployment of authentication and security mechanisms between the user and sensitive data. It is important to strike a balance between these requirements in order to motivate compliance of the users to security policies (Figure 2.3).

Accessibility is another factor that needs to be balanced in network security. In this context, accessibility describes the degree to which a network (service) is

Table 2.5 Basic security tradeoffs

Tradeoff	Security versus
Usability	Difficulty of use and mental overhead
Accessibility	Access restrictions based on location or role
Overhead	Costs on system and network resources
Economics	Monetary and manpower costs

Figure 2.3 Security systems should not forsake usability and accessibility.

accessible to as many people as possible. The famous *Metcalfe's law* states that "the value of a network is proportional to the square of the number of its users." However, securing a network often involves segmentation of the network into trust zones through access control and filtering schemes, inherently decreasing the value of the network.

As a common example, "a user can send email to anyone" means at the same time that "anyone can send an email to that user." This medium of almost instantaneous data communication makes email a killer application on the Internet, yet the unrestricted accessibility has a downside as reminded to all email users by *spam* (unsolicited and unwanted bulk emails abusing email systems) every day. The spam problem illustrates some of the characteristics of **accessibility tradeoffs** in network security: nine out of ten emails are spam today, which indicates that one end of the tradeoff spectrum is not pleasant. On the other hand, whitelists, social network messaging, or instant messaging all have limited value by Metcalfe's law as an inherent disadvantage. Hence, email spam is a good example of a network security problem that needs to be managed rather than one that can be solved once and for all.

Another important tradeoff is the **overhead** caused by security systems in terms of system and economic resources. Any defense mechanism, whether an antivirus software, firewall, or IDS, uses network and system resources such as bandwidth, memory, and CPU cycles. Furthermore, even if they are open source and free, they still require system administrator attention for maintenance and operation. The economic aspects of security play an important role in the decision making of individuals and organizations and affect the resulting security level significantly [74].

In addition to generic ones, there are many **specific tradeoffs** that need to be taken into account during the operation of defense systems. For example, a basic performance criterion for an IDS is the false-alarm rate. There exists a tradeoff between reduction in the false-alarm rate by decreasing the system sensitivity and increase in the rate of undetected intrusions (Figure 2.4). Clearly, on either extreme the IDS becomes totally ineffective. Therefore, the IDS should satisfy some upper and lower bounds on the false-alarm rate and undetected intrusions according to the specifications of the deployed network.

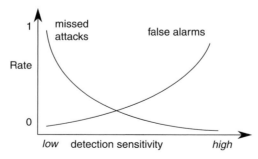

Figure 2.4 An example tradeoff in network intrusion detection.

Figure 2.5 The cycle of security risk-management.

2.4.2   Security risk-management

Security risk-management usually involves the identification of vulnerabilities, assessment of potential risks, creating a risk-management plan, and implementation of prevention or protection measures (Figure 2.5). While risk-management is well established in some fields such as finance, it is relatively young in the context of network security. Most of the current risk-management approaches focus more on organizational and management aspects of network security rather than technological ones. However, the main goal is the same: defending the network against attacks. It is also interesting to note that security risk-management reinforces the point of view that network security is a problem to be managed rather than solved once and for all through pure engineering. Chapter 6 provides a detailed discussion on the topic.

When assessing security risks, it is important to be aware of the **limits of risk models**. In finance, a recent theory called *Black swan* studies large-impact, hard-to-predict, and rare events (i.e. black swans) beyond the realm of "normal" expectations [170]. Rare but high-impact security events may also look like a black swan if the models used in risk assessment are limited and take into account only past experiences within an organization. For example, a disgruntled employee stealing and selling millions of customer records of a multinational company is clearly a rare event and may be classified as a "black swan." One way to counter such rare events is by not relying solely on past experiences but simulating virtual "what if" scenarios to assess unusual and unexpected security risks.

2.5   Discussion and further reading

This chapter, which contains a rudimentary summary of various security concepts, shows the breadth of the network security field. Unsurprisingly, there are multiple books presenting the subject from different perspectives. While reference [23] provides an excellent introductory coverage of security concepts, reference [161] focuses more on real-life lessons. Reference [25] presents intrusion detection systems (earlier version available online [26]). The recent work [50] adopts a game-theoretic and analytical approach to the security of wireless networks. There are also numerous practical network security books on specific architectures, systems, and protocols that discuss best practices and daily guidelines.

Although not discussed in detail in this work, cryptography is one of the main foundations of network security. A very useful historical perspective on cryptography can be found in reference [166]. The reference book [110], which is available online, contains a detailed and easy-to-access overview of cryptography. While topics such as key management in sensor networks is a popular research topic today, new directions such as quantum cryptography are increasingly commanding more attention.

Classical textbooks for a general overview of communication and computer networks include references [37, 93, 140, 171]. There has been an explosion of activity in the area of networking during the last decade from optical technologies and cognitive radio at the physical and networking layers to social networks and various Web technologies (e.g. AJAX, HTML 5) at the application layer. We will discuss topics such as social networks, vehicular, and wireless networks in the upcoming chapters within the context of network security and will refer to respective publications.

Part II

Security games

3   Deterministic security games

Chapter overview

1. Definitions and game model
   - components of a game: players, actions, outcomes, information
2. Intrusion detection games
   - formal decision-making when facing attacks
   - matrix game formulation
   - games with dynamic information structures
3. Sensitivity analysis
   - the effect of game parameters on the results
4. Modeling malicious behavior in social networks
   - modeling incentives for disruption and cooperation
   - interaction between collaborative and malicious players
5. Security games for vehicular networks
   - prioritization of assets and defensive actions
   - numerical analysis using a realistic dataset
6. Security games in wireless networks
   - random and multiple access channels
7. Revocation games

Chapter summary

Security games provide a quantitative framework for modeling the interaction between attackers and defenders. Security games and their equilibrium solutions are used as a basis for formal decision-making and algorithm development as well as to predict attacker behavior. The presented analytical foundation is applicable to security problems in a variety of areas ranging from intrusion detection to social, wireless, and vehicular networks.

Table 3.1 Components of a security game

Component	Description
Players	Attacker and defender
Action space	Set of attacks or defensive measures
Outcome	Cost and benefit to players for each action–reaction or game branch
Information structures	Players fully or partially observe each other's actions

3.1　Security game model

Network security is a **strategic game** played between malicious attackers who aim to compromise networks, and owners or administrators defending them. The game is played on complex and interconnected systems, where attacks exploiting vulnerabilities and defensive countermeasures constitute its moves. This metaphorical "game" over the control of the network and the associated interaction between attackers and defenders can be formally analyzed using the rich mathematical basis provided by the field of game theory. We refer to Appendix A.2 for a brief overview of noncooperative game theory.

A game-theoretic framework for defensive decision-making has a distinct **advantage over optimization** as it explicitly captures the effects of the attacker behavior in the model, in addition to those of the defensive actions. Plain optimization formulations, on the other hand, focus only on the optimization of defensive resources without taking attackers into account. Security games with their inherent multiplayer nature not only provide a basis for defensive algorithm development but can also be used to predict attacker behavior. Nonetheless, the security games discussed here adopt first a more defense-oriented point of view and focus more on defense than attacker strategies.

A security game can be defined as having four components, as summarized in Table 3.1: the players, the set of possible actions for each player, the outcome of each player interaction (action–reaction), and information structures in the game. This chapter focuses mainly on *two-player* noncooperative security games. One player is the *attacker*, which abstracts one or multiple people with malicious intent to compromise the defended system. Although different malicious individuals may have varying objectives, defining a single metaphorical "attacker" player simplifies the analysis at the first iteration. Moreover, this abstraction is often justified as the defenders only have vague (limited) information on the profile of attackers. In the case of additional information, multiple games can be defined against individual attacker types and combined within a Bayesian game. Such formulations will be discussed in detail in Chapter 5. The system administrators and security officers protecting a network are also represented by a single abstract player, a *defender*, in the game due to the fact that these individuals usually share a common goal and information set.

The action spaces of attackers and defenders are the sets of possible attacks and defensive countermeasures, respectively. The attacks often exploit vulnerabilities in the networks and use the complexity and connectivity to their advantage. The specific security games in this chapter abstract each attack as a single action even though many of them consist of a multi-stage process such as reconnaissance (e.g. collecting information about a network by scanning), preparatory actions (e.g. obtaining access to the system by exploiting a vulnerability), and the main attack of achieving a specific goal. Similarly, the defensive actions are modeled as a set of single "actions." They may vary from simply monitoring the network or systems and setting alarms to taking (automated) measures to thwart the attacks through reconfiguration or patching.

Based on the **properties of action spaces**, the security game may be classified as *finite* or *infinite*. If the set of actions is modeled as discrete choices such as $\{scan, shell\,access\}$ and $\{monitor, patch\}$, then this is a finite security game. On the other hand, if the set of actions are continuous, e.g. when each attack or defensive action is associated with a continuous intensity value, then the game is infinite. Another important distinction is the one between *actions and strategies*. A strategy is a decision rule whose outcome is an action. For example, "if a scan is detected, monitor that subnetwork more intensely" is a strategy (decision rule) where the action (control) is the intensity of monitoring of a subnetwork in the system. This distinction disappears in static, one-shot games where action and strategy are synonymous.

The outcome of a security game is quantified by the cost (payoff) values for each possible action–reaction combination of the players. The actions of the attacker and defender are mapped to a specific cost (payoff) value, representing the gain or loss of each player for each branch of the game after it is played. In certain two-player finite games, these mappings may be represented by a matrix for each player. If the loss of one player is exactly the same as the gain of the other for each and every branch of the game, then the game is said to be *zero-sum*. Otherwise, it is a *nonzero-sum* game. In the finite static case, a zero-sum game is often described by a single game matrix and called simply a *matrix game*. Similarly, a nonzero-sum finite two-player game is called a *bi-matrix game*. In a matrix security game each row and column corresponds to a specific action of the attacker and defender, respectively. Then, each matrix entry represents the cost and gain of the attacker and defender for their respective actions.

In security games, the attacker and defender often cannot fully observe each other's moves and the evolution of the underlying system. Furthermore, each player has only limited information on the opponent's specific objectives. Hence, a security game is more analogous to the strategy board games of *Risk* or *Stratego* rather than *chess*. These **information limitations** can be formally modeled through various means including (Bayesian) games with chance moves, information sets, learning schemes, or fuzzy games. This chapter focuses on full information security games as a starting point. Chapter 5 provides a detailed treatment of various limited information security games.

Nash equilibrium (NE) provides a powerful **solution concept** for security games as is the case in other applications of game theory. At NE, a player cannot improve his outcome by altering his decision unilaterally while others play the NE strategy. In a security game, the NE solution provides insights to expected attacker behavior by mapping attacker preferences to an attack strategy under a given set of circumstances. Likewise, the NE strategy of the defender can be used as a guideline on how to allocate limited defense resources when facing an attacker. In zero-sum games, the NE corresponds to **saddle-point equilibrium** (SPE), and hence in this case the two will be used interchangeably.

Rationality of players is an important underlying assumption of the class of security games considered here. However, individuals often do not act rationally in real life, partly due to limitations on observations and available information. Despite this, security games provide organizations – within the boundaries of the models – with rational guidelines for defensive strategies and principles for algorithms that can also be implemented in (semi)automated security systems. In a zero-sum setting, if the defender adopts an NE strategy, it does not matter whether the attacker is rational, since any deviation from the NE solution will decrease the cost of the defender and benefit of the attacker. Therefore, when properly implemented by the members of the organization or by security systems, NE (equivalently, SPE) solutions (in zero-sum games) represent the defense strategies against competent attackers in the worst case.

Definitions and conventions

The security games in this chapter and in the remainder of the book adopt the following definitions and conventions. The players of a static (bi)matrix security game are denoted by *Player A* or $\mathcal{P}^A$ for attackers and *Player D* or $\mathcal{P}^D$ for defenders. The finite action spaces are the set of attacks

$$\mathcal{A}^A := \{a_1, \ldots, a_{N_A}\}$$

and the set of defensive measures

$$\mathcal{A}^D := \{d_1, \ldots, d_{N_D}\}.$$

The outcome of the game is captured by the $N_A \times N_D$ game matrices G^A and G^D for the attacker (row player) and defender (column player), respectively. The entries in the matrices G^A and G^D represent the costs for players which they minimize. In the case of a zero-sum security game, i.e. when $G^A = -G^D$, the matrix

$$G := G^D = -G^A$$

is said to be the **game matrix**. In this convention, $\mathcal{P}^A$ (attackers, row player) maximizes its payoff while $\mathcal{P}^D$ (defender, column player) minimizes its cost based on the entries of the game matrix.

If a security game admits an NE solution in pure strategies, then it is denoted with a superscript star (e.g. $a_1^* \in \mathcal{A}^A$).

Define

$$p^A := [p_1, \ldots, p_{N_A}]$$

as a probability distribution on the attack (action) set $\mathcal{A}^A$ and

$$q^D := [q_1, \ldots, q_{N_D}]$$

as a probability distribution on the defense (action) set $\mathcal{A}^D$ such that $0 \leq p_i, q_i \leq 1 \ \forall i$ and $\sum_i p_i = \sum_i q_i = 1$. Then, an NE solution in mixed strategies is denoted by the pair of probability distributions (p^*, q^*). It is a well-known fact that every (bi)matrix game admits an NE solution in mixed strategies [31]. Furthermore, the pair

$$(v^{A*}, v^{D*}) = (p^* G^A q^{*T}, p^* G^D q^{*T})$$

is the NE outcome of the security game for $\mathcal{P}^A$ and $\mathcal{P}^D$, respectively. Here, $[.]^T$ denotes the transpose of a vector or a matrix.

3.2 Intrusion detection games

Intrusion detection can be defined as the process of monitoring the events occurring in a computer system or network and analyzing them for signs of intrusions. This classical definition is somewhat limited and focuses mainly on detecting attacks after they occur and reacting to them. Recent developments in the field have resulted in a more extended approach of intrusion prevention where monitoring capabilities are utilized to take preventive actions to defend against malicious attacks before or as they occur.

An intrusion detection and prevention system (IDPS) consists of three main components: *information sources*, *analysis*, and *response*. Security games in the context of intrusion detection aim to formalize decision-making processes in all three components, often with a focus on response. A detailed discussion on IDPSs can be found in Chapter 9.

3.2.1 Matrix games

The following example game (Figure 3.1), which is one of the simplest possible security game formulations, provides basic insights to security decision making and the interaction between attackers and defenders in the context of intrusion detection. Consider a network monitored by an intrusion detection system and targeted by malicious attackers. In the simplest case, the action set of the attacker, $\mathcal{P}^A$, is $\mathcal{A}^A = \{a_1, na\}$, where a_1 is launching an attack on the network and na denotes no attack. The action set of

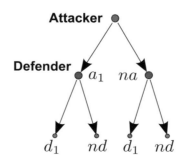

Figure 3.1 Simple intrusion detection game.

the defender, $\mathcal{P}^D$, is $\mathcal{A}^D = \{d_1, nd\}$, where d_1 is intensified monitoring and nd denotes default response, i.e. no major defensive action. Adopting a nonzero-sum formulation, the game matrices are defined as

$$G^A = \begin{matrix} (d_1) \ (nd) \\ \begin{bmatrix} \beta_c & -\beta_s \\ 0 & 0 \end{bmatrix} \begin{matrix} (a_1) \\ (na) \end{matrix} \end{matrix}, \quad G^D = \begin{matrix} (d_1) \ (nd) \\ \begin{bmatrix} -\alpha_c & \alpha_m \\ \alpha_f & 0 \end{bmatrix} \begin{matrix} (a_1) \\ (na) \end{matrix} \end{matrix}, \tag{3.1}$$

where the α's and β's are positive scalars. The parameter $-\alpha_c$ quantifies the gain of $\mathcal{P}^D$ for detecting the attack. The quantities α_f and α_m are the costs of a false-alarm and missing an attack, respectively, for the defender. The cost β_c represents the detection penalty for the attacker, $\mathcal{P}^A$, whereas $-\beta_s$ is the benefit from a successful (undetected) attack. Although missing an attack is associated with a cost for the defense, false-alarms cost nothing to the attacker, and, hence, are denoted with zero entries in the game matrices.

The *min-max* or **safe strategy** of a player guarantees a maximum cost, or so-called *safety level* regardless of the actions of the opponent. However, it provides only limited insight into the underlying problem. Here, the positive detection cost of an attack, $\beta_c > 0$, forces the attacker's safe strategy to be no attack, *na*. In cases when the cost associated with *na* is zero, for example, if the attack is executed by an anonymous remote script or botnet, both actions become equally safe for the attacker. On the other hand, the defender $\mathcal{P}^D$'s safe strategy depends on the relative values of α_f and α_m, the costs of false-alarm and missing an attack. If $\alpha_f > \alpha_m$, then $\mathcal{P}^D$ chooses not to monitor $nd1$ as a safe strategy and the opposite d_1, if $\alpha_f < \alpha_m$.

The **NE** of the security game is computed next. Notice that the game does not admit a pure NE solution. Let $0 \leq p_1 \leq 1$ and $1 - p_1$ be the probabilities for actions a_1 and *na* of $\mathcal{P}^A$, respectively. Also, let $0 \leq q_1 \leq 1$ and $1 - q_1$ be the probabilities for actions d_1 and *nd* of $\mathcal{P}^D$. In this security game the pair of mixed strategies, (p^*, q^*), is said to constitute a noncooperative NE solution, if the following inequalities are satisfied

$$
\begin{aligned}
p_1^* \left(\beta_c q_1^* - \beta_s (1 - q_1^*) \right) &\leq p_1 \left(\beta_c q_1^* - \beta_s (1 - q_1^*) \right), \\
p_1^* \alpha_m + q_1^* \left[\alpha_f - (\alpha_f + \alpha_c + \alpha_m) p_1^* \right] &\leq p_1^* \alpha_m + \\
& \quad q_1 \left[\alpha_f - (\alpha_f + \alpha_c + \alpha_m) p_1^* \right].
\end{aligned}
\tag{3.2}
$$

Note that at such an NE, the players do not have any incentive (in terms of improving their utility or cost) to unilateral deviation.

This set of inequalities can be solved by setting the coefficients of p_1 and q_1 to zero on the right-hand sides [31]. Thus, the only solution to the set of inequalities in (3.2) constitutes the unique NE of the game

$$p_1^* = \frac{\alpha_f}{\alpha_f + \alpha_c + \alpha_m} \quad \text{and} \quad q_1^* = \frac{\beta_s}{\beta_c + \beta_s}. \tag{3.3}$$

The NE solution above exhibits an interesting feature as all completely mixed NE do. While computing own NE strategy, each player pays attention only to the cost (parameters) of the opponent, and attempts to *neutralize* the opponent. Hence, the nature of the optimization (i.e. minimization or maximization) becomes irrelevant in this case.

This security game solution is the opposite of doing an optimization based only on own preferences and parameters.

In the context of intrusion detection, the NE solution (3.3) of the security game is interpreted as follows. The probability of the attacker targeting the network at NE, p_1^*, increases proportionally with α_f. A smaller false-alarm cost for the defense means the availability of extensive defense resources. This naturally discourages attacks. Similarly, an increase in α_c and α_m plays a deterrent role for the attacker. On the other hand, the probability of the defender monitoring is affected by the attacker's gain from a successful intrusion, $-\beta_s$. If $\beta_s \gg \beta_c$, then the defender is inclined to monitor with increased frequency. The penalty for the attacker getting detected may vary significantly, depending on his physical reachability. If, for example, the attackers are physically close to the network, then β_c is much larger due to the increased risk, when compared to the marginal detection cost of a script-based attack from the other side of the globe.

3.2.2 Games with dynamic information

The bi-matrix intrusion detection game is now extended by introducing a dynamic information structure and modeling attacks to multiple systems on a network. Assume a scenario where the defending player $\mathcal{P}^D$ protects two systems using only one sensor, which reports attacks to system 2 in a reliable manner, whereas there is no reliable information on attacks to system 1. Hence, $\mathcal{P}^D$ can only distinguish whether the attacker $\mathcal{P}^A$ targets system 2 at a given time instance and has to make decisions accordingly.

Such cases can be modeled using separate **information sets** in the security game (Figure 3.1). Since $\mathcal{P}^D$ can distinguish between information sets but not attacker actions within them, this is called a dynamic information game. The second group of targets and the related attacker, $\mathcal{P}^A$, actions are $\{a_2, a_{12}\}$, where a_{12} denotes targeting both systems and a_2 means targeting only system 2. The interpretation for $\{a_1, na\}$ is as before, with a_1 standing for targeting only system 1, and na meaning neither system is targeted. Correspondingly, $\{d_2, d_{12}\}$ are the actions of $\mathcal{P}^D$, at information set 2, where d_{12} means defending both systems.

The resulting **intrusion detection game** is studied recursively. The first information set has already been analyzed in the previous section. A similar analysis is repeated for the second information set in Figure 3.2. The same game parameters are used as before for simplicity. The parameter $-\alpha_c$ quantifies the gain of $\mathcal{P}^D$ for detecting the attack. The quantities α_f and α_m are the costs of a false-alarm and missing an attack, respectively, for the defender. The cost β_c represents the detection penalty for the attacker, $\mathcal{P}^A$, whereas $-\beta_s$ is the benefit from a successful (undetected) attack. Due to the existence of two separate targets, the cost terms of each target are added to each other accordingly. Hence, the following 2×2 bi-matrix game is obtained

$$G^A = \begin{bmatrix} \beta_c & -\beta_c - \alpha_f \\ \beta_c - \beta_s & 2\beta_c \end{bmatrix} \begin{matrix} (a_2) \\ (a_{12}) \end{matrix} , \quad G^D = \begin{bmatrix} -\alpha_c & -\alpha_c + \alpha_f \\ -\alpha_c + \alpha_m & -2\alpha_c \end{bmatrix} \begin{matrix} (a_2) \\ (a_{12}) \end{matrix} . \quad (3.4)$$

with column headers $(d_2)\ (d_{12})$ for both matrices.

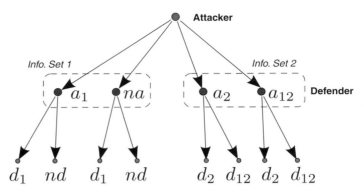

Figure 3.2 Intrusion detection game with two systems as potential targets and two information sets.

Let $\bar{p}_1$ and $1 - \bar{p}_1$ be the probabilities of $\mathcal{P}^A$ actions $\{a_2, a_{12}\}$ and $\bar{q}_1$ and $1 - \bar{q}_1$ be the respective probabilities of $\mathcal{P}^D$ actions $\{d_2, d_{12}\}$. The security strategy of $\mathcal{P}^A$ is determined using the relative values of α_c, α_f, and α_m as in the previous case. Furthermore, there is again no NE in pure strategies.

The unique **NE solution** in mixed strategies is obtained by solving the counterpart of the set of inequalities (3.3), and is given by

$$\bar{p}_1^* = \frac{\alpha_m + \alpha_f}{\alpha_m + \alpha_c + \alpha_f} \quad \text{and} \quad \bar{q}_1^* = \frac{\beta_c + \alpha_f}{\beta_c + \beta_s + \alpha_f} \ . \tag{3.5}$$

In general, it is possible to adjust the cost parameters by taking into account various factors such as relative importance of a subsystem for the organization, threat levels given the output of sensors, etc. Therefore, the given game parameters should only be considered as an example case. The equilibrium probabilities of $\mathcal{P}^A$ and $\mathcal{P}^D$ have a similar interpretation as those in the previous analysis.

Given the equilibrium solutions and costs of the bi-matrix games in each information set, $\mathcal{P}^A$ and $\mathcal{P}^D$ determine their overall **strategies**. The equilibrium strategy of $\mathcal{P}^D$, γ^{*D}, for example, is given by

$$\gamma^{*D} = \begin{cases} \begin{array}{l} d_1 \quad \text{w.p.} \ q_1^* \\ nd \quad \text{w.p.} \ 1 - q_1^* \end{array}, & \text{if in information set 1,} \\ \begin{array}{l} d_2 \quad \text{w.p.} \ \bar{q}_1^* \\ d_{12} \quad \text{w.p.} \ 1 - \bar{q}_1^* \end{array}, & \text{if in information set 2.} \end{cases}$$

This is a so-called **behavioral strategy** for $\mathcal{P}^D$, since he is "mixing" between the alternatives in each information set separately (and independently), and not "mixing" between the four actions available to him, namely $\{d_1, nd, d_2, d_{12}\}$.

The equilibrium strategy of the attacker, $\mathcal{P}^A$, depends on the outcome of the subgames associated with the respective information sets defined. For example, if

$$[\bar{p}_1^*, 1 - \bar{p}_1^*] \, G^A \, [\bar{q}_1^*, 1 - \bar{q}_1^*]^T < [p_1^*, 1 - p_1^*] \, G^A \, [q_1^*, 1 - q_1^*]^T,$$

then the equilibrium strategy γ^{*A} is

$$\gamma^{*A} = \begin{cases} a_2 & \text{w.p. } \bar{p}_1^* \\ a_{12} & \text{w.p. } 1 - \bar{p}_1^*. \end{cases}$$

The simplifying assumptions of this intrusion detection game, which allow for analytical results, can easily be extended to capture more realistic scenarios. Hence, the example game in Figure 3.1 can be scaled up arbitrarily. Although increasing complexity may prevent derivation of a closed form solution, such games can easily be solved numerically. Thus, the discussed framework becomes applicable to practical scenarios.

3.3 Sensitivity analysis

The solution of a security game is not only affected by the game model but also is a complex and nonlinear function of **game parameters**. Therefore, even if the game structure is realistic, the output of the security game crucially depends on the input values, defined for example in the game matrix. This input–output relationship, or how a variation (uncertainty) in the input of a security game affects its outcome, can be studied using *sensitivity analysis*.

In simple cases, such as the intrusion game (3.1), the sensitivity analysis can be conducted simply by taking the derivative of the NE solution with respect to entries in the game matrix to obtain sensitivity functions

$$\frac{\partial p_1^*}{\partial \alpha_f} = \frac{\alpha_c + \alpha_m}{(\alpha_f + \alpha_c + \alpha_m)^2}, \quad \frac{\partial p_1^*}{\partial \alpha_m} = \frac{\partial p_1^*}{\partial \alpha_c} = \frac{-\alpha_f}{(\alpha_f + \alpha_c + \alpha_m)^2},$$

and

$$\frac{\partial q_1^*}{\partial \beta_s} = \frac{\beta_c}{(\beta_c + \beta_s)^2}, \quad \frac{\partial q_1^*}{\partial \beta_c} = \frac{-\beta_s}{(\beta_c + \beta_s)^2}.$$

Analytical expressions such as these unfortunately cannot be derived in a majority of games. Therefore, the sensitivity analysis can be conducted numerically, for example using Monte Carlo sampling methods. The **sensitivity** of defense NE probability q_1^* to parameter β_s is computed numerically for the intrusion game (3.1) and shown in Figure 3.3 for various values of β_h.

A related important question in security games is how to determine the game parameters reliably and accurately such that the solution is useful and realistic. Definition 3.1 below from reference [31, p. 81] establishes a **strategic equivalence result** for bi-matrix security games such that any affine transformation of the game matrices does not affect the solution of the game. Therefore, as long as the input (game) values are determined accurately *relative to each other* using any method, the solution of the game is not sensitive to the absolute numbers in the game matrix. This flexibility allows for various methods ranging from expert opinions to machine learning to be used in defining game parameters. At the same time it increases the usefulness of game-theoretic models significantly for real-world applications.

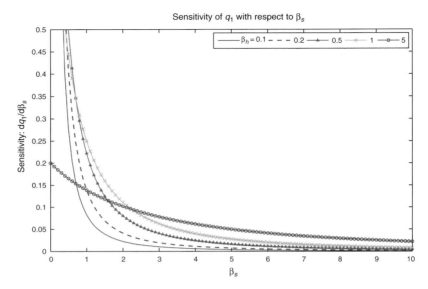

Figure 3.3 Sensitivity of NE probability q_1^* to parameter β_s for various values of β_c in the intrusion detection game.

Definition 3.1 *The bi-matrix game* $(\bar{G}^A, \bar{G}^D)$ *is said to be strategically equivalent to the original game* (G^A, G^D), *if it is obtained by the following affine transform*

$$\bar{G}_{i,j}^A = k_1 G_{i,j}^A + l_1, \quad \bar{G}_{i,j}^D = k_2 G_{i,j}^D + l_2, \quad \forall i, j,$$

where k_1, k_2 *are positive constants and* l_1, l_2 *are scalar parameters.*

3.4   Modeling malicious behavior in social networks

Incentive mechanisms and selfish behavior in social (e.g. peer-to-peer) and ad-hoc networks have been analyzed extensively in the literature using game-theoretic models, e.g. references [50, 64, 141]. However, malicious behavior in this context has received much less attention from the research community. The following simple but interesting games [174] **model malicious behavior** where the attacker aims to disrupt collaboration in the network rather than behaving selfishly to get a disproportionate share of network resources.

It is possible to analyze the behavior types of different players on social networks as part of a *behavior spectrum* that starts with cooperation at the one end and ends with malicious behavior at the other (Figure 3.4). Even when players are fully cooperative,

Figure 3.4 Graphical representation of user behavior spectrum on networks.

they can make unintentional mistakes which may affect the overall performance. The next step is selfish behavior, where players still play by the rules but selfishly aim to obtain as big a share of available network resources as possible. Finally, the players can be outright malicious and try to destroy system performance rather than obtaining any resource for themselves. Each of these cases can be captured using varying cost structures for respective types of player based on their preferences. For example, the cost function of an attacker is expected to be different from that of a collaborative player.

Consider an ad-hoc network consisting of nodes capable of deciding whether to collaborate with each other or defect. On the one hand, the nodes have to forward packets to each other in order to enable multi-hop communication. On the other hand, most of them are selfish and want to preserve their limited energy (battery). For a successful transmission between two nodes, both of them have to collaborate. This situation can be formalized in a **forwarding game** played by a set of *collaborative* players, $\mathcal{P}_i^D$, and *malicious or disruptive* ones, $\mathcal{P}_j^A$, $i,j \in \{1,2,\ldots,M\}$. The action set of each player is

$$\mathcal{A} = \{c,d\},$$

where c represents *collaboration* and d denotes *defection*. In order to describe the outcome of the game, the following parameters are defined. The value E denotes the effort (e.g. energy) spent during collaboration. The value B denotes the benefit obtained from collaborating or forwarding. It is assumed that $B > E > 0$, i.e. collaboration is more beneficial than energy preservation. In addition, define $\alpha := B - E > 0$ and $\beta := E > 0$ for notational convenience.

An example game between two collaborative players, $\mathcal{P}_1^D$ and $\mathcal{P}_2^D$, is investigated first. The game matrices of respective players are

$$G^{D1} = \begin{array}{c} (c)\ (d) \\ \begin{bmatrix} -\alpha & \beta \\ 0 & 0 \end{bmatrix} \end{array} \begin{array}{c} (c) \\ (d) \end{array}, \quad G^{D2} = \begin{array}{c} (c)\ (d) \\ \begin{bmatrix} -\alpha & 0 \\ \beta & 0 \end{bmatrix} \end{array} \begin{array}{c} (c) \\ (d) \end{array}. \tag{3.6}$$

Note that $G^{D1} = \left(G^{D2}\right)^T$ since in accordance with our notational convention $\mathcal{P}_1^D$ is the row player and $\mathcal{P}_2^D$ the column one. This nonzero-sum game admits two NE in pure strategies, (c,c) and (d,d), with respective equilibrium costs of $(-\alpha,-\alpha)$ and $(0,0)$ for the players. Since the equilibrium strategy (c,c) results in better costs for both players, it is preferable over the other(s). Formally, such equilibria are called *admissible*.

If this game is played repeatedly between these two players, an interesting phenomenon arises. If one player starts with strategy d or defects, the other player also plays d. Since (d,d) is an NE, neither player has any incentive to deviate and they are stuck with an unfavorable solution forever. A similar situation also emerges for the other NE, (c,c). However, this time the outcome is desirable for both players. *Within the given model*, this tit-for-tat (TFT) strategy, i.e. mimicking the behavior of the opponent, is the only rational option for the players unless there is an external (unmodeled) factor. Moreover, the safety strategy for both of the players is d. If both players are conservative, then the outcome of the game is the undesirable equilibrium.

When additional factors are brought into play, the players can be motivated to try more collaborative strategies to improve the outcome. For example, if the players take into account a (finite) time horizon when choosing their strategies, then they can adopt richer strategies. Such strategies involve taking a risk over the safety strategy d such as starting the game with collaboration c or randomly trying collaboration from time to time [64].

A game between a malicious attacker and a collaborative player, $\mathcal{P}^A$ and $\mathcal{P}^D$, is analyzed next. The game matrices of the respective players are

$$
G^A = \begin{matrix} (c) \ \ (d) \\ \begin{bmatrix} \alpha & \beta \\ -\beta & 0 \end{bmatrix} \begin{matrix} (c) \\ (d) \end{matrix} \end{matrix}, \quad
G^D = \begin{matrix} (c) \ \ (d) \\ \begin{bmatrix} -\alpha & 0 \\ \beta & 0 \end{bmatrix} \begin{matrix} (c) \\ (d) \end{matrix} \end{matrix}. \tag{3.7}
$$

This game admits a unique pure strategy NE, (d,d) with the associated costs of $(0,0)$. It is a desirable outcome for the attacker but not for the collaborative player. In other words, the attacker successfully disrupts the network.

Finally, both games discussed are brought together in **a mixed game** to model the case when a collaborative player is surrounded by multiple others but does not know which ones are malicious. This lack of information can be modeled within a Bayesian game framework, which in this case is a two-player nonzero-sum game with chance moves (Figure 3.5). Here, the well-behaved node, $\mathcal{P}_1^D$, plays against an opponent $\mathcal{P}_2^D$ which may be malicious. Given a number of attackers on the network, let $0 \le k \le 1$ be the probability of the neighbor, $\mathcal{P}_2^D$, being an attacker. Then, $\mathcal{P}_1^D$ plays the game (3.7) against $\mathcal{P}_2^D$ with probability k and the game (3.6) with probability $1-k$.

In order to find the NE solution(s) of this Bayesian game, the k weighted average of game matrices (3.6) and (3.7) are computed to obtain

$$
G^{D1} = \begin{matrix} (c) \ \ (d) \\ \begin{bmatrix} \alpha & 0 \\ -\beta & 0 \end{bmatrix} \begin{matrix} (c) \\ (d) \end{matrix} \end{matrix}, \quad
G^D = \begin{matrix} (c) \quad\quad (d) \\ \begin{bmatrix} -\alpha(1-2k) & \beta \\ -\beta & 0 \end{bmatrix} \begin{matrix} (c) \\ (d) \end{matrix} \end{matrix}. \tag{3.8}
$$

The strategy (d,d) is one of the NE solutions.

If the probability k satisfies

$$
k < \frac{\alpha}{2\alpha + \beta},
$$

then the collaborative strategy (c,c) is also an NE, which makes the game very similar to that between two collaborative players. Therefore, the number of malicious nodes

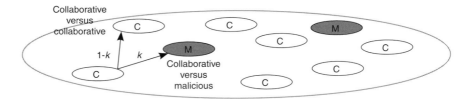

Figure 3.5 Collaborative users against malicious ones in a Bayesian game where the probability of a neighboring node being malicious is k.

has to be sufficiently high in order to disrupt communication on the network by breaking the potential trust between collaborative players. For example, on a network with M nodes, the number of attackers has to be at least $M\alpha/(2\alpha+\beta)$. The collaborative nodes might use a centralized or decentralized learning scheme using their own (and possibly others') observations to detect malicious neighbors and estimate the probability k [174]. An example scheme where the rights and credentials of malicious nodes are revoked by benign ones collaborating with each other to defend the system is discussed in Section 3.7.

3.5   Security games for vehicular networks

The study of vehicular networks (VANETs) enabling vehicle-to-vehicle and vehicle-to-infrastructure communications is an emerging research field aiming to improve transportation security, reliability, and management. **VANET security games** model the interaction between malicious attackers to vehicular networks and defense mechanisms protecting them on a road network. VANET security games take as an input various measures such as traffic density or centrality measures (e.g. betweenness centrality). Then, the objective of the game is to locate central (vulnerable) points on the road topology as potential attack targets (e.g. for jamming) and deploy countermeasures in the most effective manner. The defense system can be static in the form of roadside units (RSUs) or dynamic in the form of mobile law enforcement units. In the static case, all RSU positions are precomputed. In the dynamic case, RSU placement is adaptive to conditions in the vehicular network, such as traffic patterns or attacks detected. In both cases, the defense mechanisms are assumed to be capable of detecting attackers and rendering them ineffective.

3.5.1   Vehicular network model

A vehicular network model consists of three layers: data traffic, vehicular traffic, and road network. While the first two are dynamic, the last is naturally fixed. Each network can be formally modeled as a separate graph, yet they are closely related to each other.

The vehicles can communicate with neighboring vehicles and RSUs. The vehicles' neighbors are defined by their limited-radius (e.g. 300 m) radio coverage. The range and data rates can be modeled, for example, as circular and fixed, respectively. The framework developed also allows for more complex radio models, which along with other metrics determine the input parameters of the security game. Communications can be multi-hop and RSUs are assumed to be connected with each other. The RSUs can also help vehicle-to-vehicle communication by tunneling data.

The road network is modeled as a graph of discrete road segments. For a given map segment, e.g. a city district or rural region with a road network, the graph can be obtained by discretizing (quantizing) the roads to fixed-sized segments along their length. Then, road segments constitute the set of nodes of a road graph where the set of edges represent neighborhood relationships between the nodes. Subsequently, each segment's characteristics such as traffic density can be calculated for a given time period.

The data traffic generated and disseminated on a VANET depends on the specific scenario and applications deployed. For example, in evaluating metrics for RSU placement for a time-critical accident warning scenario, the warning messages are disseminated to all cars within the three-hops broadcast range. Hence, data traffic plays an indirect role in determining the characteristics of the vehicular traffic network. In security game formulations, the data traffic model is implicitly taken into account when defining the vehicular network, and hence in determining the payoff matrices.

3.5.2   Attack and defense model

VANET security games abstract the interaction between malicious attackers to VANETs and various defense mechanisms protecting them. The abstract model considered makes the following broad assumptions regarding the nature of possible attacks and defensive measures as well as attackers and defenders.

1. An attack causes (temporary) damage or disruption to one or more VANET applications at a certain location.
2. The attackers have some incentive for (benefit from) causing damage to VANET applications. At the same time, they incur costs such as the risk of being captured.
3. The defenders have mechanisms that are capable of detecting attacks (attackers) and rendering them ineffective (capturing attackers) with some probability, if they allocate resources to the attack location.
4. The defense systems can be static (e.g. deployed in RSUs) or dynamic (e.g. deployed in police cars).
5. The attackers and defenders have limited information on each other's objectives.
6. Both attackers and defenders deploy randomized (mixed) strategies.

The general class of attacks satisfying the assumptions above are **location-based**. One such attack is *jamming*, which disrupts all communications in a region. These attacks can be detected early by ordinary users or defensive forces if they are present at that location. Furthermore, the attackers can be identified to some extent by their location using triangulation techniques as long as the attack continues. Another class of attacks involve *bogus messages* disseminated by the attackers for disruption of traffic or for selfish aims, e.g. sending a false accident message to clear the road. These messages are restricted to their initial neighborhood first, even if they reach a broader area with time. However, the attackers will probably move away by the time the message reaches the infrastructure. Again, deployment of defensive systems at the same location provides better capabilities for checking the correctness of the messages. In addition, mobile defenses such as police cars may quickly assess the situation and physically capture the perpetrators if necessary, something beyond the capabilities of ordinary users. A third class of relevant attacks involves *Sybil attacks* where the attackers create and operate under multiple forged identities for self-protection as well as to increase the intensity of their attacks. Checking the authenticity of these identities may be resource-wise infeasible for the ordinary vehicles nearby, due to communication overheads and limited access rights. Deploying appropriate local defensive systems can help to detect the attacks early and physically identify the attackers.

3.5.3   Game formulation and numerical analysis

A realistic **traffic dataset** is used in defining the parameters of an example security game. The data consists of traces of car movements generated by a simulator [167], which was created at ETH Zurich with maps from the Swiss geographic information system (GIS). The road map of a rural region, depicted in Figure 3.6, is quantized to an 11×11 grid. The traces offer snapshots in 1 s intervals about the identity of a car, its x- and y-coordinates on the map, and a time stamp. The traffic density in the region averaged over 500 s is shown in Figure 3.7.

The specific **VANET security game** formulation is a finite *zero-sum* game where the actions spaces of the attacker and defender consist of attacking and defending a specific road segment. Each road segment corresponds in this case to a square element of an 11×11 grid obtained by uniformly discretizing the rural region map. For convenience the discrete map grid is represented as a vector of size $121 = 11 \times 11$. Hence, converting the map matrix to a vector, the action spaces of $\mathcal{P}^A$ and $\mathcal{P}^D$ are defined as

$$\mathcal{A}^A := \{a_1, \ldots, a_{121}\}$$

and

$$\mathcal{A}^D := \{d_1, \ldots, d_{121}\},$$

respectively. As an illustrative sample instance of the defined security game, consider an attacker jamming one road segment with some attack probability. In response, the defender (e.g. designer, city planner, law enforcement) allocates defense resources to

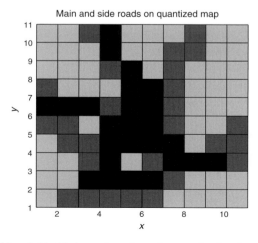

Figure 3.6 Main (black) and side (dark gray) roads on the quantized region map.

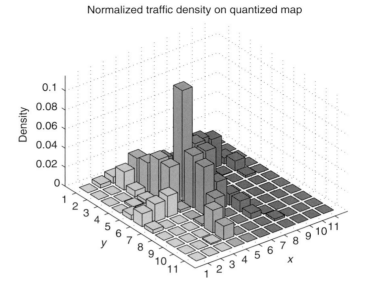

Normalized traffic density on quantized map

Figure 3.7 Vehicle density on the quantized region map.

the same or another road segment according to its own strategy. The outcome of this game instance is then determined by the corresponding entry in the game matrix.

The zero-sum game matrix maps player actions on the road segment to outcomes for $\mathcal{P}^A$ and $\mathcal{P}^D$. The game matrix entries can be a function of the importance of each road segment (as characterized by, for example, the traffic density), the risk of detection (gain from capture) for the attacker (defender), as well as other factors. As an example, the game matrix, G, of this game is defined as:

$$G = [G_{i,j}]_{121 \times 121} := \begin{cases} C(i), \text{ if } i \neq j \\ r, \text{ if } i = j, \ \forall i, j \end{cases} \tag{3.9}$$

where $C(j)$ is defined as vehicle density on a road segment j, which can be computed through averaging over time interval $[0, T]$:

$$C(j) = \frac{1}{T} \sum_{t=1}^{T} \sum_{i} \delta(i, j, t).$$

The indicator function δ is defined as:

$$\delta(i, j, t) := \begin{cases} 1, \text{ if vehicle } i \text{ is on road segment } j \text{ at time } t \\ 0, \text{ else.} \end{cases}$$

The fixed scalar r represents the risk or penalty of capture for the attacker (benefit for defender), if the defender allocates resources to the location of the attack, i.e. the same square on the map. It is set to $r = -0.1$ here, indicating a large risk for $\mathcal{P}^A$ relative to the gain C.

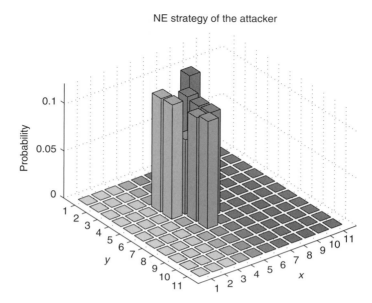

Figure 3.8 Equilibrium attack probabilities of the vehicular network security game shown on the rural region map.

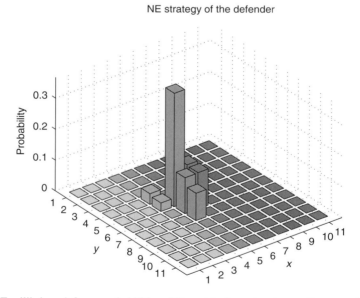

Figure 3.9 Equilibrium defense probabilities of the vehicular network security game shown on the rural region map.

The zero-sum vehicular network security game defined is solved using standard methods. There is no pure-strategy equilibrium and the game admits a unique (saddle-point) equilibrium solution in mixed strategies. The equilibrium strategies are shown on the region map in Figure 3.8 for the attacker, $\mathcal{P}^A$, and Figure 3.9 for the defender, $\mathcal{P}^D$. A big penalty for $\mathcal{P}^A$ (higher diagonal values in the game matrix) leads to diversification in attack probabilities instead of narrowly focusing on the most valuable locations. While the attacker chooses a few squares with almost equal probability, the defender probabilities on those same squares show more variability. This is a result of the penalty (or risk) of capture for the attacker being uniform, whereas the defender losses are more diverse as they are proportional to the traffic density, e.g. disruptions affecting a larger population.

3.6   Security games in wireless networks

In wireless networks, medium access control (MAC) is necessary for coordinating multiple transmissions originating from different nodes to achieve reliable and efficient communication. Various random and multiple access approaches have been developed to achieve this goal. The ALOHA protocol and code division multiple access (CDMA) are two well-known examples of random and multiple access schemes, respectively.

This section analyzes the interactions between selfish and malicious nodes at the MAC layer using security games. While the selfish nodes aim to transmit to a common receiver and maximize their throughput, the malicious nodes try to disrupt communication by simultaneous transmissions and jamming. A security game framework is presented that describes not only selfish behavior but also jamming and DoS at the MAC layer.

For a finite number of transmitter nodes, random access schemes have been extensively studied as a cooperative throughput optimization problem. An alternative is noncooperative operation, in which selfish nodes select the transmission probabilities to optimize their individual performance objectives. Similarly, cooperative and noncooperative formulations have been thoroughly studied for interference-limited systems such as CDMA, especially in the context of power management, where individual transmissions create interference for nearby nodes resulting in a closely coupled environment.

The noncooperative game formulations allow modeling of malicious behavior in addition to selfishness in MAC. Different from selfish nodes who try to maximize own throughput, the malicious nodes pursue destructive objectives such as jamming the packet transmissions of others even when it does not improve own transmissions. Such jamming and DoS attacks may have significant detrimental effects on the reliability and efficiency of the wireless channel. Consequently, malicious behavior in wireless networks has been studied with increasing intensity [153, 154, 192].

Two specific access control schemes are considered within the context of security games in wireless networks. One of them is a slotted random access channel similar to slotted ALOHA where only one node can use the channel at a given time slot. The other one is an interference-limited multi-access scheme where the nodes choose their transmission power levels in order to maximize their signal-to-interference-plus-noise-ratio (SINR) which is correlated with throughput. Additional factors such as energy are also taken into account, especially considering the often battery-limited nature of the nodes.

It is also assumed that the malicious nodes follow the random access protocols "to hide in the crowd" and avoid short and bursty transmissions to minimize their detection risk.

NE provides, as before, a sound solution concept in wireless security games. The NE solutions indicate optimal attack and defense strategies for noncooperative random access systems. Additionally, pricing schemes and other incentive mechanisms have been investigated in the literature to improve the throughput or other aggregate properties of the system. In particular, the possible performance loss has been evaluated if selfish nodes defect by becoming malicious in their transmission strategies, where the results are compared with a cooperative equilibrium outcome that optimizes the total weighted utility sum in random access [154].

The players of the security games studied in this section are either selfish or malicious. An interesting question is, what happens if non-malicious nodes or users collaborate against malicious attackers and jointly implement defensive strategies such as revocation of malicious players' certifications? An example game considering such a scenario is presented subsequently in Section 3.7.

3.6.1 Random access security games

The random access game models a wireless network with multiple nodes transmitting packets to a common receiver over a shared random access channel. Each transmitter is assumed to have saturated queues with uninterrupted availability of packets at any time slot and infinite buffer capacity for simplicity. The random access channel is a slotted system, in which each packet transmission takes one time slot. Any simultaneous transmission or collusion results in packet loss, which corresponds to a *classical collision channel*.

Let p_i denote the transmission probability of a node i from the set of $\mathcal{N}$ transmitters. A packet of node i is successfully received with probability $\prod_{j \neq i}(1 - p_j)$, if the other nodes $j \neq i$, $j \in \mathcal{N}$ do not transmit in the same time slot. The average throughput of the channel for successful transmissions is chosen to be one for simplicity and without any loss of generality. The reward for any successful packet transmission of node i is quantified by the value $r_i > 0$. Since the nodes are mobile, it is natural to assume that they are battery limited and take into account the amount of energy spent for transmissions. We introduce the parameter $e_i \geq 0$ to represent the average transmission energy cost of a node i per time slot.

The transmitting nodes in set $\mathcal{N}$ constitute at the same time the players of the random access game. The players (nodes) decide on their transmission probabilities $p = [p_1, \ldots, p_N]$, where $0 \leq p_i \leq 1 \ \forall i$ and N is the cardinality of the set $\mathcal{N}$ or the number of players. Each node is associated with a cost function, J_i, that quantifies the above-discussed positive and negative factors affecting its decision. The set of players can be divided into two nonoverlapping sets of selfish nodes $\mathcal{P}^D \subset \mathcal{N}$, who still follow the rules, and malicious nodes, $\mathcal{P}^A \subset \mathcal{N}$, such that $\mathcal{P}^D \bigcup \mathcal{P}^A = \mathcal{N}$.

A selfish node $i \in \mathcal{P}^D$ chooses the transmission probability p_i given the transmission probabilities of other nodes in order to minimize the individual cost function, J_i^D, that reflects the difference between the throughput reward and the cost of transmission energy. The cost function for selfish nodes is defined as

$$J_i^D(p) := p_i e_i - r_i p_i \prod_{j \neq i} (1 - p_j) \ \forall i \in \mathcal{P}^D \subset \mathcal{N}. \tag{3.10}$$

The malicious nodes, unlike selfish ones, are motivated by the "reward" for disrupting or jamming the transmissions of other nodes rather than their own throughput. Hence, the cost function for a malicious node j is defined as

$$J_j^A(p) := p_j e_j - c_j p_j \sum_{k \in \mathcal{P}^D} p_k \prod_{l \neq j,k} (1 - p_l) \ \forall j \in \mathcal{P}^A \subset \mathcal{N}, \tag{3.11}$$

where $c_j > 0$ is the reward for blocking random access of a selfish node. The second term is the sum of all such rewards obtained by playing a role in jamming selfish nodes. Note that if the transmission of a selfish node is already blocked by another malicious or selfish node, then the malicious node j does not receive any reward.

The player optimization problems for both selfish and malicious nodes are

$$\min_{p_i} J_i^D(p) \ \forall i \in \mathcal{P}^D, \text{ and } \min_{p_j} J_j^A(p) \ \forall j \in \mathcal{P}^A, \tag{3.12}$$

respectively.

It is illustrative to consider first a random access game with two players, $\mathcal{N} = \{1,2\}$, and analyze the problem of two selfish but not malicious noncooperative transmitters. Then, the impact of one malicious transmitter on another selfish node is evaluated. Finally, the case of an arbitrary number of selfish and malicious transmitters is discussed.

Two selfish players
When both players are selfish, the respective cost functions, as a special case of the one in (3.10) to (3.11), are

$$J_1(p) := p_1 e_1 - r_1 p_1 (1 - p_2),$$

and

$$J_2(p) := p_2 e_2 - r_2 p_2 (1 - p_1).$$

An analysis of this game shows that for some values of the parameters, there are multiple NE solutions, whereas for some other values, the NE solution is unique. This result is summarized in the following theorem.

Theorem 3.2 *The random access game with two selfish players admits the following NE strategies (p_1^*, p_2^*) under the respective conditions.*

1. If $0 < e_1 \leq r_1$ and $0 < e_2 \leq r_2$, there exist three NE:

$$p_1^* = 1, \ p_2^* = 0; \ p_1^* = 0, \ p_2^* = 1; \ p_1^* = 1 - \frac{e_2}{r_2}, \ p_2^* = 1 - \frac{e_1}{r_1}.$$

2. *If $e_2 = 0$ and $e_1 < r_1$, there exists a continuum of NE:*

$$p_1^* = 1, \ p_2^* \in \left[1 - \frac{e_1}{r_1}, 1\right],$$

and if $e_1 = 0$ and $e_2 < r_2$, there exists likewise a continuum of NE:

$$p_1^* \in \left[1 - \frac{e_2}{r_2}, 1\right], \ p_2^* = 1.$$

3. *If $e_1 > r_1$ or $e_2 > r_2$ or both, the NE is unique:*
   (a) $p_1^* = p_2^* = 0$, if $e_1 > r_1$ and $e_2 > r_2$.
   (b) $p_1^* = 0$, $p_2^* = 1$, if $e_1 > r_1$ and $e_2 < r_2$.
   (c) $p_1^* = 1$, $p_2^* = 0$, if $e_1 < r_1$ and $e_2 > r_2$.

Proof The NE solutions of the random access game are in the square $0 \leq p_1 \leq 1$, $0 \leq p_2 \leq 1$. Consider first NE strictly inside the square, so-called *inner* NE. Since both J_1 and J_2 are linear in p_1 and p_2, respectively, every inner NE (p_1^*, p_2^*) has to have the property that $J_1 (p_1, p_2^*)$ is independent of p_1 and $J_2 (p_1^*, p_2)$ is independent of p_2, which leads to $p_1^* = 1 - e_2/r_2$, $p_2^* = 1 - e_1/r_1$. This solution is inner provided that $0 < e_1 \leq r_1$, $0 < e_2 \leq r_2$. Clearly, one can also allow $e_1 = r_1$ and/or $e_2 = r_2$ in this solution which by inspection is also an NE, though not inner. Likewise for $e_1 = 0$ and $e_2 = 0$, which are covered by case (b). Now, any other NE solution candidate will have to be at the corners of the square. These cases are covered in parts (a) and (c) of the theorem, which follow directly from the definition of NE. □

According to Theorem 3.2, the NE strategies may not be unique for certain values of systems parameters. For instance, if $0 < e_1 < r_1$ and $0 < e_2 < r_2$, then there exist three NE strategies (p_1^*, p_2^*), two of them at the corners $(1,0)$, $(0,1)$ and the other one inner $(1 - e_2/r_2, 1 - e_1/r_1)$. The NE solutions are depicted in Figure 3.10.

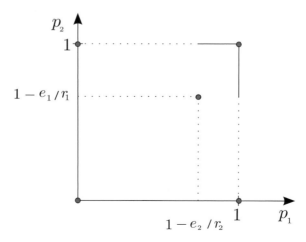

Figure 3.10 NE strategies of the random access game from Theorem 3.2.

It is worth noting that for the classical collision channels, the envelope of the throughput rate under NE strategies coincides with the boundary of the achievable throughput region, if $\sum_{i=1}^{2} e_i/r_i = 1$.

One selfish player and one malicious player

Consider now the case when the second player is a malicious attacker. Then, the respective cost functions of the players, as special cases of those in (3.10) and (3.11), are

$$J_1(p) := p_1 e_1 - r_1 p_1 (1 - p_2),$$

and

$$J_2(p) := p_2 e_2 - c p_2 p_1.$$

This game again admits multiple and even a continuum of NE solutions, depending on the parameter values.

Theorem 3.3 *The random access game with one selfish player 1 and a malicious player 2 admits the following NE strategies (p_1^*, p_2^*) under the respective conditions.*

(i) $p_1^* = \dfrac{e_2}{c}$, $p_2^* = 1 - \dfrac{e_1}{r_1}$, if $0 < e_1 \leq r_1$ and $0 < e_2 \leq c$.

(ii) $p_1^* = p_2^* = 0$, if $e_1 > r_1$.

(iii) $p_1^* = 0$, $p_2^* \in \left[1 - \dfrac{e_1}{r_1}, 1\right]$, if $e_1 \leq r_1$ and $e_2 = 0$.

(iv) $p_1^* = 1$, $p_2^* = 0$, if $e_1 < r_1$ and $e_2 > c$.

(v) $p_1^* \in \left[\dfrac{e_2}{c}, 1\right]$, $p_2^* = 1$, if $e_1 = 0$ and $e_2 \leq c$.

Proof The proof is similar to that of Theorem 3.2. Note that the solution in (i) here is the inner NE.   □

The noncooperative NE strategies may not be unique, depending on the reward and cost parameters, as observed also for the case of two selfish transmitters. The NE solutions are depicted in Figure 3.11.

It is interesting to note that the *inner* NE strategies p_2^* of player 2 are the same, regardless of whether node 2 is selfish or malicious. In other words, malicious operation of node 2 only changes the inner equilibrium strategy of the selfish node 1 to be attacked. If node 2 is selfish, the throughput rate λ_1^s of node 1 is $p_1^*(1 - p_2^*)$, where p_1^* and p_2^* are given in Theorem 3.2, respectively, such that

$$\lambda_1^s = \left(1 - \frac{e_2}{r_2}\right) \frac{e_1}{r_1}. \tag{3.13}$$

On the other hand, if node 2 is malicious, the throughput rate λ_1^m of node 1 is changed to

$$\lambda_1^m = \frac{e_1 e_2}{c r_1}, \tag{3.14}$$

where the respective values of p_1^* and p_2^* are taken from Theorem 3.3.

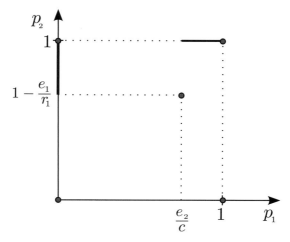

Figure 3.11 NE strategies of the random access game from Theorem 3.3.

Player 2 can reduce the throughput rate of node 1 by switching to malicious operation, i.e. the DoS attack of node 2 is successful if, and only if,

$$\lambda_1^m < \lambda_1^s \Rightarrow c > \frac{e_2}{1 - e_2/r_2}. \tag{3.15}$$

Consequently, the reward for jamming transmissions of node 1 should be high enough for node 2 to compensate the increase in energy cost and the decrease in individual throughput rate.

It is interesting to note that the selfish behavior of benign players brings a certain degree of robustness to the overall system. Unlike the collaborative case discussed in Section 3.4, the effect of malicious nodes is much more limited, partly because there is less to destroy in a selfish environment than in a collaborative one. As the basic analysis above shows, under some conditions the malicious behavior fails even to distinguish its effect from the selfish one. This phenomenon can be interpreted as a boundary between selfish and malicious behavior in the broad spectrum that starts with collaboration and ends at malicious attacks.

Multiple selfish and malicious players

The results of Theorems 3.2 and 3.3 are extended to the cases of N selfish nodes and N selfish plus one malicious node, respectively, for the inner strategy NE.

Consider N selfish transmitter nodes (players) with the properties described in the game model of Section 3.6.1. The following theorem extends the results of Theorem 3.2 to N selfish players for inner NE strategies.

Theorem 3.4 *The random access game with N selfish players admits the following inner NE strategies*

$$p_i^* = 1 - \frac{r_i}{e_i} \left(\prod_{j=1}^N \frac{e_j}{r_j} \right)^{\frac{1}{N-1}}, \; if\, 0 < e_i < r_i, \; i = 1, \ldots, N.$$

The corresponding NE throughput rate of node i is

$$\lambda_i^* = \frac{e_i}{r_i} - \left(\prod_{j=1}^{N} \frac{e_j}{r_j} \right)^{\frac{1}{N-1}}. \tag{3.16}$$

Proof The counterpart of the argument used in the proof of Theorem 3.2 for inner NE yields the set of equations

$$\prod_{j\neq i} \left(1 - p_j^*\right) = \frac{e_i}{r_i}, \quad i = 1,\dots,N. \tag{3.17}$$

Simple algebraic manipulations lead to

$$\left(1 - p_i^*\right) \frac{e_i}{r_i} = \left(1 - p_j^*\right) \frac{e_j}{r_j} \ \forall i,j,$$

which indicates a solution of the form

$$p_i^* = 1 - \frac{r_i}{e_i} \alpha$$

for some α. Substituting it back into (3.17) yields the inner NE transmission probabilities.

The inner NE throughput rate of a node i is

$$\lambda_i^* = p_i^* \prod_{j\neq i} \left(1 - p_j^*\right) = p_i^* \frac{e_i}{r_i},$$

which follows directly from (3.17). Substituting the term p_i^* with the above obtained inner NE transmission probability completes the proof. □

Consider now an additional malicious node $\mathcal{A}$ with the objective of blocking the random access of the other N selfish nodes. For the classical collision channel studied here, only one transmission is sufficient to block the random access of all other selfish nodes, i.e. independent transmissions of multiple malicious nodes increase the energy costs without extra benefits in terms of reducing the throughput rate of target nodes. If there are multiple malicious nodes that can cooperate with each other, they should form a coalition such that only one node transmits at a time to minimize the energy costs. Therefore, the single malicious node case provides insights into more general scenarios.

The counterparts of the cost functions (3.10) and (3.11) are respectively

$$J_i^D(p) := p_i e_i - r_i p_i (1 - p_m) \prod_{j\neq i} (1 - p_j), \quad i = 1,\dots,N \tag{3.18}$$

and

$$J^A(p) := p_m e_m - c p_m \sum_{j=1}^{N} p_j \prod_{k\neq j} (1 - p_k), \tag{3.19}$$

where p_m is the transmission probability and e_m the energy cost of the malicious node.

The NE transmission strategies can be characterized using a similar analysis to before. However, the set of equations is more complex, which hinders the derivation of simple analytical expressions for equilibrium transmission probabilities and throughput rates. The inner NE strategies satisfy

$$p_i^* = 1 - \frac{r_i}{e_i} \left(\prod_{j=1}^{N} \frac{e_j}{r_j} \right)^{\frac{1}{N-1}} \left(\frac{1}{1 - p_m^*} \right)^{\frac{1}{N-1}}, \quad i = 1, \ldots, N$$

and

$$p_m^* = 1 - \frac{r_m}{e_m} \left[\sum_{j=1}^{N} \frac{e_j}{r_j} - \left(\prod_{j=1}^{N} \frac{e_j}{r_j} \right)^{\frac{1}{N-1}} \right].$$

When the selfish nodes are symmetric in terms of their energy costs and rewards $e_i = e_s$ and $r_i = r_s$ $\forall i$, the relationship between the equilibrium transmission probabilities of selfish, p_s^*, and malicious players, p_m^*, simplify to

$$p_s^* = \frac{1}{N} \frac{e_m \, r_s}{e_s \, c} (1 - p_m^*).$$

This equation has multiple intuitive interpretations. The equilibrium solution of selfish players, p_s^*, is

- proportional to the energy cost of malicious players and throughput reward of selfish players;
- inversely proportional to the equilibrium transmission probability of the malicious node;
- inversely proportional to the number of selfish nodes, energy cost of selfish players, and the reward of the malicious node.

Similar observations can also be made for the equilibrium solution of the malicious player, p_m^*. All of these relationships are in line with intuitive expectations stemming from the game context.

3.6.2 Interference limited multiple access security games

The previous section discussed security games played over random access channels, specifically, a slotted ALOHA model where only one node can use the channel at a given time slot. This section presents multiple access security games within an interference-limited multi-access scheme where the nodes choose their transmission power levels in order to maximize their SINR that is correlated with their throughput.

Following a treatment similar to those in Section 3.6.1, consider a security game between transmitters of two possible (selfish or malicious) types. Define $P_i \geq 0$ and $e_i \geq 0$ as the transmission power level and the corresponding energy cost (per unit power) of a node i, respectively. Each node independently chooses the power P_i for transmitting to a common receiver in order to minimize the individual expected cost J_i. The SINR value achievable by a selfish node i is

$$\gamma_i = \frac{h_i P_i}{\frac{1}{L} \sum_{j \neq i} h_j P_j + \sigma^2}, \qquad (3.20)$$

where $h_{(.)}$ is the respective channel gain, L is the processing gain, and σ^2 is the background noise. The throughput reward for a selfish node i is $f_i(\gamma_i)$, which is an increasing function of the SINR value γ_i. For simplicity, consider a reward function linear in SINR such that:

$$f_i(\gamma_i) = \gamma_i. \tag{3.21}$$

An alternative formulation, where $f_i(\gamma_i) = \log(1 + \gamma_i)$, is discussed in reference [153].

The performance objectives of the players incorporate *throughput* rewards based on achieved SINR levels, transmission *energy* costs, and malicious interests such as rewards obtained from *jamming*. The nodes also incur transmission energy costs proportional to their power level, eP. Then, the cost function minimized by a selfish node is

$$J_i(P) = e_i P_i - \gamma_i = e_i P_i - \frac{h_i P_i}{\frac{1}{L}\sum_{j\neq i} h_j P_j + \sigma^2}, \tag{3.22}$$

where $P = [P_1, \ldots, P_N]$ is the transmission power vector of all nodes. Such games among selfish players only have been studied extensively in the literature [15, 59, 157].

Unlike selfish ones, malicious nodes receive rewards for jamming others (in terms of decreasing their performance or SINR) rather than improving their own SINR levels. Thus, the cost function of a malicious node is defined as

$$J_i(P) = e_i P_i + \sum_{j\in\mathcal{P}^D} \gamma_j, \tag{3.23}$$

where $\mathcal{P}^D$ is the set of selfish nodes that the malicious one targets. As before, the set of malicious nodes is denoted by $\mathcal{P}^A$ and $\mathcal{P}^D \bigcup \mathcal{P}^A = \mathcal{N}$, which is the set of all nodes. It is natural to assume here that malicious nodes do not have any incentive of interfering with or degrading each other's transmissions. Notice additionally that without energy costs, a game between one malicious node and one selfish node will be a zero-sum game of throughput balancing.

It is illustrative to consider, as in Section 3.6.1, first a multiple access game between two selfish players, $\mathcal{N} = \{1, 2\}$, and subsequently analyze the game between one malicious and one selfish transmitter.

Two selfish transmitters

Theorem 3.5 *The unique NE strategies (transmission power levels) for two selfish transmitters on an interference-limited multiple access channel are*

$$P_i^* = \frac{L}{h_i}\left(\frac{h_j}{e_j} - \sigma^2\right), \ j \neq i, \ if \ h_i \geq \sigma^2 e_i, \ i = 1, 2,$$

$$P_i^* = 0, \ if \ h_i < \sigma^2 e_i, \ i = 1, 2, \tag{3.24}$$

$$P_i^* = 0, P_j^* \to \infty, \ if \ h_i < \sigma^2 e_i, \ h_j > \sigma^2 e_j, \ j \neq i.$$

Proof The individual constrained optimization problem for each transmitter $i = 1, 2$ is $\min_{P_i \geq 0} J_i(P)$. Define the Lagrangian

$$L_i(P_1, P_2) = J_i(P_1, P_2) - \lambda_i P_i, \quad i = 1, 2, \tag{3.25}$$

where $\lambda_i \geq 0$ is a Lagrange multiplier corresponding to the inequality constraint in the individual optimization problem. The corresponding Karush–Kuhn–Tucker (KKT) conditions are then

$$\frac{\partial L_i(P_1, P_2)}{\partial P_i} = 0, \ P_i \geq 0, \ \lambda_i \geq 0, \ \lambda_i P_i = 0, \ i = 1, 2. \tag{3.26}$$

These necessary conditions are also sufficient for optimality, since the cost function $J_i(P_1, P_2)$ and inequality constraint $P_i \geq 0$ are continuously differentiable and convex functions of P_i. The equilibrium strategies (3.24) follow from applying the KKT conditions (3.26) separately to each cost function $J_i(P_1, P_2)$, $i = 1, 2$, with constraint $P_i \geq 0$, where the costs of selfish nodes are given by (3.22). □

The performance measure for the transmitting selfish nodes is the achievable SINR value. The SINR values for the two selfish nodes $i = 1, 2$ at the NE are

$$\gamma_i = \begin{cases} e_i P_i^*, & \text{if } h_j > \sigma^2 e_j, \ j \neq i, \\[2mm] \dfrac{h_i P_i^*}{\sigma^2}, & \text{otherwise.} \end{cases} \tag{3.27}$$

One malicious transmitter and one selfish transmitter

Theorem 3.6 *The unique NE strategies (transmission power levels) for a selfish transmitter 1 and a malicious transmitter 2 on an interference-limited multiple access channel are*

$$P_1^* = \frac{L}{h_2} \frac{e_2 h_1}{(e_1)^2}, \ P_2^* = \frac{L}{h_2} \left(\frac{h_1}{e_1} - \sigma^2 \right), \ \text{if } h_1 \geq \sigma^2 e_1, \tag{3.28}$$

$$P_1^* = 0, \ P_2^* = 0, \ \text{if } h_1 < \sigma^2 e_1.$$

Proof The equilibrium strategies (3.28) follow from applying the KKT conditions (3.26) separately to each objective function $J_i(P_1, P_2)$, $i = 1, 2$, with constraint $P_i \geq 0$, where the costs $J_1(P_1, P_2)$ and $J_2(P_1, P_2)$ are given by (3.22) and (3.23), respectively. □

The equilibrium SINR level of selfish node 1 is given by (3.27) with P_1 from (3.28). The malicious attack of node 2 is more successful in reducing the SINR of selfish node 1 compared to the alternative selfish behavior of node 2 only under the following conditions. Under the assumption of $h_i \geq \sigma^2 e_i$, $i = 1, 2$, ensuring nonzero transmission power levels, the malicious node is effective if, and only if,

$$\frac{h_2}{e_2} > \sigma^2 + \frac{e_2}{h_2} \left(\frac{h_1}{e_1} \right)^2,$$

i.e. if h_1 is sufficiently small and e_1 large. Otherwise, the malicious attack fails compared to the selfish operation, which is sometimes referred to as the "windfall of malice." A similar situation has also been observed in Section 3.6.1. If both transmitters are malicious, they do not receive any reward from interfering with each other and the NE strategies are $P_1 = 0$ and $P_2 = 0$.

In the **general case of multiple selfish and malicious nodes**, the NE solutions are generally difficult to characterize explicitly with analytical expressions. In such cases, first the existence (and possibly uniqueness) of equilibrium solutions needs to be established. Then, the solutions can be computed numerically or through iterative algorithms.

3.7   Revocation games

In settings where there is no immediately available central defense system, e.g. in ad-hoc, vehicular, or decentralized social networks, it makes sense for the benign (defending) players to collaborate against malicious ones and jointly implement defensive strategies such as revocation of malicious players' rights and certifications. This section presents revocation games, which formalize a revocation approach to handling misbehavior, and are a variant of security games.

The revocation approach facilitates "local" and fast response to misbehavior by allowing benign players, who follow the rules, collaboratively to denounce and punish their misbehaving "neighbors" even in rapidly changing environments. The neighborhood relationships and location may be defined through physical characteristics (e.g. wireless range) of the system such as the case in ad-hoc networks as well as through different means, e.g. in online social networks. Revocation constitutes an alternative to reputation or trust-based schemes which require long-term monitoring and keeping of state information as well as more stationary settings. On the other hand, both approaches are distributed in nature and rely on localized algorithms that do not require participation of all nodes at all times.

Revocation schemes aim to remove attackers from the system. During a revocation decision, the amount of interaction between nodes and state information of a given attacker are limited. A typical application of revocation involves fast and local annulment of credentials, such as a compromised key issued by a central but "distant" certificate authority (CA). This kind of local revocation can be particularly useful when the central authority is not immediately available to revoke the misbehaving node (e.g. when the CA is offline) or when it cannot detect the misbehavior in time, especially when the latter does not involve a key compromise (e.g. sending bogus information).

The underlying system conforms to the following basic **assumptions** regarding the nature of the malicious attackers and the detection capabilities of defending nodes. The malicious nodes have the same communication capabilities and credentials as the benign ones. The attacks aim to disrupt the network by, for example, disseminating false information for fun or personal gain. It is assumed that the benign (defending) nodes can detect such attacks by virtue of possessing special detection capabilities. Such nodes can then act as initiators and participants of revocation games. As a simplification, the detection system of the participating nodes is assumed to function perfectly and misbehavior by the malicious nodes is always correctly classified. In other words, all participants correctly identify the malicious node, and hence the revocation game is more about whether the defending nodes will converge to the correct decision of revoking the accused node. Each information piece (e.g. message), even false ones, needs to be associated with a verifiable identity, which constitutes the basis for the revocation system. Information without verifiable identity is simply ignored by the nodes. In the system considered, *Sybil attacks*, where the attacker impersonates another node, are difficult to implement and do not pose a significant issue. This assumption is justified, for example, in a public key infrastructure with an offline certificate validation process.

3.7.1 Revocation game model

A revocation game is played among the nodes in a neighborhood relationship assuming a certain stationary period of time. Let $\mathcal{N}$ be the set of all nodes in the neighborhood, $\mathcal{P}^A \subset \mathcal{N}$ the malicious, and $\mathcal{P}^D \subset \mathcal{N}$ the defending ones. Each revocation game is then played among the n defending players (nodes) $\mathcal{P}^D$ against a single malicious one $M \in \mathcal{P}^A$.

Each player $i \in \mathcal{P}^D$ has three possible actions, $\mathcal{A}_i = \{A, V, S\}$, in the revocation game against the malicious node $M \in \mathcal{P}^A$. First, the player can **abstain**, A, from the local revocation procedure. This action means that the player is not willing to contribute to the local revocation procedure and instead expects other players or eventually the CA to revoke the attacker. Second, the player i can participate by casting a **vote**, V, against a detected attacker [55]. It is assumed, as a simplification, that a majority vote is required to revoke an attacker in the game. Finally, following the protocol suggested in reference [117], the player can commit **self-sacrifice**, denoted by the action S, and invalidate both its current identity (the pseudonym it currently uses) and the identity of the attacker. This action essentially finishes the game by revoking the attacker immediately but usually comes at a high cost to the defender. A revocation game is illustrated in Figure 3.12.

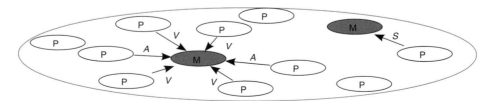

Figure 3.12 Mobile nodes (P) may collaboratively revoke the credentials of nearby malicious ones (M) through voting or by resorting to self-sacrifice.

The costs associated with the actions defined above for a defending node i can be defined as

$$J_i(\mathcal{A}_i, k) = \begin{cases} (1-k)c, & \text{if } \mathcal{A}_i = A, \\ v + (1-k)c - kb, & \text{if } \mathcal{A}_i = V, \\ c_s - B, & \text{if } \mathcal{A}_i = S, \end{cases} \qquad (3.29)$$

where the parameter k takes the value one if the revocation is successful after the game and zero otherwise. In addition, the scalar parameters v and b denote the cost and benefit of voting, respectively, c the cost of attack, c_s the cost of self-sacrifice, and B the benefit of self sacrifice. All parameters are assumed to be non-negative. The cost structure (3.29) is chosen to be broad with multiple parameters to capture various scenarios. Notice that the parameter k provides the coupling in the revocation game. Without k, the game would have been a trivial one.

Based on the timing of the game play, a sequential and concurrent (static) variant of the defined revocation game will be studied next. Subsequently, various extensions, open issues, and future research directions will be discussed.

3.7.2   Sequential revocation games

In sequential revocation games each player makes a decision sequentially (one-by-one) while observing the current state of the game. Only a single round of the game is considered as the underlying system may not be stationary over long time intervals. The order of play is assumed to be random. For each player i, the quantity n_i denotes the number of remaining nodes who have not yet played, and n_r the number of remaining votes required to revoke the attacker by voting. Both of these state that information is available to the players at their turn of play.

To analyze this sequential game, which can also be expressed in an extensive form, one can use the concept of the NE. However, the NE concept without any refinement is somewhat loose when it is applied to extensive-form games because, first, it is generally nonunique, and second, some of the NE could predict outcomes that are not *credible* for some players, i.e. these outcomes are unreachable because the players will not play, out of self-interest, according to the computed NE path. Hence, the stronger concept of *subgame-perfect equilibrium* is a more suitable solution concept for sequential revocation games. The strategy s is a subgame-perfect equilibrium of a finite extensive-form game, if it is an NE of any subgame defined by the appropriate subtree of the original game. The game studied here has – by definition – a finite number of stages.

The existence of subgame-perfect equilibria can be checked by applying the *one-deviation property*. This property requires that there exists no single stage in the game in which a player i can gain by deviating from her subgame-perfect equilibrium strategy while conforming to it at other stages. Hence, at a subgame-perfect equilibrium of a finite extensive-form game, the one-deviation property holds.

Consider a simplified variant of the cost structure (3.29) in the sequential revocation game. Specifically, let $B = b = 0$, $c_s = 1$, and $v < 1$. Then, the subgame-perfect equilibrium strategies of a player i are characterized in the following theorem.

Theorem 3.7 *In the sequential revocation game, for any given values of n_i, n_r, $v < 1$, and c, the strategy of player i that results in a subgame-perfect equilibrium is:*

$$\mathcal{A}_i^* = \begin{cases} A, & if\ [c < v]\ or\ [(c > 1)\ and\ (n_i \geq 1)]\ or\ [(v < c < 1)\ and\ (n_i \geq n_r)] \\ V, & if\ [(v < c < 1)\ and\ (n_i = n_r - 1)] \\ S, & if\ [(c > 1)\ and\ (n_i = 0)] \end{cases}$$

Proof The existence of subgame-perfect equilibria in the revocation game can be obtained by applying the technique of *backward induction*, which is also called "Zermelo's algorithm" in dynamic programming. Backward induction works by eliminating suboptimal actions, i.e. yielding higher costs than the other actions in the same subtree and at the same stage of the game tree, beginning at the leaves of the extensive-form game tree. The obtained path (sequence of actions) in the game tree defines the backward induction solution and any strategy profile that realizes this solution is a subgame-perfect equilibrium [67].

Here, the one-stage-deviation principle is used to prove that deviating from each of the strategies in the theorem under the corresponding conditions will not result in a gain.

First, assume that $c < v$, i.e. voting is more expensive than enduring the attack-induced cost. If at any stage, player i deviates from the strategy A, then playing V or S would result in a cost of v or 1, respectively. In both cases, the cost is higher than c, since $v < 1$.

If $v < c < 1$ and $n_i \geq n_r$, i.e. voting is less expensive than the attack-induced cost and the number of remaining players is higher than the required number of voters, then playing S or V would result in a cost of 1 or v, respectively. These costs are greater than 0 and the attacker will be revoked anyway, since $v < c$. Hence, the player i cannot gain by deviating from the action A.

Another case that makes action A the best-response is when the attack-induced cost c is bigger than 1, the cost of self-sacrifice, and the number of remaining players is higher than 1 ($n_i \geq 1$), i.e. the attacker will be revoked by another player anyway. The proof is similar to the previous cases.

Next, assume that $v < c < 1$, $n_i = n_r - 1$, and player i that is supposed to play V according to strategy $\mathcal{A}_i^*$ above, deviates in a single stage. If it plays S or A, it loses one or c, respectively, both bigger than v. In both cases, the player cannot gain by deviating from V.

Finally, if $c > 1$, $n_i = 0$, and the player deviates from S by playing A or V, then the player's cost will be c or $c + v$, respectively. Both costs are greater than one and deviation results in a loss for the player.

Thus, a deviation from any action $\mathcal{A}_i^*$ under the corresponding conditions results in a loss for the deviating player i which completes the proof that the strategy $\mathcal{A}_i^*$ leads to a subgame-perfect equilibrium. □

Theorem 3.7 essentially states that, since the attacker cost is fixed, the only objective of the players is to remove the attacker, regardless of the actual game stage of this happening. Thus, the revocation decision is left to the last players, either by voting or by self-sacrifice, whichever induces less cost. For example, a node plays S only if the attack-induced cost is higher than the cost of self-sacrifice and if it is the last player in the sequential game.

The solution in Theorem 3.7 is not robust to changes in the system. For example, if some of the last players move out of the system before their turn to play, then a revocation decision cannot be reached. To overcome this limitation, a variable cost version of the revocation game is defined next.

Define a sequential game with variable costs where $c_j = j\alpha$, where $1 \leq j < n$ is the stage of the game, and $\alpha > 0$ is the stage cost of attack. Let the attack cost at the final stage of the revocation game grow infinitely, $c_n = \infty$, if the attacker is not revoked. Furthermore, assume that $v < \alpha$. The subgame-perfect equilibrium strategies in the defined variable cost sequential game are characterized in the following theorem.

Theorem 3.8 *In the sequential revocation game with variable costs, for any given values of n_i, n_r, $v < \alpha$, and α, the strategy of player i that results in a subgame-perfect equilibrium is:*

$$
\mathcal{A}_i^* = \begin{cases}
A, & \text{if } \left[\left(1 \leq n_i < \min\left\{n_r - 1, \frac{1}{\alpha}\right\}\right) \text{ and } (v + (n_r - 1)\alpha < 1)\right] \\
& \quad \text{or } \left[\left(1 \leq n_i < \frac{1}{\alpha}\right) \text{ and } (v + (n_r - 1)\alpha > 1)\right] \\
V, & \text{if } \left[(n_i \geq n_r - 1) \text{ and } (v + (n_r - 1)\alpha < 1)\right] \\
S, & \text{otherwise}
\end{cases}
$$

Proof The proof follows an argument similar to that of Theorem 3.7. If there are not enough voters but at least another player can self-sacrifice, $1 \leq n_i < \min\left\{n_r - 1, \frac{1}{\alpha}\right\}$ and self-sacrifice is more expensive than voting, $v + (n_r - 1)\alpha < 1$, then deviating from playing A has a cost of one. This is due to the fact that playing S is the only possible option in this case as the number of voters is insufficient. Hence, the player cannot gain by a deviation from action A.

If $1 \leq n_i < \frac{1}{\alpha}$ and $v + (n_r - 1)\alpha > 1$, deviating from playing A will cause player i a cost of one if it plays S and a cost of $v + (n_r - 1)\alpha$ if it plays V. Hence, the player again does not gain by deviating from A.

If there are enough voters, $n_i \geq n_r - 1$, and voting is less expensive than self-sacrifice, $v + (n_r - 1)\alpha < 1$, then deviating from playing V by choosing A or S costs the player $n_i\alpha$ or one, respectively. In both cases, the cost will be greater than $v + (n_r - 1)\alpha$, assuming that v is negligible. Hence, the player does not gain by one-stage deviation.

The explicit condition for playing S is:

$$\alpha > 1 \text{ or}$$
$$(n_i < n_r - 1) \text{ and } \left((n_i = 0) \text{ or } \left(n_i \geq \frac{1}{\alpha}\right)\right) \text{ and } (v + (n_r - 1)\alpha < 1) \text{ or}$$
$$(n_i = 0) \vee \left(n_i \geq \frac{1}{\alpha}\right)) \wedge (\alpha < 1 < v + (n_r - 1)\alpha).$$

If the attack-induced cost is more expensive than self-sacrifice, $\alpha > 1$, and player i deviates from strategy S, it can play V or A. If it plays V, it loses $v + (n_r - 1)\alpha > 1$ and if it plays A, it loses $n_i\alpha > 1$.

If the current player is the last one, $n_i = 0$, or $n_i \geq \frac{1}{\alpha}$ (alternately $n_i\alpha \geq 1$, which means that the cost of abstaining is higher than the cost of self-sacrifice), and voting is cheaper than self-sacrifice but there are not enough voters, $v + (n_r - 1)\alpha < 1$ and $n_i < n_r - 1$, deviating from S by playing A would result in a cost of $n_i\alpha$ if $n_i \geq \frac{1}{\alpha}$ or ∞ if $n_i = 0$ as the attacker will not be revoked. Playing V would not lead to revocation (as there are not enough voters) and hence result in a cost of $v + n_i\alpha$ if $n_i \geq \frac{1}{\alpha}$ or ∞ if $n_i = 0$. Hence, deviation from S would not pay off. A similar argument can be made if $n_i = 0$ or $n_i \geq \frac{1}{\alpha}$, and $\alpha < 1 < v + (n_r - 1)\alpha$: both voting and abstaining are more expensive than self-sacrifice, and hence deviation from S would increase the cost.

Based on the arguments above, one-stage-deviation from the subgame-perfect equilibrium strategy $\mathcal{A}_i^*$ degrades the deviating player's payoff, which completes the proof. $\qquad\square$

An interpretation of the result of Theorem 3.8 is that, in contrast to the game with fixed costs, the players are more concerned about quickly revoking the attacker as their costs increase with time. Hence, under appropriate conditions, they will begin the revocation process by voting or self-sacrifice in the early stages of the game rather than waiting until the last opportunity.

3.7.3 Static revocation games

An alternative to the sequential revocation game is a static version where all nodes take actions simultaneously. One of the main features of such a static revocation game is that the players do not need to know others' decisions before taking their own, which obviously shortens the revocation game duration significantly and decreases the need for keeping state information during the process. Both of these are issues that are important, especially in ephemeral networks where stationary time windows are usually quite short.

Consider an n-player static revocation game with the action space $\mathcal{A} = \{A, V, S\}$, cost structure as in (3.29), and game parameters as defined before. Let, in addition, $n_v < n$ be the number of votes required to revoke a malicious player. The NE solutions of this game are characterized in the following theorem.

Theorem 3.9 *The static revocation game admits at least one NE such that the malicious node is revoked under one of the following conditions:*

1. $B \geq c_s$
2. $B < c_s$ and $b < v$ and **either**
    a. $B - c_s > b - v > -c$ or
    b. $b - v > B - c_s > -c$.

Proof The proof can be structured into four parts all of which follow directly from the definition of the NE.

First, if $B = c_s$ and $b > v$, in other words the payoff for voting is strictly greater than for self-sacrifice, then all players voting $\mathcal{A}_i^* = V\ \forall i$ is the unique NE strategy as all players are better off voting. At this unique equilibrium, the malicious node is revoked for $n > n_v$ by definition.

Second, if $B = c_s$ and $b < v$, i.e. a strictly lower payoff for voting than self-sacrifice suggests that the latter strategy is always preferred for revocation by at least one participating player. In this case, the NE solutions belong to the set of all combinations with at least one self-sacrifice and all others abstain. Any of the NE revokes the malicious node.

Third, if $B < c_s$, $b < v$, and $B - c_s > b - v > -c$, then the NE belongs to the set of all combinations with exactly one self-sacrifice and $n - 1$ abstentions. Even though both payoffs are negative, if self-sacrifice is still better than voting in terms of costs, and the revocation is performed by one player, then it is in the best interests of all the other players to abstain. Again, any of the NE revokes the malicious node. Notice that the self-sacrificing node has no incentive to deviate from this equilibrium, for if it also abstains, then its cost will be higher than self-sacrifice due to the malicious node not revoked in that case.

Finally, if $B < c_s$, $b < v$, and $b - v > B - c_s > -c$, then the NE are all strategies that have either one self-sacrifice with $n - 1$ abstentions as before or n_v votes with $n - n_v$ abstentions. In any of these NE solutions, the malicious node is clearly revoked. This follows again directly from the definition of NE as in the previous cases. Notice that, for the case of voting, the revocation is performed by the strict minimum number of voters, n_v. $\square$

One issue with the results in Theorem 3.9 is that there are multiple NE solutions under many different conditions. If the game is played in one shot such that the players act concurrently, then making a choice between the feasible NE is difficult for the players. This problem can be circumvented by adopting dynamic voting schemes [142] or using other characteristics of the players such that the symmetry between equilibrium solutions is broken and an ordering among them is obtained. While doing this, it is preferable to keep the information exchange between players at a minimum and avoid introducing global state variables as in trust-based systems.

3.8   Discussion and further reading

Although there are many books on game theory, two specific ones, references [31] and [136], provide substantial background to follow all game-theoretic models and analysis in this book. Other relevant books on game theory include references [95, 135]; see Appendix A.2 for additional references. The recent reference [132] provides an overview of game theory more from a computer science perspective, and discusses its applications to a wide variety of subjects. A brief introduction to noncooperative game theory is in Appendix A.2.

Although this book focuses on the NE as the main solution concept, partly due to its simplicity, there are ongoing efforts to find alternatives to address issues arising especially when there are multiple equilibria. Correlated equilibrium [135] by Aumann and regret minimization [72] constitute two representative examples among many.

An extensive overview of intrusion detection systems can be found in reference [26]. Earlier articles [5, 6] complement and extend the material in Section 3.2. There are multiple works worth mentioning on security games but these could not be included due to space limitations. The thesis [56] presents a different perspective to security games. Additional formulations have been proposed in references [70, 71], and [73]. An application of intrusion detection games to sensor networks is analyzed in reference [1]. The impact of malicious players on system efficiency in the context of distributed computing is investigated in reference [119], where security games are played over graphs.

Incentive mechanisms and selfish behavior in social (e.g. peer-to-peer) and ad-hoc networks have been analyzed extensively in the literature using game-theoretic models, e.g. references [50, 64, 174], which provide the basis for the brief summary in Section 3.4 and contain a more detailed treatment. Other recent works on the subject include references [108, 109] which present a graph-theoretic formulation.

An introductory discussion on the security of vehicular networks is in reference [84]. A game-theoretic approach has also been presented in reference [50]. The articles [14, 46], upon which Section 3.5 is based, extend the discussion on the subject.

Jamming in wireless networks was first introduced in reference [192] and has subsequently been analyzed extensively in the literature [50]. Among these are a dynamic game formulation of jamming [102], correlated jamming on MIMO Gaussian fading channels [86, 105], and key establishment using frequency hopping [168]. The material in Section 3.6 is a revised version of the treatment in reference [153, 154] by Sagduyu *et al.*, where a more extended discussion of the subject can be found.

Revocation games in Section 3.7 are based on the articles [40, 141, 142], which extend the models and results presented. They have introduced a set of protocols implementing the revocation games analyzed, and discussed dynamic voting schemes for cost and social welfare optimization. A related study on the subject is reference [117].

4   Stochastic security games

Chapter overview

1. Markov security games
   - games played on a probabilistic state space
   - solving Markov games using an extension of MDPs
2. Stochastic intrusion detection game
   - random emergence of system vulnerabilities
3. Security of interconnected systems
   - the effect of individual units on others
   - states modeling attack steps and compromises
4. Malware filter placement game
   - algorithms under dynamic routing
   - simulations and illustrative example

Chapter summary

Stochastic or Markov security games extend the deterministic security game framework of Chapter 3 through the utilization of probability theory to model the unknown and uncontrollable parameters in security problems. Although they are mathematically more complex, due to their mathematical sophistication Markov games enable a study of the interaction between attackers and defenders in a more realistic way. Moreover, the dynamic nature of the underlying game parameters, interdependencies, and external factors are captured within the stochastic framework.

4.1   Markov security games

Probabilistic approaches have been used extensively to model the faults, errors, and failures of networked systems in the context of reliability and dependability. However, a study of security incidents differs from traditional reliability analysis. Unlike unplanned (random) faults and errors in the system, security compromises are caused by malicious attackers with specific goals. An increasing number of studies have recently focused on quantifying operational security and relating security assessment to the reliability domain [155] through probabilistic models as well as adopted game-theoretic approaches.

Stochastic or Markov[1] **security games** depart from the deterministic security games of Chapter 3 in several aspects. The stochastic (Markov) models considered here capture the complexities and unknown properties of the underlying networked system better than deterministic counterparts. Hence, a more realistic depiction of attacker versus system interaction is obtained. Another major difference is the dynamic nature of the current model. The static security games of Chapter 3 are played repeatedly over time in a myopic manner. On the other hand, the players here optimize their strategies, taking into account future (discounted) costs, which allows them to refine their own strategies over a time horizon. Markov security games rely on Markov decision processes (MDPs) as a theoretical foundation for the development and analysis of player strategies.

Unlike their deterministic counterparts, stochastic games are played between the attacker, $\mathcal{P}^A$, and the defender, $\mathcal{P}^D$, on a **state space** representing the environment of the game. A *state* may be an operational mode of the networked system such as which units are operational, active countermeasures, or whether parts of the system are compromised.

In the adopted stochastic model, the states evolve probabilistically according to a defined stochastic process with the Markov property. A Markov property holds naturally in many systems and provides a nice simplification for others. The resulting stochastic process is parameterized by player actions enabling the effect of player decisions to be modeled on the networked system properties. For example, the probability of detection of a specific attack is a function of attacker behavior, e.g. intensity of attack or whether the system is targeted, as well as the amount of monitoring resources allocated by the defender.

In addition to providing the defense system guidelines for countermeasures and resource allocation, stochastic security games aim to analyze the **behavior of rational attackers**. Within the framework of stochastic security games, attacker behavior is represented as a probability distribution over possible attacks (actions) in each state. Attacker strategies can be derived under various assumptions and for different scenarios resulting in NE and other solutions.

At the same time, the stochastic state space model of the network provides a basis for analyzing its security properties. Based on the attacker strategies, the transition

---

[1] The two terms, Markov and stochastic, will be used interchangeably in this book when qualifying the described class of games.

probabilities between states can be computed for various cases. Hence, a network security state diagram is obtained that explicitly incorporates attacker behavior.

Markov games, which have been studied extensively by the research community in recent years, and their variants constitute the basis of stochastic security games. The players in security games are direct adversaries, which can be modeled as zero-sum games in most cases. Hence, zero-sum Markov games are the main focus of this chapter. Multi-agent MDPs, such as dynamic programming methods derived from MDPs, are utilized for solving these types of game.

Nonzero-sum Markov games, on the other hand, pose a serious challenge in terms of convergence of solutions due to the nonuniqueness of NE at each stage. Various methods have been suggested to overcome these convergence problems, such as generalizations of NE to correlated equilibria, choosing asymmetric learning rates for players to prevent synchronization, or cooperation schemes. Despite these efforts, there is still no unified theory that is easily applicable to solve nonzero-sum stochastic security games.

When limitations are imposed on information available to players in stochastic security games, they can adopt various learning schemes. Such games are said to be of **limited information** due to the fact that each player observes the other players' moves and the evolution of the underlying system only partially or indirectly. The players refine their own strategies online in such cases by continuously learning more about the system and their adversaries. Consequently, they base their decisions on limited observations by using, for example, *Q-learning* methods. Such limited information security games will be discussed in Chapter 5.

4.1.1   Markov game model

As a basis for stochastic security games, consider a two-player ($\mathcal{P}^A$ versus $\mathcal{P}^D$) zero-sum **Markov game** played on a finite state space, where each player has a finite number of actions to choose from. As in Chapter 3, the action space of attacker, $\mathcal{P}^A$, is defined as $\mathcal{A}^A := \{a_1, \ldots, a_{N_A}\}$ and constitutes the various possible attack types. Similarly, the action set of defensive measures for $\mathcal{P}^D$ is $\mathcal{A}^D := \{d_1, \ldots, d_{N_D}\}$. The environment of the networked system is captured by a finite number of environment states, $S = \{s_1, s_2, \ldots, s_{N_S}\}$.

It is assumed that the **states** evolve according to a discrete-time finite-state Markov chain which enables the utilization of well-established analytical tools to study the problem. Then, the state transitions parameterized by player actions are determined by the mapping

$$\mathcal{M} : S \times \mathcal{A}^A \times \mathcal{A}^D \to S. \tag{4.1}$$

Let $p^S := [p_1, \ldots, p_{N_p}]$ be a probability distribution on the state space S, where $0 \le p_i^S \le 1 \ \forall i$ and $\sum_i p_i^S = 1$. Then, in the discrete and finite case considered here, the mapping $\mathcal{M}$ can be represented by an $N_S \times N_S$ *transition (or Markov) matrix* $M(a,d) = [M_{i,j}(a,d)]_{N_S \times N_S}$, parameterized by $a \in \mathcal{A}^A, d \in \mathcal{A}^D$ such that

$$p^S(t+1) = M(a,d)\, p^S(t), \tag{4.2}$$

where $t \geq 1$ denotes the stage of the repeated stochastic security game.

The mapping $\mathcal{M}$ in (4.1) can be alternatively parameterized by the states to obtain as many zero-sum game matrices $G(s)$ as the number of states, $s \in S$, each of dimension $N_A \times N_D$. In other words, given a state $s(t) \in S$ at a stage t, the players play the zero-sum game

$$G(s(t)) = [G_{a,d}(s(t))]_{N_A \times N_D}. \tag{4.3}$$

For example, if $\mathcal{P}^A$ takes action a_4 and $\mathcal{P}^D$ the action d_2 when in state s_3, then the outcome of the game is $G_{4,2}(3)$ (gain of $\mathcal{P}^A$ and loss of $\mathcal{P}^D$).

The definition in (4.2) represents the most general case for the types of Markov game considered. The **state transition matrix** M does not have to depend on all actions of the players. For example, in cases when the actions of $\mathcal{P}^D$ do not have an effect on the state evolution, the transition matrix does not have to be parameterized by them resulting in $M(a)$, $a \in \mathcal{A}^A$. Likewise, if neither player has an effect on state transitions, then (4.2) simply becomes $p^S(t+1) = M p^S(t)$. On the other hand, as long as the Markov game is not trivial, the zero-sum game matrices for each state, $G(s)$, remain as before.

The **strategies of the players** are state dependent and are extensions of those in Chapter 3. The strategy of $\mathcal{P}^A$,

$$p^A(s) := \left[p_1^A(s), \ldots, p_{N_A}^A(s) \right],$$

is defined as a probability distribution on the attack (action) set $\mathcal{A}^A$ for a given state $s \in S$ and the one of $\mathcal{P}^D$ is

$$p^D(s) := \left[p_1^D(s), \ldots, p_{N_D}^D(s) \right],$$

a probability distribution on the defense (action) set $\mathcal{A}^D$, such that $0 \leq p_i^A, p_i^D \leq 1 \; \forall i$ and $\sum_i p_i^A = \sum_i p_i^D = 1$. The *mixed* strategies, unlike pure ones, ensure that there exists a saddle-point equilibrium at each stage of the matrix game, $G(s)$ by Corollary 2.3 in reference [31, p. 28], which also states that the saddle-point value in mixed strategies is unique.

4.1.2 Solving Markov games

There is a close relationship between the solution methods of Markov games and MDPs. If one of the players in a stochastic security game adopts a fixed strategy, then the game degenerates to an optimization problem for the other player and the Markov game turns into an MDP. Consequently, the methods for solving MDPs, such as value iteration, are directly applicable with slight modifications to Markov games.

For the zero-sum Markov game formulation in Section 4.1.1, the defending player $\mathcal{P}^D$ aims to minimize own *aggregate cost*, $\bar{Q}$,[2] while facing the attacker $\mathcal{P}^A$ who tries to maximize it. The reverse is true for the player $\mathcal{P}^A$ due to the zero-sum nature of

---

[2] The superscript D is dropped to simplify the notation.

the game. Hence, it is sufficient to describe the solution algorithm for only one player. To avoid duplication, the rest of the analysis focuses on the defensive player, $\mathcal{P}^D$. The game is played in discrete-time (stages) over an infinite time horizon. As in MDP, the aggregate cost of the defending player $\mathcal{P}^D$ at the end of a game is the sum of all realized stage costs discounted by the scalar factor $\alpha \in [0,1)$ and given by

$$\bar{Q} := \sum_{t=1}^{\infty} \alpha^t G_{a(t),d(t)}(s(t)), \quad a(t) \in \mathcal{A}^A, d(t) \in \mathcal{A}^D, s(t) \in \mathcal{S}, \tag{4.4}$$

where $G_{a(t),d(t)}(s(t))$ is an entry in the stage game matrix $G(s(t))$ defined in (4.3).

The player can theoretically choose a different strategy $p^A(s(t))$ at each stage t of the game to minimize the final realized cost $\bar{Q}$ in (4.4). Fortunately, this complex problem can be simplified significantly. First, it can be shown that a stationary strategy $p^A(s) = p^A(s(t)) \forall t$ is optimal, and hence there is no need to compute a separate optimal strategy for each stage. Second, the problem can be solved recursively using dynamic programming (DP) to obtain the stationary optimal strategy (solving a zero-sum matrix game at each stage). Notice that, unlike MDP, the optimal strategy can be mixed, i.e. stochastic for each state s. A basic overview of DP in discrete- and continuous-time is provided in Appendix A.3.

At a given stage t, the Q-value or optimal t-stage cost, $Q_t(a,d,s)$, can be computed iteratively using DP recursion

$$Q_{t+1}(a,d,s) = G_{a,d}(s) + \alpha \sum_{s' \in \mathcal{S}} M_{s,s'}(a,d)$$
$$\times \min_{p^D(s')} \max_a \sum_{d \in \mathcal{A}^D} Q_t(a,d,s') p_d^D(s'), \tag{4.5}$$

for $t = 0, 1, \ldots$, and a given initial condition Q_0. Here, the Q values are defined over not only states as in MDP but also player actions.

This recursion can be equivalently written in functional form

$$T_t(Q)(a,d,s),$$

by defining the mapping $\mathcal{T} : \mathcal{A}^A \times \mathcal{A}^D \times \mathcal{S} \to \mathcal{A}^A \times \mathcal{A}^D \times \mathcal{S}$. The mapping $\mathcal{T}$ can be interpreted as the optimal cost function for the one-stage problem that has stage cost $G(s)$. Furthermore, the iteration (DP algorithm) provably converges to the optimal cost function

$$Q^*(a,d,s) := \lim_{t \to \infty} T_t(Q)(a,d,s), \quad \forall a,d,s. \tag{4.6}$$

It is possible to write the counterpart of Bellman's equation for the Markov game by splitting (4.5) into two parts:

$$V(s) = \min_{p^D(s)} \max_a \sum_{d \in \mathcal{A}^D} Q_t(a,d,s) p_d^D(s) \tag{4.7}$$

and

$$Q_t(a,d,s) = G_{a,d}(s) + \alpha \sum_{s' \in \mathcal{S}} M_{s,s'}(a,d) V(s'), \quad t = 1, \ldots \tag{4.8}$$

The strategy $p^D(s) \, \forall s$ obtained by solving (4.7) is the *minimax* strategy with respect to Q. It can be computed by $\mathcal{P}^D$ for any state s by solving the linear program

$$\min_{p^D(s)} V(s)$$

$$\text{s.t.} \quad \sum_{d \in \mathcal{A}^D} Q_t(a,d,s) p_d^D(s) \geq V(s), \, \forall a \in \mathcal{A}^A, \tag{4.9}$$

$$\sum_d p_d^D = 1, \, p_d^D \geq 0, \, \forall d \in \mathcal{A}^D.$$

This linear program above is a generalization of the min operator for MDP.

The fixed points of equations (4.7) and (4.8), V^* and Q^*, lead to the optimal minimax solution for the defender. One would also obtain the corresponding mixed strategy of the attacker (in the saddle-point equilibrium) by interchanging the positions of min and max in (4.7), this time the maximization being over $p^A(s)$ and minimization over d. This does not change the values of V^* and Q^*, since the game has a saddle-point in mixed strategies.

The Bellman equation to obtain Q^* (for either players) is $Q^* = T(Q^*)$. There are multiple ways to solve the Bellman equation. One of them is the algorithm described by (4.7) and (4.8) together with (4.9), as captured in Algorithm 4.1 below, which is also known as *value iteration* or *successive approximation*. An alternative method is policy iteration. However, value iteration is often preferred over its alternatives due to its scalability.

Algorithm 4.1 Value iteration algorithm

1: Given arbitrary $Q_0(a,d,s)$ and $V(s)$
2: **repeat**
3:    **for** $a \in \mathcal{A}^A$ and $d \in \mathcal{A}^D$ **do**
4:       Update V and Q according to (4.7) and (4.8)
5:    **end for**
6: **until** $V(s) \to V^*$, i.e. $V(s)$ converges.

An alternative stochastic game formulation

An alternative formulation of zero-sum stochastic games is given in reference [136, Chap. V.3]. There, a stochastic game is said to consist of "game elements" which are the counterparts of zero-sum game matrices $G(s)$ for each state $s \in \mathcal{S}$ defined in Section 4.1.1. The term

$$q_{ij}(a,d), \quad a \in \mathcal{A}^A, d \in \mathcal{A}^D, i,j \in \mathcal{S},$$

denotes the probability of having to play the j-th game element (state) when currently in element i under the given actions of the players. Hence, these "state transition" probabilities approximately correspond to the entries of the Markov transition matrix $M(a,d)$ in (4.2).

This zero-sum stochastic game model of reference [136] differs from that in Section 4.1.1 in one important aspect. It introduces a nonzero termination probability to the stochastic game such that the probability of infinite play is zero and all expected costs are finite even without a discount factor. In other words, there exists a nonzero transition probability $q_{i,0}(a,d) > 0$ to a state 0 at which the game terminates regardless of the actions of the players. Consequently, even if the discount parameter is chosen to be one, $\alpha = 1$, the aggregate cost remains finite.

It is possible to state the counterpart of value iteration equations (4.7) and (4.8) for this stochastic game as

$$V(s) = \min_{p^D(s)} \max_a \sum_{d \in \mathcal{A}^D} Q_t(a,d,s) p_d^D(s) \qquad (4.10)$$

and

$$Q_t(a,d,s) = G_{a,d}(s) + \sum_{s' \in S} q_{s,s'}(a,d) V(s'), \quad t = 1,\dots \qquad (4.11)$$

Again, the value iteration algorithm asymptotically converges to the unique fixed points V^* and Q^*, which constitute the optimal minimax solution for the player, due to

$$\sum_{j \in S} q_{ij}(a,d) < 1 \quad \forall i,a,d.$$

4.2   Stochastic intrusion detection game

The stochastic security game framework described in the previous section opens the door to more comprehensive and realistic intrusion detection games than that investigated in Section 3.2. As an illustrative example, a stochastic intrusion detection game is defined and studied, where vulnerabilities of individual systems monitored on a network are modeled using a controlled Markov chain.

A single system is considered first, where the state space consists of two states

$$S = \{v, nv\},$$

representing vulnerability to attacks and being not vulnerable at a given time instance. The state transition probabilities of the Markov chain capture the discovery or exploitation probability of vulnerabilities affecting the system as well as the probability of the system being upgraded or patched such that the vulnerability is removed. These probabilities naturally depend on the actions of the defender. If the defender allocates additional resources to the system for maintenance, then the probability of new vulnerabilities emerging decreases while one of their removal increases. In contrast, a neglected system is more vulnerable to attacks and remains so with high probability.

The stochastic intrusion detection game is defined as a zero-sum Markov game between the attacker, $\mathcal{P}^A$, and the defender $\mathcal{P}^D$. As in the game in Section 3.2, the action set of $\mathcal{P}^A$ is $\mathcal{A}^A = \{a, na\}$, where a corresponds to launching an attack on the system and na denotes no attack. The corresponding action set of $\mathcal{P}^D$ is

$$\mathcal{A}^D = \{d, nd\},$$

where d is intensified monitoring and maintenance, and nd denotes default defensive level. Given the state space $S = \{v, nv\}$, the probability of being in a specific state is given by the vector

$$p^S := \left[p_v^S, p_{nv}^S \right], \quad 0 \le p_v^S, p_{nv}^S \le 1 \ \forall i \text{ and } p_v^S + p_{nv}^S = 1.$$

Then, the state probabilities evolve according to

$$p^S(t+1) = M(\cdot)p^S(t),$$

where $t \ge 1$ denotes the stage of the game. The state transition or Markov matrices conditioned on the actions of $\mathcal{P}^D$, n, and nd, are denoted by $M(n)$ and $M(nd)$, respectively. Similarly, the game matrices conditioned on the states are $G(nv)$ and $G(v)$.

In addition to the stochastic model of the underlying environment (in this case the system vulnerability), the stochastic intrusion detection game differs from the previous one in Section 3.2 with its dynamic nature. The players here optimize their strategies not only with respect to instantaneous costs but also taking into account future discounted costs given the expected evolution of the game. The Markov game model (Section 4.1.1) and its solution (Section 4.1.2) provide the necessary theoretical foundation for the analysis and computation of player strategies.

Numerical example

The game introduced above is now analyzed numerically for a set of example parameters. Let

$$M(d) = \begin{bmatrix} 0.9 & 0.8 \\ 0.1 & 0.2 \end{bmatrix}, M(nd) = \begin{bmatrix} 0.1 & 0.2 \\ 0.9 & 0.8 \end{bmatrix}. \tag{4.12}$$

The diagrams (a) and (b) in Figure 4.1 depict the state transition matrices $M(d)$ and $M(nd)$, respectively.

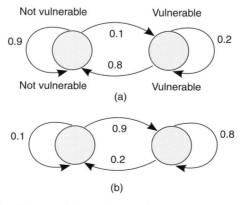

Figure 4.1 State transition diagram of the stochastic intrusion detection game for (a) the defensive action, $M(d)$, and (b) default level, $M(nd)$.

The game matrices are chosen as

$$G(nv) = \begin{bmatrix} -5 & 10 \\ 1 & 0 \end{bmatrix}, \; G(v) = \begin{bmatrix} -3 & 11 \\ 2 & 0 \end{bmatrix}. \qquad (4.13)$$

Notice that the values in $G(v)$ indicate a higher gain and loss for $\mathcal{P}^A$ and $\mathcal{P}^D$, respectively, when the system is vulnerable, when compared to those in $G(nv)$. Furthermore, they are consistent with the parametric static game version in (3.1).

The game defined is solved using the Algorithm 4.1, where α is chosen as 0.3 to represent heavy discounting of future costs. Just after three iterations, the Q values converge to their final values with an accuracy of 0.01,

$$Q(nv) = \begin{bmatrix} -4.3 & 10.1 \\ 1.7 & 0.1 \end{bmatrix}, \; Q(v) = \begin{bmatrix} -2.9 & 11.7 \\ 2.1 & 0.7 \end{bmatrix}. \qquad (4.14)$$

The corresponding V values (4.7) are $[1.1, 1.7]$. The resulting equilibrium strategies of $\mathcal{P}^A$ and $\mathcal{P}^D$ in the stochastic intrusion detection game are depicted in Figure 4.2. The strategies obtained by solving the two static games (4.13) separately are shown in Figure 4.3 for comparison.

The stochastic intrusion detection game discussed illustrates Markov security games and relevant solution methods through a simple and easy-to-understand example. Real-world intrusion detection systems monitor and defend large-scale networks with many interconnected components. Stochastic security games can easily model such systems within the presented mathematical framework as will be illustrated in the next section. The state space can be extended to capture not only multiple systems on the network

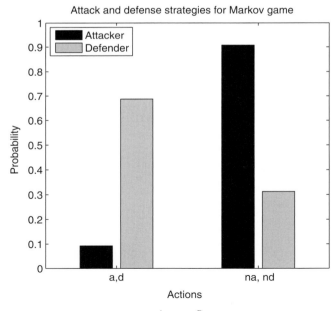

Figure 4.2 Equilibrium strategies of players $\mathcal{P}^A$ and $\mathcal{P}^D$ in the stochastic intrusion detection game.

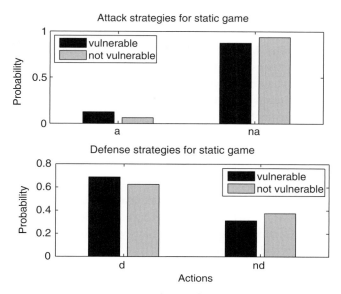

Figure 4.3 Equilibrium strategies of players $\mathcal{P}^A$ and $\mathcal{P}^D$ in the zero-sum intrusion detection games (4.13).

but also their interdependencies. In addition, the effects of defender and attacker actions can be incorporated into state transition probabilities. On the other hand, when using stochastic security games, the well-known state explosion problem should be watched out for. This can often be addressed using hierarchical decomposition and other state reduction techniques. Furthermore, the value iteration algorithm for solving the Markov security games involves mainly linear programming, which is very scalable.

4.3 Security of interconnected systems

Interdependencies among individual systems play a significant role in network security. Stochastic (Markov) models, which have been successfully used in dependency analysis, can be extended and applied to the network security domain. Malicious attacks, security compromises, and defensive actions along with (random) fault and recovery processes form a single stochastic state-based model. Thus, the stochastic security game framework presented in the previous section provides a basis for investigation of attacker and defender interactions on a networked system with interconnected components.

Each **state** in the stochastic model may represent an operational mode of the network based on subsystem characteristics, whether specific nodes are compromised or vulnerable, defensive processes, attack stages, etc. The number and definition of states is an important part of security modeling and involves a tradeoff between representativeness and complexity. Models with extreme fine granularity may become not only computationally complex but also problematic during parameterization.

The **transition probabilities** between states capture not only intrinsic random events of the system such as errors and failures but also attacks and defensive actions. Hence, the controlled Markov model considered encompasses attack graphs, where each state transition corresponds to a single atomic step of a penetration, along with the impact of attacks and defense measures on the system. This facilitates the modeling of unknown attacks in terms of generic state transitions.

The **parameters** underlying the transition probabilities constitute one of the most important components of the model. There are well-established procedures for obtaining accidental failure and repair rates as part of traditional dependability analysis. However, quantifying the effects of attacker and defender actions can be challenging. Possible solutions to address this problem include collecting the opinions of security experts, using empirical data such as historical attack data obtained from honeypots, or results of controlled experiments. In any case, parameter uncertainty is hard to circumvent. Therefore, conducting a sensitivity analysis as in Section 3.3 is crucial for reliability of the results and in order to test their robustness with respect to parameter deviations.

4.3.1   Analysis of an illustrative example

Application of the stochastic security game framework of Section 4.1 to an interconnected system is best demonstrated through analysis of a simple example network. Consider a portion of a corporate network with three important elements: the web server hosting various web-based applications (system 1), the file server hosting the corporate database (system 2), and the administrator accounts (system 3), as depicted in Figure 4.4.

The system states

$$S = \{000, 000^*, 001, 010, \ldots, 111\}$$

are defined based on whether any element is compromised, 1, or not, 0. The states are enumerated in the order given such that 000 is state 1, 000^* is state 2, and 001, 010, 100, 110, 111 correspond to states $3, 4, 5, 6, 7$, respectively. In this model, the state 010 means the web server (system 2), and state 011 means that both the web server and administrator account (systems 2 and 4) are compromised. The state 000^* represents the state when no system is compromised but the attacker has gained valuable information about the

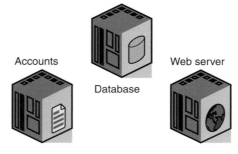

Figure 4.4 Visual illustration of a corporate network.

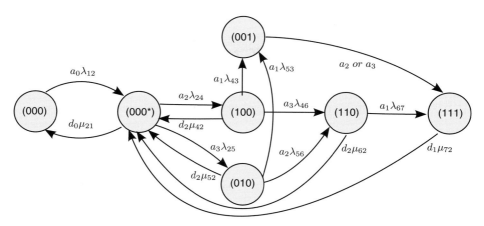

Figure 4.5 Finite state diagram of the example network with system states $\mathcal{S}$.

system through scanning. The attacker $\mathcal{P}^A$ has the actions $\mathcal{A}^A = \{a_0, a_1, a_2, a_3\}$, which represent scanning the network a_0, compromising an administrator account a_1, the web server a_2, and file server a_3, respectively. Similarly, the defender $\mathcal{P}^D$ has the options $\mathcal{A}^D = \{d_0, d_1, d_2\}$ denoting patching the systems, d_0, resetting administrator passwords, d_1, and checking both file and web servers, d_2, respectively.

Figure 4.5 shows the finite state diagram of the example network. Here, the transition probabilities λ's indicate the success probability of an individual attack. The parameters μ's denote the system recovery probabilities, e.g. due to routine maintenance or specific defensive actions. The probability of the system staying in a given state is defined as γ. Then, the Markov matrices similar to those in (4.12) are computed between the states in the diagram for each given action–reaction pair of $\mathcal{P}^A$ and $\mathcal{P}^D$. For example, the transition probabilities between states 000 and 000* given $\mathcal{P}^A$ and $\mathcal{P}^D$ actions, a_0 and d_0, are:

$$p^S(000 \rightarrow 000^*, a_0) = \frac{\lambda_{12}}{\lambda_{12} + \gamma_1},$$

$$p^S(000^* \rightarrow 000, a_0, d_0) = \frac{\mu_{21}}{\lambda_{12} + \mu_{21} + \gamma_2}.$$

The Markov model illustrated is quite versatile and provides a basis for different types of analysis in addition to the equilibrium one of the associated stochastic game. They are summarized in Table 4.1. The game matrices of the stochastic security game (game analysis in the table) are defined for each state, seven in total for this example, in a way similar to those in (4.13). Subsequently, the game can be solved using methods discussed in Section 4.1.2, for example, using the value iteration algorithm. The worst-case analysis entails setting all attack probabilities to one and optimizing the defender strategy using an MDP. In the best case, $\mathcal{P}^A$ adopts the equilibrium strategy whereas the defense uses all the measures available at the same time without worrying about the cost of these actions. The dependency analysis, on the other hand, considers only accidental failures in the system that are independent of attacker actions. Finally, notice that the

Table 4.1 Security analysis types

Analysis	Description
Worst case	Network under attack on all fronts, i.e. all possible attacks simultaneously
Best case	Full defensive measures against rational attackers playing equilibrium strategy
Game analysis	Both $\mathcal{P}^A$ and $\mathcal{P}^D$ play the equilibrium strategies from stochastic game
Dependency analysis	Only accidental failures are considered without any malicious attacks

Markov model also captures multi-stage attacks with the simple two-stage example of "scan first and then attack" depicted here.

4.3.2   Linear influence models

This section presents two applications of a linear interdependency model based on the concept of linear influence networks [114]; one for the relationship between security assets and one for vulnerabilities. Both of these terms are used in a broad sense such that security assets may refer to system nodes or business assets in the security context, whereas vulnerabilities may also mean threats and potential attacks. Each influence network is represented by a separate weighted directed graph.

Independent versus effective value of security assets

The term *security asset* refers to a node in a complex and interconnected system that plays a security related role. The network of assets is modeled as a weighted directed graph $G_s = \{\mathcal{N}, \mathcal{E}_s\}$ where $\mathcal{N}$ is the set of nodes (assets) with cardinality n, and the set of edges $\mathcal{E}_s$ represents the influence among the nodes. The weight of each edge $e_{ij} \in \mathcal{E}_s$ is denoted by a scalar w_{ij} that signifies the influence of node i on node j, $i, j \in \mathcal{N}$. The resulting *influence matrix* is then defined as

$$W_{ij} := \begin{cases} w_{ij} & \text{if } e_{ij} \in \mathcal{E}_s \\ 0 & \text{otherwise,} \end{cases} \tag{4.15}$$

where $0 < w_{ij} \le 1 \; \forall i, j \in \mathcal{N}$ and $\sum_{i=1}^{n} w_{ij} = 1$, $\forall j \in \mathcal{N}$. The entry $w_{jj} = 1 - \sum_{i=1, i \ne j}^{n} w_{ij}$ is the self-influence of a node on itself.

Let the vector

$$x = [x_1, x_2, \dots, x_n]$$

quantify the value of *independent security assets*, i.e. nodes in $\mathcal{N}$. The vector

$$y = [y_1, y_2, \dots, y_n]$$

representing the *effective security assets* takes into account the influence of nodes on each other's value. Both are related to each other by the one-to-one linear influence mapping

$$y = W x. \tag{4.16}$$

Since W is a stochastic matrix, i.e. $\sum_{i=1}^{n} w_{ij} = 1, \forall j \in \mathcal{N}$, the aggregate value of all the effective security assets is equal to the sum of the value of all independent security assets, which follows from

$$\sum_{i=1}^{n} y_i = \sum_{i=1}^{n} \sum_{j=1}^{n} w_{ij} x_j = \sum_{j=1}^{n} \sum_{i=1}^{n} w_{ij} x_j = \sum_{j=1}^{n} x_j \sum_{i=1}^{n} w_{ij} = \sum_{j=1}^{n} x_j. \tag{4.17}$$

The influence matrix W thus signifies the redistribution of security asset values. The value of an independent security asset or node i is redistributed to all nodes in the network that have an influence on i, including itself.

If a node is compromised, then the node itself and all the edges connected to it will be removed from the graph. Hence, the security loss of the network will be the node's effective security asset value instead of its independent one. Conversely, if a node is secured, it regains its original influence on other nodes. In either case, the entries of the influence matrix W are normalized to keep it stochastic.

Notice that this linear model of independent and effective security assets can be easily applied to a variety of security scenarios including the illustrative example in the previous section.

Influence model for vulnerabilities

The linear influence network model is now utilized to capture the interdependencies between security vulnerabilities in a network, where the vulnerabilities of a security asset (node) influence other nodes. For example, in a corporate network, if a workstation is compromised, the data stored in this computer can be exploited in attacks against other workstations; these latter computers will thus become more vulnerable to intrusion. Such an interdependency can be captured using the linear influence network model.

Let the nodes $\mathcal{N}$ of the weighted directed graph $\mathcal{G}_v = \{\mathcal{N}, \mathcal{E}_v\}$ represent the security assets as before. The edges $\mathcal{E}_v$ now denote the amount of influence between the nodes in terms of vulnerabilities. Consequently, the *vulnerability matrix* is defined as

$$V := \begin{cases} v_{ij}, & \text{if } e_{ij} \in \mathcal{E}_v \\ 1, & \text{if } e_{ii} \\ 0, & \text{otherwise,} \end{cases} \tag{4.18}$$

where $0 \leq v_{ij} \leq 1$ quantifies the vulnerability of node i due to node j as a result of interdependencies in the system. The self-influence is defined to be one, $v_{ii} = 1, \forall i$, which can be interpreted as the default level of vulnerability of a node independent of others. Define in addition the aggregate influence on node i from all other nodes as

$$v_i = \sum_{j=1}^{n} v_{ij}.$$

Note that $v_i \geq 1$ as $v_{ii} = 1 \; \forall i$. Unlike the model for security assets in the previous section, the influence matrix here is not normalized, and hence not stochastic.

In this vulnerability influence model, the more connected a node is, the more vulnerabilities it has due to outside influences. This is very much in line with the discussion in Chapter 2 and the argument "the most secure computer is the one totally disconnected from the network."

Define next the probability that node i is compromised as

$$c_i(a,d,s) := \max\left\{c_i^b(a,d,s)\, v_i, 1\right\},$$

where $0 \leq c_i^b(a,d,s) \leq 1$ is the baseline or standalone probability of compromise of node i. Here, in accordance with the model in Section 4.1.1, $s \in S$ is the given system state, $a \in \mathcal{A}^A$ the current attack, and $d \in \mathcal{A}^D$ the defensive action taken.

Numerical example
As an illustrative example, we now analyze numerically a stochastic security game for the network with three nodes in Section 4.3.1 (Figure 4.4) using the alternative solution method in Section 4.1.2. Similar to the example in Section 4.3.1, the system states are defined as

$$S = \{000, 001, 010, \ldots, 111\}.$$

The states indicate whether any node is compromised one or not zero, and are enumerated in the order given such that 000 is state 1, 001 is state 2, and 010, 100, 011, 101, 110, 111 correspond to states $3, 4, 5, 6, 7$, respectively.

The influence matrix quantifying the redistribution of security asset values among the nodes is chosen to be

$$W := \begin{pmatrix} 0.8 & 0.2 & 0 \\ 0 & 0.6 & 0 \\ 0.2 & 0.2 & 1 \end{pmatrix}. \tag{4.19}$$

The independent security asset values of the individual nodes are $x := [8, 8, 15]$. The vulnerability matrix modeling the interdependencies between the nodes is

$$V = \begin{pmatrix} 1 & 0.1 & 0 \\ 0.4 & 1 & 0 \\ 0.1 & 0.3 & 1 \end{pmatrix}.$$

The parameters in the example are chosen as follows. The probability that an independent defended node j gets compromised is $p_d^{(j)} = 0.2$ whereas this probability increases to $p_n^{(j)} = 0.4$ if not defended. The probability that the system goes to state 000 is $p_r^{(1)} = 0.7$ while other transition probabilities are chosen to be $p_r^{(\cdot)} = 0.2$. Finally, the probability that the game (attacks) ends is $p_e = 0.3$.

For example, suppose the system is at S_1 $(0,0,0)$. The next state could be one in $\{000, 001, 010, 100\}$ depending on the Attacker and Defender actions. The Attacker's pure strategies are $\mathcal{A}^A = \{a_1, a_2, a_3\}$, and $\emptyset$, which mean to attack node 1, node 2, node 3, or do nothing, respectively. Similarly, the Defender's pure strategies are $\mathcal{A}^D = \{d_1, d_2, d_3\}$, and $\emptyset$.

If the Attacker attacks node 1 and the Defender defends it, then the possible next states are $\{s_1 = 000, s_4 = 100\}$ and

$$G_{11}(1) = c_1(1,1,1)\,y(1)$$
$$q_{11}(1,1) = (1 - c_1(1,1,1))(1 - p_e),$$
$$q_{14}(1,1) = c_1(1,1,1),$$
$$q_{1j}(1,1) = 0 \;\forall j \neq 1,4,$$

where $c_1(1,1,1) = p_d^{(1)} \sum_k V_{1,k}$ denotes the probability that node 1 is compromised at state 000 and has *full* defensive support, as defined in the previous section. Here, the actual probability that the game ends at this state is $p_e^{(1)} = (1 - c_1(1,1,1))\,p_e > 0$.

On the other hand, if the Attacker attacks node 1 and the Defender defends node 2, we have

$$G_{12}(1) = c_1(1,2,1)\,y(1)$$
$$q_{11}(1,2) = (1 - c_1(1,2,1))(1 - p_e),$$
$$q_{14}(1,2) = c_1(1,2,1),$$
$$q_{1j}(1,2) = 0 \;\forall j \neq 1,4,$$

where $c_1(1,2,1) = p_n^{(1)} \sum_k V_{1,k}$ is the probability that node 1 is compromised at state 000 and has *no* defensive support.

Next, suppose that the system is at $s_4 = 100$, i.e. node 1 is compromised. The next state could be one of $\{000, 100, 101, 110\}$. The Attacker's pure strategies are a_2, a_3, and $\emptyset$, which mean to attack node 2, node 3, or do nothing, respectively. Similarly, the Defender's pure strategies include d_2, d_3, and $\emptyset$. If the Attacker attacks node 2 and the Defender defends it, then

$$G_{22}(4) = c_2(2,2,4)\,y(4)$$
$$q_{47}(2,2) = c_2(2,2,4),$$
$$q_{41}(2,2) = (1 - c_2(2,2,4))\,p_r^{(4)},$$
$$q_{44}(2,2) = (1 - c_2(2,2,4))\left(1 - p_r^{(4)} - p_e^{(4)}\right),$$
$$q_{4j}(2,2) = 0 \;\forall j \neq 1,4,7,$$

where $c_2(2,2,4) = p_d^{(4)} \sum_k V_{2,k}$ is the probability that node 2 is compromised at state $s_4 = 100$ while defended.

The other entries can be calculated in a similar way. Then, the resulting stochastic security game can be solved using the alternative method in Section 4.1.2. The computed optimal strategies of the Attacker and the Defender are depicted in Figures 4.6(a) and 4.6(b), respectively. These strategies can be interpreted as a guideline for the players to allocate their resources in the security game.

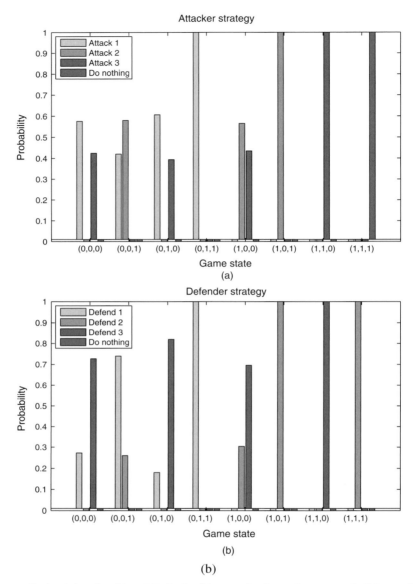

Figure 4.6 Optimal Attacker (a) and Defender (b) strategies for each state $s \in \mathcal{S}$ of the stochastic security game computed using the alternative method in Section 4.1.2.

4.4   Malware filter placement game

The malware detector and filter placement, which will be referred to here as the *filter placement problem*, investigates detector and filter placement algorithms within given hardware and network constraints. The specific objective here is to study optimal placement and activation schemes taking into account the actions of malicious attackers based on the stochastic game framework of Section 4.1.1.

Malware filters are network security measures that are implemented on or next to the network elements, instead of hosts, in order to enforce certain network-wide security policies. Hence, the deployment of detection capabilities are analyzed as an integral part of the network itself. Malware filtering is closely related to passive network monitoring, where packets are sampled with the intention of finding out more about flows in the network as well as identifying malicious packets. While monitors are only about observing, the filters react to security problems directly by dropping (some of the) potentially malicious packets. Both the monitors and filters often run on dedicated hardware in order to handle high traffic volumes. The amount of traffic on the link monitored also affects the type of pattern recognition algorithms that can be deployed. While only simple algorithms can be run on backbone gigabit links, more sophisticated stateful inspection can be used on lesser utilized edge links.

Malware filters cannot be fully deployed or activated on all links all the time due to restrictions on capacity, delay, and energy. The last limitation is especially relevant in the case of ad-hoc networks where the nodes often have limited energy. The identification of the strategic points for deploying active network monitors can be computationally complex due to the presence of multiple tradeoffs involved. The filter placement problem also has more general implications than security-related ones due to its importance within the context of network management. It is studied using standard optimization techniques in Chapter 7 through multiple formulations and utilizing betweenness centrality algorithms originating from social network analysis, while omitting the effect of malicious attackers' actions.

The filter placement problem is investigated in this section adopting a game-theoretic approach, more specifically the stochastic game framework of Section 4.1.1. In the worst-case scenario, the attackers attempting to gain unauthorized access to a target system residing in the network or to compromise its accessibility through distributed denial of service (DDoS) attacks are expected to have complete knowledge of the internal configuration of the network such as routing states or detector locations. Thus, the attackers should be seen as rational and intelligent players who respond to defensive actions by choosing different targets or routes to inject the malware. Due to this presumed adaptive behavior of the opponent, dynamic defensive measures should be considered that take into account the actions of the attackers. Otherwise, the filter placement and activation algorithms may yield suboptimal results as a consequence of the attackers circumventing them through intelligent routing.

The specific stochastic security game here utilizes the notion of a sampling budget, where the sampling effort can be distributed arbitrarily over the links of the network. This approach is useful in architectures where the routers have built-in sampling capabilities. The framework considered is also applicable to the deployment of dedicated hardware devices for detecting and filtering malicious packets. The formulation also encompasses frequent routing state changes on the network, as is the case in, for example, ad-hoc networks, which are modeled explicitly by the routing state transition matrix and strategies of the players.

In practice, one of the main benefits of using stochastic security games for solving the filter placement problem is the quantitative framework enabling dynamic and adaptive filter placement without manual intervention. Furthermore, the equilibrium solution ensures that, assuming guaranteed local detection rates at the links, the global detection and filtering rate will never fall below a certain level. This is an important feature for the protection of critical infrastructures.

4.4.1   Stochastic game formulation

The underlying network for filter placement is represented by an undirected graph (V, E), where V is the set of vertices (e.g. computers or servers) and E the set of network links. Let $v_i v_j$ denote the link between nodes v_i and v_j. The attacker $\mathcal{P}^A$ controls directly or indirectly a subset of vertices $V_A \subseteq V$, which, for example, may belong to a botnet. On the other hand, the defender $\mathcal{P}^D$ protects a set of targets $V_T \subseteq V$, where it is assumed that $V_A \cap V_T = \varnothing$ without any loss of generality. In some cases, the vertices may represent entire subnetworks with multiple hosts, while in others only a single networked device. In other words, a node $v \in V_A$ may be either a single client or a subnet from which even multiple distinct attackers operate, yet are mathematically treated as a single attacker, $\mathcal{P}^A$.

Adopting the stochastic security game framework of Section 4.1.1, the filter placement problem is formulated as a zero-sum finite Markov game, where each player has a finite number of attack or defensive actions to choose from. The attacker's action space is defined as $\mathcal{A}^A := V_A \times V_T = \{a_1, a_2, \ldots, a_{max}\}$, representing the attack routes available to the attacker from a certain node $v_i \in V_A$ from which the attack is initiated to the target nodes $v_j \in V_T$. On the other hand, the action space of the defense is taken as $\mathcal{A}^D = \{d_1, d_2, \ldots, d_{max}\} \subseteq E$, which is the set of links on which the malware filters can be deployed or activated. The actions of $\mathcal{P}^D$ are restricted to the network core under direct control, i.e. the set $\{v_i v_j \in E$ such that $v_i, v_j \notin (V_A \cup V_T)\}$. This is, for example, the case when $\mathcal{P}^D$ represents an Internet service provider (ISP). For simplicity of analysis, $\mathcal{P}^A$ is taken to choose a single source-target vertex pair and $\mathcal{P}^D$ to deploy a filter at a given time. This assumption is immediately relaxed when extending the analysis from pure to mixed strategies.

The players interact on a network consisting of a set of nodes whose routing tables may change randomly at discrete time instances with a predefined probability, which models load balancing on the network. In addition, the framework allows for analysis of static routing and imperfectly functioning (defective) detectors under static routing. Each of these constitute the underlying stochastic system on which the players interact. Focusing on the dynamic load-balancing case, each possible routing configuration defines the space of routing states, $S = \{s_1, s_2, \ldots, s_{max}\}$. The exact mathematical definition of the states depends on the routing protocol employed in the underlying real-life network. In *session routing*, for example, for each source–sink pair $(v_i, v_j) \in \mathcal{A}^A$ there is a distinct path on which the packets are routed. On the other hand, in architectures that are not flow-based, all packets with the same target (Internet protocol address) arriving at a certain router will be forwarded to the same next router. Similarly, the failures of the detectors can be modeled probabilistically to obtain a set of detector states similar to the routing ones. The transition probabilities between the states s for each case are denoted by M and are independent of player actions unlike the case in the previous section. Then, the probability of being in a specific state, $p^S := [p_1^S, p_2^S, \ldots, p_{max}^S]$, evolves according to $p^S(t+1) = M p^S(t)$, where $t \geq 1$ denotes the stage of the game.

Each player is associated with a set of costs that are not only a function of the other players' actions but also the state of the system. The costs of $\mathcal{P}^D$ and $\mathcal{P}^A$ are captured by the game matrices $G(s)$, where $s \in \mathcal{S}$ for a given state s. The cost of $\mathcal{P}^A$ simply represents the number (value) of detected malware packets, in other words the cost of a failed attack attempt. The $\mathcal{P}^D$ benefits from such a situation (approximately) equally. In the reverse situation, the cost represents missed malware packets, i.e. a successful attack which benefits the attacker. Thus, the game is zero-sum. Each player knows its own cost at each stage of the game as well as the (state) transition probabilities between the routing configurations. Hence, the stochastic security game can be solved using the methods of Section 4.1.2.

4.4.2   Simulations

Setup

The filter placement game described above is illustrated now on the example network shown in Figure 4.7, which is composed of eleven nodes. Two systems (a_1 and a_2) are controlled by the attacker $\mathcal{P}^A$ and may be used to launch an attack on any of the three target systems (t_1, t_2, t_3). Two of these target systems are connected to the same access router. Using the notational convention of the previous section, the action space of $\mathcal{P}^A$ is then $\{a_1t_1, a_1t_2, a_1t_3, a_2t_1, a_2t_2, a_2t_3\}$, where a_it_j corresponds to the attack from a_i on t_j in the attacker's action space. These attacks are enumerated for notational convenience to define the $\mathcal{P}^A$ action set as $\mathcal{A}^A = \{a_1, a_2, \ldots, a_6\}$.

The action space of $\mathcal{P}^D$ consists of all the links on the network where a filter may be placed or activated by the defense. Placement of a filter directly before an

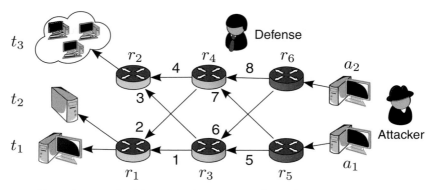

Figure 4.7 Example network with two attack points $V_A = \{a_1, a_2\}$ and three target systems $V_T = \{t_1, t_2, t_3\}$. The candidate links for filter placement are labeled with numbers.

attacked node or access router (r_5, r_6) is generally suboptimal due to lack of information on the exact location of attackers and targets. Hence, the action space of $\mathcal{P}^D$ is $\{r_1 r_3, r_1 r_4, r_2 r_3, r_2 r_4, r_3 r_5, r_3 r_6, r_4 r_5, r_4 r_6\}$, which is enumerated as $\mathcal{A}^D = \{d_1, \ldots, d_8\}$ as shown in Figure 4.7.

The game matrices $G(s)$ are defined numerically next. Let $p(s_l, a_i)$ denote the path of the packets of attack $a_i = a_m t_n \in \mathcal{A}^A$ traveling from a_m to t_n under routing configuration s_l. In addition, let $\mathcal{P}^D$ deploy a filter at $d_j \in \mathcal{A}^D$. Then, the entry $g_{i,j}(s_l)$ in game matrix $G(s)$ is defined as

$$g_{i,j}(s_l) = \begin{cases} c, \text{ if } d_j \in p(s_l, a_i) \\ -c, \text{ otherwise} \end{cases}, \qquad (4.20)$$

where c is a positive constant. This means at the same time that the benefit of $\mathcal{P}^D$ is c if the current attack uses a route on which a filter is deployed and $-c$ in case it traverses the network undetected (due to lack of sniffers on its path or detector failures).

The state space of the game consists of the possible routing configurations on the network. For each attack $a_i \in \mathcal{A}^A$, there are exactly two distinct paths from the attacker to the target system. Once a packet has traveled two hops and arrives at one of the routers r_3 or r_4, there is only one possible path to t_i. Therefore, the routing decision has to be made after the first hop. In real-life networks, it is possible that, even though each attack path has the same length (three hops), packets arriving at the ingress nodes r_5, r_6 will not be routed over the same outgoing link. For example, this is the case for flow-based resource reservation architectures or multi-protocol label switching (MPLS) domains often encountered in quality of service (QoS)-aware architectures. For the sake of reducing the number of states and to preserve the simplicity of the example, all packets arriving at node r_i will be routed over the same outgoing link $r_i r_j$. However, this routing configuration changes from time to time for load-balancing purposes or due to failures.

Consequently, four possible environment states $S = \{ll, lr, rl, rr\}$ corresponding to the four routing configurations are defined in the example network. As a notational convention, for example, $s_3 = rl$ denotes the configuration where all packets arriving at r_5 will be routed to the right and all packets arriving at r_6 will be routed to the left. The colloquial terms "left" and "right" hereby refer to the intuitive graphical interpretation arising from Figure 4.7. The transition probabilities between states follow directly from the underlying routing architecture of the network.

Results

The simulations are run on the Network Security Simulator (NeSSi) [51], a realistic packet-level network simulation environment which allows for detailed examination and testing of security related network algorithms. NeSSi is implemented in the Java programming language and built upon the Java-based intelligent agent componentware (JIAC) framework. JIAC agents are used within the simulator for modeling and implementing the network entities such as routers, clients, and servers. The front-end of NeSSi consists of a graphical user interface that allows the creation of arbitrary network topologies and is used here to create the one in Figure 4.7. The communication between clients, servers, and routers takes place by realistic IPv4 packet transmission, which is implemented using communication services between the agents. Built upon the network layer, the simulator implements TCP and UDP protocols[3] on the transport layer. At the application layer, the hypertext transfer protocol (HTTP) and simple mail transfer protocol (SMTP) protocols[4] are emulated faithfully. Thus, the results obtained through the simulation tool are applicable to real IP networks. Moreover, different types of routing protocol encountered in real-life IP networks, static as well as dynamic, are supported in NeSSi. As part of NeSSi, a monitoring/filtering agent is implemented which can be deployed on a set of links. This means that once one or multiple packets containing a previously generated signature, indicating malicious content, traverses a sniffed link, an alert is issued and/or those packets are filtered. In addition, since the TCP/IP stack is faithfully emulated, the captured traffic can also be written to a common dump file format for later post-processing with standard inspection tools.

Each simulation consists of a period of 1000 time steps in which the players update their actions, i.e. $\mathcal{P}^A$ sends a malware packet with a known signature over a chosen attack path and $\mathcal{P}^D$ deploys the filter at a link. Here, the time interval between the steps is not exactly specified but assumed to be long enough to satisfy the information assumptions made earlier. The malware packets sent on the network simulated in NeSSi are real UDP packets and are captured by a realistic filter implementation using pattern matching algorithms and a malware signature database to filter out the packets. The routing configuration on the network changes every two time steps in accordance with the state transition probabilities. In the simulations, it is observed that the time average of the routing states matches well with the theoretical invariant distribution.

In the simulations, first both $\mathcal{P}^A$ and $\mathcal{P}^D$ use the equilibrium strategy computed from the stochastic security game defined. The performance of the optimal filter placement strategy of $\mathcal{P}^D$ is shown in Figure 4.8(a). This result is compared to the case when $\mathcal{P}^D$ activates the filters randomly with a uniform distribution, which is depicted in Figure 4.8(b). The total number of malware packets captured in this case is, as expected, smaller than in the previous scenario. In addition, the opposite scenario is simulated where $\mathcal{P}^A$ selects a uniformly random strategy whereas $\mathcal{P}^D$ uses the equilibrium strategy. Unsurprisingly, the attacker is observed to be at a disadvantage and more packets are filtered out than in both of the previous simulations.

4.5   Discussion and further reading

The Markov game framework adopted in this chapter is mainly based on references [98, 195], and a discussion on value function approximation can be found in reference [91]. Reference [36] provides a detailed overview of Markov decision

---

[3] TCP and UDP are two of the main transport protocols on the Internet.
[4] HTTP is one of the core protocols on the Web whereas SMTP is the main standard for electronic mail.

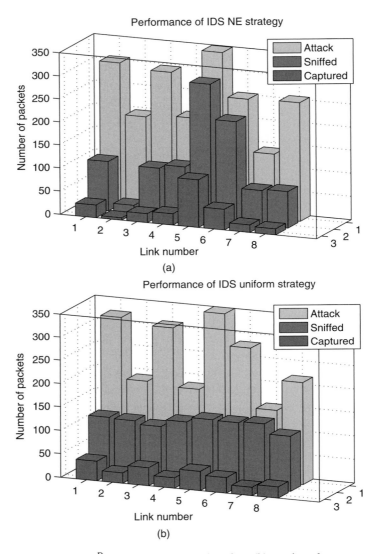

Figure 4.8 Performance of $\mathcal{P}^D$ under *optimal* (a) and *uniform* (b) monitor placement strategies on the example network from Figure 4.7 under optimal attacks and dynamic routing.

processes, which Markov games take as a basis and extend upon. A basic overview of DP in discrete- and continuous-time is provided in Appendix A.3. An alternative approach to stochastic games presented in Section 4.1.2 is from reference [136], where instead of an infinite horizon and discount factor α, a nonzero probability for terminating the game is introduced. In the security game setting, this can be interpreted as the attacker ceasing to attack the system with some nonzero probability due to an external reason.

The stochastic intrusion detection game in Section 4.2 is a reinterpretation of the framework introduced in reference [8]. Section 4.3 summarizes some of the contributions in reference [155] and adopts them to the Markov game framework presented. The concept of linear influence networks [114] and their applications to security [129] are further elaborated upon in Section 4.3.2.

The malware filter placement game in Section 4.4 is based on [159], and provides an example application of the Markov game framework presented.

5 Security games with information limitations

Chapter summary

In many applications, the players, attackers, and especially defenders do not have access to each other's payoff functions. They adjust their strategies based on estimates of opponent's type or observations of opponent actions. However, these observations may not be accurate due to imperfect "communication channels" that connect the players, such as sensors with detection errors. Moreover, there may be inaccuracies in player decisions and actions as a result of, for example, actuator errors. The Bayesian game approach and fictitious play are utilized to analyze such security games with information limitations. Illustrative examples on intrusion detection and wireless security games are provided.

5.1 Bayesian security games

In many security scenarios, the defenders and malicious attackers have limited information about each other. One of the main restrictions on available information in the security domain is the defenders' limited observation capabilities. This can be, for example, due to imperfections in (intrusion) detection systems. Even if the defender has an accurate estimate of attacker preferences, the limitations on detection capabilities have to be taken into account as a factor in defensive decisions. Similarly, the attackers can exploit their knowledge of imperfect detection when choosing their targets. All these considerations on limited observation and detection can be formalized within the framework of **Bayesian security games**.

Bayesian games model lack of information about the properties of players in a non-cooperative game using a probabilistic approach. In a Bayesian game, the players are usually assumed to be one of many specific types. A special *nature* player is introduced to the game which assigns a predetermined probability distribution to each player and type combination, which constitutes its fixed strategy. Subsequently, the original players compute their own strategies by taking into account each possible player-type combination weighted by the predetermined probability distribution.

Bayesian security games can utilize the described probabilistic approach to model the imperfect detection capabilities of security defense systems. For example, a virtual sensor network used for detection of malicious activity by the defender can be defined as the nature player. Then, by extending the security game framework of Chapter 3, both the attacker $\mathcal{P}^A$ and defender $\mathcal{P}^D$ can take into account the detection probability of the sensors, i.e. the fixed strategy of the nature player, when determining their best-response strategies. The resulting Bayesian intrusion detection game as well as a Bayesian extension of the wireless security (jamming) in Section 3.6 are discussed next as two representative Bayesian security games.

5.1.1 Bayesian intrusion detection game

Sensors and imperfect detection

The defense systems often include a (virtual) sensor network in order to collect information and detect malicious attacks, which can be represented by the nature player in the Bayesian security game model. A virtual sensor network is defined as a collection of autonomous hardware and/or software agents that monitors the system and collects data, e.g. for detection purposes. The sensors report possible intrusions or anomalies occurring in a subsystem using common techniques such as signature comparison, pattern detection, or statistical analysis. The data from individual sensor nodes is often correlated to be able to observe overall trends on the network. The virtual sensors can be classified according to their functionality as host-based or network-based [26]. While host-based ones collect data directly from the system they reside on, e.g. by observing log files and system calls, the network-based sensors obtain data from network traffic by analyzing packets and headers. A graphical depiction of an intrusion detection and response system with a network of virtual sensors is shown in Figure 5.1. Section 9.1.1

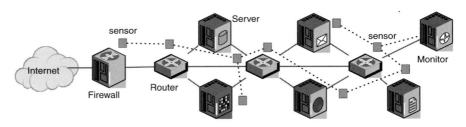

Figure 5.1 Intrusion detection system with a network of virtual sensors.

contains further information on (virtual) sensors and their role in intrusion detection systems.

Consider a distributed network of sensors

$$\mathcal{S} := \{s_1, s_2, \ldots, s_{N_S}\},$$

similar to the one depicted in Figure 5.1. The defended system monitored for signs of malicious behavior can be represented as a set of subsystems

$$\mathcal{T} := \{t_1, t_2, \ldots, t_{N_T}\},$$

which may be targeted by the attacker. The subsystems in $\mathcal{T}$ may represent the operating system, middleware, applications, or parts of the network as well as (business) processes distributed over multiple hosts. Define the augmented set of attacks and anomalies over a target system $\mathcal{T}$ as

$$\mathcal{A}^A := \{a_1, a_2, \ldots, a_{N_A}\} \cup \{na\},$$

where *na* represents "no attack." In this context, the generic term "attack" is associated with two specific attributes: a target subsystem, $t \in \mathcal{T}$, and a threat or anomaly. Hence, the set of attacks $\mathcal{A}^A$ is defined as the cross-product of the set $\mathcal{T}$ and the set of documented threats and detectable anomalies targeting it.

Given the set of attacks $\mathcal{A}^A$ thus defined, the linear mapping $\bar{P} : \mathcal{A}^A \to \mathcal{A}^A$ describes the relationship between the actual attacks and the output of the sensor network $\mathcal{S}$. Specifically, the matrix

$$\bar{P} := [\bar{P}_{ij}]_{N_A \times N_A}, \text{ where } 0 \leq \bar{P}_{ij} \leq 1, \ \ \forall i, j \in \{1, \ldots, N_A\}, \tag{5.1}$$

represents whether or not an attack is correctly reported. The entry $\bar{P}_{ij}$ of the matrix denotes the probability of attack i being reported as attack j. If $i \neq j$, then the sensor network confuses one attack for another. Such misreporting is quite beneficial for the attacker. In the case of $j = na$, $\bar{P}_{ij}$ is the probability of failing to report an existing attack. Similarly, if $i = na$ and $j \neq na$, then $\bar{P}_{ij}$ is the probability of false alarm for attack j. Thus, the matrix $\bar{P}$ describes the fixed strategy of the nature player (virtual sensor network).

Game formulation in extensive form

Based on the model in the previous section, a finite Bayesian intrusion detection game is defined within the security game framework of Chapter 3. The finite action spaces of the attacker $\mathcal{P}^A$ and the defender $\mathcal{P}^D$ are $\mathcal{A}^A := \{a_1, \ldots, a_{N_A}\}$ and $\mathcal{A}^D := \{d_1, \ldots, d_{N_D}\}$, respectively. Notice that the attacker's action space coincides with the set of attacks previously defined. The defense responses may vary from a simple alert, which corresponds to a passive response, to reconfiguration of the sensors and limiting access of users to the system, which are active responses. The special action "no response" is also included in the response set $\mathcal{A}^D$.

Introduce $p^A := [p_1, \ldots, p_{N_A}]$ as a probability distribution on the attack (action) set $\mathcal{A}^A$ and $q^D := [q_1, \ldots, q_{N_D}]$ as a probability distribution on the defense (action) set $\mathcal{A}^D$ such that $0 \leq p_i, q_i \leq 1 \; \forall i$ and $\sum_i p_i = \sum_i q_i = 1$. The fixed strategy of the nature player, equivalently the detection probabilities of the sensor network, is captured by the matrix $\bar{P}$ defined in (5.1). Thus, given an attack probability p^A, the output vector of the sensor network is given by $p^A \bar{P}$.

As an illustrative example, consider the following Bayesian security game with a single system $\mathcal{T} = \{t\}$. The action space of the attacker consists of a single detectable attack and "no attack," $\mathcal{A}^A = \{a\} \cup \{na\}$. The defender actions are limited to "set an alert" or "do nothing" denoted by $\mathcal{A}^D := \{d, nd\}$. The probability distributions over these sets, p^A and p^D, are defined accordingly. A representation of this game in extensive form is depicted in Figure 5.2, where $\{r, nr\}$ represent the sensor network reporting an attack and not reporting, each with a certain probability of error. The cost values for the defender $\mathcal{P}^D$ and the attacker $\mathcal{P}^A$ are $\left[(R_1^A, R_1^D), \ldots, (R_8^A, R_8^D) \right]$. They can be chosen to reflect specific network security tradeoffs and risks.

The example Bayesian security game in Figure 5.2 can be further explained by describing a specific scenario step by step that corresponds to following a path from left to right in accordance with the order of players' actions. The lower left branch in the figure labeled a indicates an attack by the attacker(s) to the system. The sensor network detecting this attack is represented by the r branch. Finally, given the information from the sensor network, the defender decides in branch d to take a predefined response action. The outcome of this scenario is quantified by a cost to the attacker and benefit to the defender.

A numerical example of this particular game is shown in Figure 5.3 and solved using the GAMBIT software [173]. According to the conventions of GAMBIT software, the

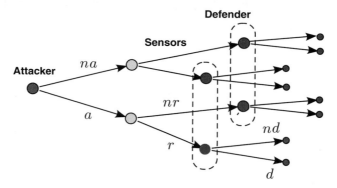

Figure 5.2 Example Bayesian security game in extensive form.

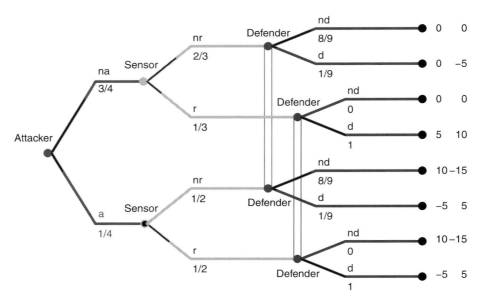

Figure 5.3 Example Bayesian security game and its numerical solution obtained using the GAMBIT software.

outcomes shown are not costs but payoffs which the players maximize. This game does not admit any NE solution in pure strategies. However, a unique NE is numerically computed in mixed strategies and shown in Figure 5.3, which also corresponds to the unique solution in behavioral strategies.

At the NE, the attacker targets the system with a probability $1/4$. A reason for this low probability is the discouraging effect of the sensor network's capability of correct detection with probability $2/3$ if there is no attack and $1/2$ when there is. Note that there are two information sets for the defender based on the reporting of sensors: one indicating an attack and one for no attack. The NE strategy of the defender, given the information by the sensor network, is "do nothing" *nd* with probability $8/9$ if no attack is reported, and a response *d* with probability 1 if an attack is reported. The defender in this case has arguably more to lose by ignoring the attack and acts despite relatively less reliable reporting by the sensor network. Unsurprisingly, the NE strategies of the players are also very much dependent on the outcome payoffs of the game [6] as well as the detection probabilities $\bar{P}$ of the sensor network. Therefore, it is crucial that the payoff values in the game correctly reflect the tradeoffs in the system at hand.

Although the extensive form of the Bayesian security game provides a detailed visualization of the interaction between the players, it also has some disadvantages. One drawback is its scalability (or lack thereof). The strategy spaces of the attacker and defender may grow substantially for a more comprehensive analysis of a larger system. Another difficulty is the choice of the payoff values, which have to be determined separately for each branch of the game tree. This process may become tedious and inaccurate for a large system. In order to overcome these limitations, a continuous-kernel version of the Bayesian intrusion detection game is introduced next.

Continuous-kernel game formulation

An alternative to the formulation in the previous section is the continuous-kernel Bayesian security game. While it is more abstract in nature, it provides a model with a much smaller number of parameters.

In the continuous-kernel formulation, the action space of the attacker is

$$\mathcal{A}^A := \left\{ u^A \in \mathbb{R}^{N_A} : u_i^A \geq 0, \ i = 1, \ldots, N_A \right\},$$

and the one of the defender is

$$\mathcal{A}^D := \left\{ u^D \in \mathbb{R}^{N_D} : u_i^D \geq 0, \ i = 1, \ldots, N_D \right\}.$$

The fixed strategy of the sensor network is given again by (5.1). For notational convenience, a matrix P is defined by multiplying the diagonal terms of the matrix $\bar{P}$ with -1 to obtain

$$P := [P_{ij}] = \begin{cases} P_{ij} = -\bar{P}_{ij} & \text{if } i = j \\ P_{ij} = \bar{P}_{ij} & \text{if } i \neq j \end{cases}. \tag{5.2}$$

The parameters of the player cost functions are defined as follows: the non-negative vector $c^D := \left[c_1^D, \ldots, c_{N_A}^d \right]$ represents the cost of each attack for the defender, whereas $c^A := \left[c_1^A, \ldots, c_{N_A}^A \right]$ quantifies the gain of the attacker from the attack, if it is successful. The non-negative matrix

$$Q := [Q_{ij}]_{N_A \times N_A}, \text{ where } Q_{ij} \geq 0, \ \forall i, j \in \{1, \ldots, N_A\}$$

with $Q_{ii} > 1 \ \forall i$ models the vulnerability of a specific subsystem to various attacks. Similarly, the matrix

$$R := [R_{ij}]_{N_A \times N_D}, \text{ where } R_{ij} \in \{0, 1\} \ \forall i \in \{1, \ldots, N_A\} \text{ and } \forall j \in \{1, \ldots, N_D\}$$

with entries of *ones* and *zeros* correlates defender response actions with the attacks. The vectors $\alpha := [\alpha_1, \ldots, \alpha_{N_D}]$ and $\beta := [\beta_1, \ldots, \beta_{N_A}]$ are the cost of the response and the cost of carrying out an attack for the defender and the attacker, respectively. The relative cost ratio of false-alarms, detection, and deception are captured by the scalar parameter γ.

Based on the defined parameters, a specific quadratic cost function is introduced for the defender J^D and the attacker J^A, as

$$J^D(u^A, u^D, P) := \gamma [u^A]^T P R u^D + [u^D]^T \text{diag}(\alpha) u^D + c^D (Q u^A - R u^D), \tag{5.3}$$

and

$$J^A(u^A, u^D, P) := -\gamma [u^A]^T P R u^D + [u^A]^T \text{diag}(\beta) u^A + c^A (R u^D - Q u^A), \tag{5.4}$$

where $[\cdot]^T$ denotes the transpose of a vector or matrix, and $\text{diag}(x)$ is a diagonal matrix with the diagonal entries given by the elements of the vector x.

The specific structure of the cost functions J^D and J^A aims to model various aspects of the intrusion detection game. The first terms of each cost function, $\gamma [u^A]^T P R u^D$ and $-\gamma [u^A]^T P R u^D$, represent the cost of false alarms and the benefit of detecting the attacker for the defender as well as the cost of capture and the benefit of deception for the attacker, respectively. Notice that if the cost functions consisted of only this part, then the game would have been a zero-sum one.

The second terms $[u^D]^T \text{diag}(\alpha)u^D$ and $[u^A]^T \text{diag}(\beta)u^A$ quantify the cost of defensive measures and attacks, respectively. Depending on the scenario, this term may reflect the resource usage costs or consequences of a defender action. For the attacker, it may represent the cost of resources needed to carry out the attack. The third terms, $c^D (Qu^A - Ru^D)$ and $c^A(Ru^D - Qu^A)$, capture the actual cost or benefit of a successful attack. False alarms and detection capabilities of the sensor network at a given time are incorporated into the values of the matrix P. In the ideal case of the sensor network functioning perfectly, i.e. no false alarms and 100 percent detection, the matrix $-P$ is equal to the identity matrix, $I = \text{diag}([1,\ldots,1])$.

For notational convenience, define the vectors

$$\theta^D(c^D, R, \alpha) := [(c^D R)_1/(2\alpha_1), \ldots, (c^D R)_{N_D}/(2\alpha_{N_D})]$$

and

$$\theta^A(c^A, Q, \beta) := [(c^A Q)_1/(2\beta_1), \ldots, (c^A Q)_{N_A}/2\beta_{N_A}] .$$

The reaction functions of the attacker and defender are obtained by minimizing the respective strictly convex cost functions (5.3) and (5.4). Hence, they are uniquely given by

$$u^D(u^A, P) = [u_1^D, \ldots, u_{N_D}^D]^T ,$$

and

$$u^A(u^D, P) = [u_1^A, \ldots, u_{N_A}^A]^T ,$$

respectively, where

$$u^D(u^A, P) = [\theta^D - \gamma[\text{diag}(2\alpha)]^{-1} R^T P^T u^A]^+ \tag{5.5}$$

and

$$u^A(u^D, P) = [\theta^A + \gamma[\text{diag}(2\beta)]^{-1} P \bar{Q} u^D]^+ . \tag{5.6}$$

Here, the function $[\cdot]^+$ maps all of its negative arguments to zero.

It is desirable for the defender to configure the virtual sensor network such that all possible threats are covered. It is also natural to assume a worst-case scenario where for each attack (type) targeting a subsystem, there exists at least one attacker who finds it beneficial to attack. Hence, the following holds in many practical cases: $u_i^A > 0 \, \forall i$ or $u_j^I > 0 \, \forall j$.

A simple metric quantifying the performance of the sensor network is called "detection quality" or dq. For each attack, $a \in \mathcal{A}^A$, the detection quality of the monitoring sensor network is defined as

$$dq(i) := \frac{\bar{P}_{ii}}{\sum_{j=1}^{N_D} \bar{P}_{ij}}, \quad i = 1, \ldots, N_A.$$

Within the context of the intrusion detection game defined, the pair (u^{D*}, u^{A*}) is said to be the NE strategies of the attacker and defender, if it satisfies $u^{D*} = \arg\min_{u^D} J^D(u^{A*}, u^D, P)$ and $u^{A*} = \arg\min_{u^A} J^A(u^{D*}, u^D, P)$. The following theorem establishes the existence of a unique NE solution in pure strategies as well as providing a complete analytical characterization of the NE solution.

Theorem 5.1 *There exists a unique NE solution to the defined Bayesian intrusion detection game. Furthermore, if*

$$\gamma < \min \left(\frac{\min_i \theta^D}{\left[\max_i \left(\operatorname{diag}(2\alpha) \right)^{-1} R^T P^T \theta^A \right]^+}, \frac{\min_i \theta^A}{\left[\max_i \left(\operatorname{diag}(2\beta) \right)^{-1} (-P) R \theta^D \right]^+} \right),$$
(5.7)

then the NE is an inner solution, $u^{D*}, u^{A*} > 0$, and is given by

$$u^{A*} = (I+Z)^{-1} \cdot \left[\theta^A + \gamma [\operatorname{diag}(2\beta)]^{-1} PR\theta^D \right]$$
(5.8)

and

$$u^{D*} = (I+\bar{Z})^{-1} \cdot \left[\theta^D - \gamma [\operatorname{diag}(2\alpha)]^{-1} R^T P^T \theta^A \right],$$
(5.9)

where

$$Z := \gamma^2 [\operatorname{diag}(2\beta)]^{-1} PR [\operatorname{diag}(2\alpha)]^{-1} R^T P^T,$$

$$\bar{Z} := \gamma^2 [\operatorname{diag}(2\alpha)]^{-1} R^T P^T [\operatorname{diag}(2\beta)]^{-1} PR,$$

and I is the identity matrix.

Proof The existence of the NE in the game follows from the facts that the objective functions are strictly convex, they grow unbounded as $|u| \to \infty$, and the constraint set is convex [31, p. 174].

In order to show the uniqueness of the NE, let $\overline{\nabla}$ be the pseudo-gradient operator, defined through its application on the cost vector $J := [J^D \ J^A]$, as

$$\overline{\nabla} J := \left[\nabla_{u_1^D}^T J^D \cdots \nabla_{u_{N_D}^D}^T J^D \ \nabla_{u_1^A}^T J^A \cdots \nabla_{u_{N_A}^A}^T J^A \right]^T.$$
(5.10)

Further define $g(u) := \overline{\nabla} J$ where $u := [u^D \ u^A]$. Let $G(u)$ be the Jacobian of $g(u)$ with respect to u,

$$G(u) := \begin{pmatrix} \alpha_1 & 0 & 0 & | & & & \\ 0 & \ddots & 0 & | & & \gamma[PR]^T & \\ 0 & 0 & \alpha_{N_D} & | & & & \\ - & - & - & | & - & - & - \\ & & & | & \beta_1 & 0 & 0 \\ & -\gamma[PR] & & | & 0 & \ddots & 0 \\ & & & | & 0 & 0 & \beta_{N_A} \end{pmatrix}_{(N_A+N_D) \times (N_A+N_D)},$$
(5.11)

where the matrix $[PR]$ is of size $N_A \times N_D$. Define the symmetric matrix $\bar{G}(u) := \frac{1}{2}(G(u) + G(u)^T)$. It immediately follows that $\bar{G}(u) = \mathrm{diag}([\alpha\ \beta])$ is positive definite. Thus, due to the positive definiteness of the Hessian-like matrix $\bar{G}(u)$, the NE solution is unique [4, 147]. Note that this result does not use the condition (5.7) on γ.

Next, it is shown that the unique NE solution is indeed an inner one, under the condition (5.7) on γ, by obtaining its analytical description. Substitute u^D in (5.6) with the expression in (5.5), which leads to the fixed-point equation $u^{A*} = u^A(u^D(u^{A*}, P), P)$, such that

$$u^{A*} = \theta^A + \left[\mathrm{diag}\left(\frac{2\beta}{\gamma}\right)\right]^{-1} \mathrm{PR}\theta^D - \mathrm{diag}\left(\frac{\gamma}{2\beta}\right) \mathrm{PR} \left[\mathrm{diag}\left(\frac{2\alpha}{\gamma}\right)\right]^{-1} \mathrm{R}^T\mathrm{P}^T u^{A*}. \quad (5.12)$$

Solving for u^{A*} yields (5.8), where the inverse exists because Z is non-negative definite. The equilibrium solution u^{D*} in (5.9), on the other hand, can be derived by simply substituting for u^{A*} from (5.8) into (5.5). It is then straightforward to show that if (5.7) holds, then both $u^{D*}, u^{A*} > 0$, hence the NE is strictly positive and inner. Moreover, there cannot be a boundary solution due to the uniqueness of the NE, thus completing the proof. □

5.1.2   Bayesian games for wireless security

Random access

It is often the case that the players of wireless security (jamming) games do not exactly know the types of other player, i.e. whether they are malicious or selfish, but have their own estimates. The Bayesian security games provide a framework for analyzing these more realistic scenarios, which extend the wireless security games of Section 3.6.

In a two-player wireless security game between a selfish (defender) S node and a malicious M one, let ϕ_i denote the probabilistic belief of a player (node) $i \in \{S, M\}$ that the opponent $j \neq i$ is **selfish**. In other words, $0 \leq \phi_M \leq 1$ is the probability that the opponent is selfish as believed by the malicious player. Likewise, $0 \leq \phi_S \leq 1$ denotes the probability of the opponent being selfish for the selfish player S.

The game model is the same as that in Section 3.6.1, where a synchronous slotted system with *collision channels* is assumed such that more than one simultaneous transmission fails. Define $p_i \in [0, 1]$ as the transmission probability and $e_i \in (0, 1)$ as energy cost (per transmission) of node $i \in \{S, M\}$. Here, the *selfish* node S receives a unit throughput reward for successful transmission, i.e. $r = 1$. Then, the cost of the selfish node, as a special case of (3.10), is

$$J_S(p) := p_S e_S - p_S(1 - p_j), \ j \in \{S, M\}. \quad (5.13)$$

Assume that a *malicious* node M incurs a unit reward, if the opponent is selfish *and* successfully jammed at the given time slot, i.e. $c = 1$. The cost function of the malicious node, as a special case of (3.11), is

$$J_M(p) := \begin{cases} p_M e_M - p_M p_S, & \text{if the opponent is selfish} \\ p_M e_M, & \text{if the opponent is malicious.} \end{cases} \quad (5.14)$$

The resulting expected costs of players under the given probabilistic beliefs about the opponents are

$$J_S(p, \phi_S) := p_S e_S - \phi_S p_S (1 - p_{\tilde{S}}) - (1 - \phi_S) p_S (1 - p_M), \qquad (5.15)$$

for the selfish player where $p_{\tilde{S}}$ is the transmission probability of a selfish opponent, and

$$J_M(p, \phi_M)) := p_M e_M - \phi_M p_M p_S, \qquad (5.16)$$

for the malicious player.

The following theorem characterizes the Bayesian NE strategies of malicious and selfish players of the two-player wireless security (jamming) game defined. The results follow directly from a straightforward but illustrative application of the definition of NE.

Theorem 5.2 *The Bayesian NE strategies of selfish and malicious players, p_S and p_M, with cost functions (5.13) and (5.14), respectively, in the wireless security (jamming) game are uniquely given by*

1.
$$p_S^* = 1, \; p_M^* = 0, \; if \, \phi_S < 1 - e_S, \, \phi_M < e_M, \qquad (5.17)$$

2.
$$p_S^* = \frac{1 - e_S}{\phi_S}, \; p_M^* = 0, if \, \phi_S > 1 - e_S, \, e_M \phi_S > \phi_M (1 - e_S), \qquad (5.18)$$

3.
$$p_S^* = \frac{e_M}{\phi_M}, \; p_M^* = \frac{1 - e_S - e_M \frac{\phi_S}{\phi_M}}{1 - \phi_S}, \; if \, \phi_S < e_S + e_M \frac{\phi_S}{\phi_M} < 1, \, \phi_M > e_M. \qquad (5.19)$$

Proof (1) The expected cost of the selfish node (5.15) is minimized by transmitting with probability $p_S = 1$, if $(1 - \phi_S)(1 - p_M) > e_S$. Note that the possibility of a selfish opponent is preferable over a malicious one as indicated by the cost structure, and hence does not have an effect on the given condition in the worst case. Given that $p_S = 1$, the expected utility of the malicious node (5.16) is minimized by waiting, i.e. $p_M = 0$, if $\phi_M < e_M$, or by transmitting, if $\phi_M > e_M$. However, $J_S(p_S, 1)$ is not minimized by $p_S = 1$ (therefore the strategy pair $p_S = 1$ and $p_M = 1$ does not yield an NE), whereas $J_S(p_S, 0)$ is minimized by $p_S = 1$, if $1 - \phi_S > e_S$, and thus the NE strategy pair in (5.17) follows.

The cost of a selfish node $J_S(p_S, p_M)$ can also be minimized by waiting, $p_S = 0$, if $(1 - \phi_S)(1 - p_M) < e_S$. Given that $p_S = 0$, the cost of the malicious node $J_M(p_S, p_M)$ is minimized by waiting only, $p_M = 0$. However, $J_S(p_S, 0)$ cannot be minimized by $p_S = 0$. Therefore, the strategy $p_S = 0$ does not yield an NE.

(2) The cost $J_S(p_S, p_M)$ is indifferent to p_S, if $p_S = \frac{1 - e_S}{\phi_S}$. Given that $p_S = (1 - e_S)/\phi_S$, the cost of the malicious node $J_M(p_S, p_M)$ is minimized by $p_M = 0$, if $e_M \phi_S > \phi_M (1 - e_S)$, so that the NE strategy (5.18) follows. Notice that for $p_S = (1 - e_S)/\phi_S$, the strategy $p_M = 1$ cannot yield any NE, since $J_S(p_S, 1)$ is minimized only by $p_S = 0$ provided that $e_S > \phi_S$ and $J_M(0, p_M)$ cannot be minimized by $p_M = 1$.

(3) Consider the strategies such that selfish and malicious nodes are indifferent to p_S and p_M, respectively, in order to minimize J_S and J_M. From $J_S(1, p_M) = J_S(0, p_M)$ and $J_M(p_S, 1) = J_M(p_S, 0)$, the equilibrium strategy in (5.19) subject to $0 \leq p_S \leq 1$ and $0 \leq p_M \leq 1$ follows.   □

The throughput rates corresponding to the equilibrium strategies (5.17) to (5.19) in the theorem are

$$\lambda_S = 1, \; \lambda_S = \frac{1 - E_S}{\phi_S}, \; \text{and} \; \lambda_S = \frac{E_M}{\phi_M} \left(\frac{E_S + E_M \frac{\phi_S}{\phi_M} - \phi_S}{1 - \phi_S} \right),$$

respectively. Decreasing ϕ_M or hiding own identity, i.e. convincing the malicious node that the opponent is not selfish, is beneficial for the selfish node. On the other hand, the malicious player benefits from a high ϕ_S, i.e. the opponent considering the malicious node to be a selfish one.

Interference limited multiple access

Consider now an interference limited multiple access variant of the wireless security game discussed above. The first player is again selfish and believes that the second one is also selfish with probability ϕ_1. The second player can indeed be selfish or malicious; however, regardless of his type he knows that the first player is selfish.

The game model is similar to that in Section 3.6.2. Let P_1 and h_1 be the transmission power level and the channel gain of the first (selfish) player. Define P_{2S}, P_{2M} as the power levels and e_S, e_M as the energy costs of the second player for the selfish and malicious cases, respectively. The channel gain is h_2 in both cases. Let, for simplicity, e_S also be the energy cost for the first player. The cost functions are the same as in (3.22) and (3.23) for the selfish and malicious nodes, respectively.

The NE solutions of the interference limited multiple access wireless security game defined are given in the following theorem.

Theorem 5.3 *Consider an interference limited multiple access wireless security game between a selfish transmitter (player 1) and a second one of unknown type (player 2). The first player estimates the probability of the second one being selfish as $0 \leq \phi_1 \leq 1$ while the second player knows the type of the first one. Then, the Bayesian NE strategies of the game are*

$$P_1^* = \frac{L}{h_1} \left(\frac{h_2}{e_S} - \sigma^2 \right), \tag{5.20}$$

$$P_{2S}^* = \frac{L}{h_2} \left[\frac{h_1 \phi_1}{\left(e_S - \frac{(1 - \phi_1) h_1}{\frac{h_2}{L} P_{2M} + \sigma^2} \right)} - \sigma^2 \right]^+, \tag{5.21}$$

$$P_{2M}^* = \left[\sqrt{\frac{L h_1 P_1^*}{h_2 e_M}} - \frac{L \sigma^2}{h_2} \right]^+, \tag{5.22}$$

if $h_2 \geq \sigma^2 e_S$, where $[x]^+ = \max(x, 0)$.
Otherwise, that is if $h_2 < \sigma^2 e_S$, then

$$P_1^* = \left(\left[e_S - \frac{\phi_1 h_1}{\sigma^2} \right]^+ \right)^{-2} \frac{e_M L h_1}{h_2} (1 - \phi_1)^2, \qquad (5.23)$$

$P_{2S}^* = 0$ *and P_{2M}^* is given by (5.22) with P_1^* from (5.23).*

Proof The result is obtained by applying the KKT optimality conditions separately to the objective functions of the players under the constraints $P_1, P_{2S}, P_{2M} \geq 0$. □

It is interesting to note that if this game is played repeatedly, then the first player can learn the type of the second player by observing its transmission power level unless $P_{2S} = P_{2M}$. Then, it can adjust its power level accordingly. Subsequently, the second player can deduce this information (that the first one learned its type) by observing the change in power level of the first player. Hence, the formulation turns into one of the full information games in Section 3.6.2, depending on the type of the second player.

5.2 Security games with observation and decision errors

In security games, if the game matrices are known to both attackers and defenders, then each player can compute his own best-response and the NE strategies that minimize expected loss. However, this assumption of knowing each other's payoff values is generally not valid in practice. The players can utilize various learning schemes such as fictitious play (FP) in such cases to learn the opponent's (mixed) strategies and compute their own best-response against them.

In an FP game, the players observe all the actions to estimate the mixed strategy of the opponents. At each stage, the players update their estimates and play the pure strategy that is the best-response (or generated based on the best-response) to the current estimate of the other's mixed strategy. An underlying assumption here is the perfect observability of player actions.

In security games, the observations made by the players are usually not accurate, which is an important factor to be taken into account. In practice, the sensor systems that report an attack are imperfect and the actions of the defender, e.g. whether or not countermeasures are deployed, is hard to observe for the attacker. Hence, there always exist a positive *miss probability* (false negative rate) and a positive *false alarm probability* (false positive rate) for both players. Furthermore, the players may make decision errors from time to time due to irrationality or to the channels carrying their commands being error prone.

The subsequent subsections investigate scenarios where the attackers and defenders of FP security games make decisions under incomplete information which means not only lack of knowledge about each other's payoff values and preferences but also uncertainty on the decisions and observations of actions.

5.2.1   Game model and fictitious play

Consider a security game between an attacker $\mathcal{P}^A = \mathcal{P}^1$ and a defender $\mathcal{P}^D = \mathcal{P}^2$ similar to those in Chapter 3. The action spaces of the players are $\mathcal{A}^1 = \mathcal{A}^A$ and $\mathcal{A}^2 = \mathcal{A}^D$, which have the cardinalities $N_1 = N_A$ and $N_2 = N_D$, respectively. For *notational convenience*, let us introduce

$$p_i := [p_{i1}, \ldots, p_{iN_i}]^T$$

as the probability distribution on the action set of player $\mathcal{P}^i$ with cardinality N_i, such that $0 \leq p_{ij} \leq 1 \; \forall j$ and $\sum_j p_{ij} = 1$, $i = 1, 2$.

The security game to be introduced differs from a regular matrix game in the sense that the utility functions of players include an entropy term H to randomize their own strategy, which can be interpreted as a way of concealing their true strategy. Accordingly, the utility[1] function U_i of player $\mathcal{P}^i$ is given by

$$U_i(p_i, p_{-i}) := p_i^T G^i p_{-i} + \tau_i H(p_i), \; i = 1, 2, \tag{5.24}$$

where G^i is defined for notational convenience equal to negative of the game matrix of attacker $\mathcal{P}^A$, $G^1 = -G^A$, and to the transpose of negative of the game matrix of defender $\mathcal{P}^D$, $G^2 = \left(-G^D\right)^T$. The entropy function, H, is defined as

$$H(p_i) := -\sum_{j=1}^{N_i} p_{ij} \log(p_{ij}) = -p_i^T \log(p_i),$$

and τ_i represents how much player $\mathcal{P}^i$ wants to randomize own actions. Here, as standard in game theory literature, the index $-i$ is used to indicate those of other players, or the opponent in this case.

In this static security game, the presence of the weighted entropy term $\tau_i H(p_i)$ leads to distinguishing between classical and stochastic FP. If $\tau_1 = \tau_2 = 0$, then the game leads to *classical FP*, where the best-response mapping of players can be set-valued. If $\tau_i > 0$, however, the best-response has a unique value and the game leads to *stochastic FP* [164]. This section exclusively focuses on stochastic FP due to its relevance to security games.

The *best-response* mapping of player $\mathcal{P}^i$ is the function

$$\beta_i(p_{-i}) = \arg\max_{p_i} U_i(p_i, p_{-i}). \tag{5.25}$$

For the specific cost function (5.24) and if $\tau_i > 0$, the best-response is unique and is given by the soft-max function:

$$\beta_i(p_{-i}) = \sigma\left(\frac{G^i p_{-i}}{\tau_i}\right), \tag{5.26}$$

---

[1] This subsection makes a departure from the overall convention of cost minimization adopted in the book and formulates the player problem as one of utility maximization.

where the vector-valued soft-max function σ is defined as

$$\sigma_j(x) = \frac{e^{x_j}}{\sum_{j=1}^{N_i} e^{x_j}}, \ j = 1, \ldots, N_i. \tag{5.27}$$

The range of the soft-max function is in the interior of the probability simplex and $\sigma_j(x) > 0 \ \forall j$. The NE of the game (p_1^*, p_2^*), if it exists, can be written as the fixed point of the best-response mappings:

$$p_i^* = \beta_i \left(p_{-i}^* \right), \ i = 1, 2. \tag{5.28}$$

Fictitious play with entropy softening

In a repeated play of the static security game, each player selects an integer action $a_i \in \mathcal{A}^i$ according to the mixed strategy p_i at a given time instance. The (instant) payoff for player $\mathcal{P}^i$ is $v_{a_i}^T G^i v_{a_{-i}} + \tau_i H(p_i)$, where v_j, $j = 1, \ldots, N_i$, denotes the j-th vertex of the probability simplex. For example, if $N_i = 2$, then $v_1 = [1 \ 0]^T$ for the first action and $v_2 = [0 \ 1]^T$ for the second action. For a pair of mixed strategies (p_1, p_2), the utility function in (5.24) can be computed using

$$U_i(p_i, p_{-i}) = E \left[v_{a_i}^T G^i v_{a_{-i}} \right] + \tau_i H(p_i). \tag{5.29}$$

Using the static game description above, the discrete-time FP is defined as follows. Suppose that the game is repeated at times $k \in \{0, 1, 2, \ldots\}$. The empirical frequency $q_j(k)$ of player $\mathcal{P}^j$ obtained through observations is

$$q_j(k+1) = \frac{1}{k+1} \sum_{l=0}^{k} v_{a_j}(l). \tag{5.30}$$

Using induction, the following recursive relation is derived:

$$q_j(k+1) = \frac{k}{k+1} q_j(k) + \frac{1}{k+1} v_{a_j}(k) = \frac{k}{k+1} q_j(k) + \frac{1}{k+1} \beta_j(q_i(k)), \tag{5.31}$$

where $q_i(k)$ is the empirical frequency of player $\mathcal{P}^i$. Similarly, at each time step k, the player $\mathcal{P}^i$ picks the best-response to the empirical frequency of the opponent's actions using $\beta_i(q_j(k))$. Hence, in stochastic FP each player $i = 1, 2$, carries out the steps in Algorithm 5.4 below.

Algorithm 5.4 Fictitious play algorithm

1: Given game matrix G^i
2: **for** $k \in \{0, 1, 2, \ldots\}$ **do**
3:     Update the empirical frequency of the opponent, q_{-i}, using (5.31).
4:     Compute the best-response mixed strategy $\beta_i(q_{-i}(k))$ using (5.26).
5:     Randomly generate an action $a_i(k)$ according to the best-response mixed strategy $\beta_i(q_{-i}(k))$, such that the expectation $E[a_i(k)] = \beta_i(q_{-i}(k))$.
6: **end for**

Alternatively, a **continuous-time version of FP** can be derived by approximating the discrete-time recursions in (5.30) and (5.31) [163, 164]. The recursive equations to update the empirical frequencies in stochastic discrete-time FP are

$$q_i(k+1) = \frac{k}{k+1}q_i(k) + \frac{1}{k+1}\beta_i(q_{-i}(k)).$$

Introduce now $\Delta := 1/k$ to write the above as

$$q_i(k+\Delta k) = \frac{k}{k+\Delta k}q_i(k) + \frac{\Delta k}{k+\Delta k}\beta_i(q_{-i}(k)). \tag{5.32}$$

Let $t = \log(k)$ and $\tilde{q}_i(t) = q_i(e^t)$. Then, for $\Delta > 0$ sufficiently small,

$$q_i(k+\Delta k) \simeq q_i(e^{\log(k)+\Delta}) = \tilde{q}_i(t+\Delta).$$

Again for sufficiently small $\Delta > 0$, $\frac{k}{k+\Delta k} \simeq 1 - \Delta$ and $\frac{\Delta k}{k+\Delta k} \simeq \Delta$. Hence, the recursion (5.32) can be rearranged to

$$(\tilde{q}_i(t+\Delta) - \tilde{q}_i(t))/\Delta = \beta_i(\tilde{q}_{-i}(t)) - \tilde{q}_i(t).$$

Now letting $\Delta \to 0$ and using p_i for $\tilde{q}_i$ results in

$$\dot{p}_i(t) = \frac{dp_i}{dt} = \beta_i(p_{-i}(t)) - p_i(t), \quad i = 1,2, \tag{5.33}$$

which is the counterpart of (5.31) in continuous time, known as continuous-time FP.

The following theorem, which is a variant of Theorem 3.2 of [164] for general $\tau_1, \tau_2 > 0$, establishes the convergence of the continuous-time FP for the case when players have two actions each (i.e. a 2×2 bi-matrix game).

Theorem 5.5 *Consider a two-player two-action continuous-time fictitious play with $(L^T G^1 L)(L^T G^2 L) \neq 0$, where G^i are the game matrices of players $\mathcal{P}^i$ $i = 1, 2$ and $L := (1, -1)^T$. The solutions of this continuous-time FP (5.33) satisfy*

$$\begin{aligned}\lim_{t\to\infty}(p_1(t) - \beta_1(p_2(t))) &= 0\\ \lim_{t\to\infty}(p_2(t) - \beta_2(p_1(t))) &= 0,\end{aligned} \tag{5.34}$$

where $\beta_i(p_{-i})$, $i = 1, 2$, are given in (5.26).

5.2.2   Fictitious play with observation errors

In order to illustrate the effects of observation errors, define the two-player two-action security game depicted in Figure 5.4, which is similar to those in Chapter 3, e.g. (3.1), except that the players here maximize their utility instead of minimizing cost values. The actions of the attacker $\mathcal{P}^1 = \mathcal{P}^A$ are $\mathcal{A}^A = \{a, na\}$ and of the defender $\mathcal{P}^2 = \mathcal{P}^D$ are $\mathcal{A}^D = \{d, nd\}$. The corresponding game matrices of the players are

$$G^1 = -G^A = \begin{array}{cc} & \begin{array}{cc}(d) & (nd)\end{array}\\ \begin{pmatrix} a & b \\ c & d \end{pmatrix} & \begin{array}{c}(a)\\ (na)\end{array} \end{array}, \quad G^2 = \left(-G^D\right)^T = \begin{array}{cc} & \begin{array}{cc}(a) & (na)\end{array}\\ \begin{pmatrix} e & g \\ f & h \end{pmatrix} & \begin{array}{c}(d)\\ (nd)\end{array} \end{array}, \tag{5.35}$$

where the parameters a, b, c, d, e, f, g are all positive. *Note that we depart here from the cost minimization as well as row-column player conventions, and players maximize their utilities instead of minimizing their costs.* The observation error (channel) matrices are given by

$$C_1 = \begin{pmatrix} 1 - \alpha & \gamma \\ \alpha & 1 - \gamma \end{pmatrix}, \quad C_2 = \begin{pmatrix} 1 - \varepsilon & \mu \\ \varepsilon & 1 - \mu \end{pmatrix}. \tag{5.36}$$

If defender $\mathcal{P}^2$ observes the strategy of the attacker, $p_1 = [p_{11} \, p_{12}]$, perfectly, then the resulting utility function of the defender is

$$U^2(p_1) = \begin{cases} e p_{11} + g p_{12}, & \text{if } \mathcal{P}^2 \text{ plays } d \\ f p_{11} + h p_{12}, & \text{if } \mathcal{P}^2 \text{ plays } nd \end{cases} \tag{5.37}$$

The utility function of the attacker ($\mathcal{P}^1$) can be defined similarly.

However, in this game, the players do not know each others' game matrices (only their own) and observe the actions of the opponent with the error probabilities shown in Figure 5.4. When the observation error probabilities are known to the players, then the following basic result holds.

Proposition 5.6 *Consider a discrete-time two-player fictitious play with imperfect observations as depicted in Figure 5.4. Let $\bar{q}_i$ be the observed frequency and q_i be the empirical frequency of player i. Then,*

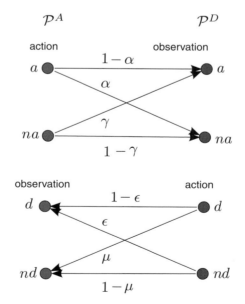

Figure 5.4 Observation errors between players are modeled as a binary channel with error probabilities $0 \le \alpha, \gamma, \varepsilon, \mu < 0.5$.

$$\text{Prob}\left(\lim_{k \to \infty} \overline{q}_i(k) = C_i q_i\right) = 1, i = 1, 2, \tag{5.38}$$

where $C_i, i = 1, 2$, are defined in (5.36). In other words, the observed frequency, $\overline{q}_i(k)$ converges almost surely to $C_i q_i$ as $k \to \infty$.

The properties of fictitious play with observation errors are further studied below using a Bayesian setting and an error compensation scheme.

Bayesian game interpretation

It is possible to analyze the security game introduced and the associated fictitious play with observation errors as a Bayesian game. Consider first the order of play $\mathcal{P}^1$-$\mathcal{P}^2$ where the defender P^2 observes attacker P^1's actions with observation errors as depicted in Figure 5.5. The formulation and discussion below also apply to the reverse order of play $\mathcal{P}^2$-$\mathcal{P}^1$.

The extensive form of the game in Figure 5.5 strongly resembles the Bayesian intrusion detection game of Section 5.1.1. The imperfect observations are captured here as actions of the *Nature player* as in Section 5.1.1. The fixed strategy of this Nature player is the same as the error probabilities in Figure 5.4. The information sets I and II in Figure 5.5 precisely correspond to the observations of the defender, P^2. While the defender knows which information set it is in, it cannot distinguish between the two nodes in each information set with certainty: given an observation, it can only tell if there is an attack or not with some probability of error.

The utility of the defender $\mathcal{P}^2$ given a particular information set $\{I, II\}$ and the fixed strategy of the nature player $p^n := [\alpha, 1 - \alpha, \gamma, 1 - \gamma]$, is

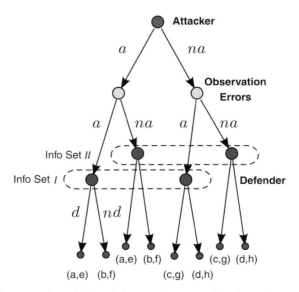

Figure 5.5 Bayesian game formulation of the security game with observation errors.

$$U^2(p_1, p^n) = \begin{cases} \dfrac{ep_{11}(1-\alpha)+gp_{12}\gamma}{p_{11}(1-\alpha)+p_{12}\gamma}, & \text{if } \mathcal{P}^2 \text{ plays } d \text{ in info. set } I \\[2ex] \dfrac{ep_{11}\alpha+gp_{12}(1-\gamma)}{p_{11}\alpha+p_{12}(1-\gamma)} & \text{if } \mathcal{P}^2 \text{ plays } d \text{ in info. set } II \\[2ex] \dfrac{fp_{11}(1-\alpha)+hp_{12}\gamma}{p_{11}(1-\alpha)+p_{12}\gamma} & \text{if } \mathcal{P}^2 \text{ plays } nd \text{ in info. set } I \\[2ex] \dfrac{fp_{11}\alpha+hp_{12}(1-\gamma)}{p_{11}\alpha+p_{12}(1-\gamma)} & \text{if } \mathcal{P}^2 \text{ plays } nd \text{ in info. set } II \end{cases} \tag{5.39}$$

Based on this utility function, the player can determine own best-response. The utility of the attacker, $\mathcal{P}^1$, is defined in a similar way using $p^n := [\varepsilon, 1-\varepsilon, \mu, 1-\mu]$ as the strategy of the nature player representing the observation errors of the attacker.

The following proposition links the studied FP and its Bayesian game interpretation. Although the result is not surprising, it serves as a theoretical basis for Algorithm 5.8 below, where the players compensate for the effects of observation errors.

Proposition 5.7 *If the observation error probabilities of an FP are known to its players, then at each stage, the best-response based on the distribution of the information sets in the corresponding Bayesian game is also the best-response based on the true empirical frequency.*

Proof If the defender $\mathcal{P}^2$ plays d, then its expected utility in the Bayesian game, obtained directly from (5.39), is

$$U^2(p_2, q) = q(I)\frac{ep_{11}(1-\alpha)+gp_{12}\gamma}{p_{11}(1-\alpha)+p_{12}\gamma} + q(II)\frac{ep_{11}\alpha+gp_{12}(1-\gamma)}{p_{11}\alpha+p_{12}(1-\gamma)},$$

where $q(I)$ and $q(II)$ are the probabilities of the information sets I and II, respectively. By definition of the game, these probabilities are

$$q(I) = p_{11}(1-\alpha)+p_{12}\gamma, \quad q(II) = p_{11}\alpha+p_{12}(1-\gamma).$$

It immediately follows that

$$U^2 = ep_{11}(1-\alpha)+gp_{12}\gamma+ep_{11}\alpha+gp_{12}(1-\gamma) = ep_{11}+gp_{12}.$$

Otherwise, if the defender plays nd, then the expected utility can be obtained similarly,

$$U^2 = fp_{11}+hp_{12}.$$

Both expected utilities are exactly the same as those in (5.37). A similar argument can also be made for the attacker, $\mathcal{P}^1$. Thus, at each stage, the best-responses of the players of the Bayesian game are the same as the ones based on the true empirical frequency.

$\square$

Algorithm 5.8 Fictitious play with known observation errors

1: Given game G^i and observation error (channel) matrices C_i, $i = 1, 2$ in (5.36),
2: **for** $k \in \{0, 1, 2, \ldots\}$ **do**
3:     Update the observed frequency of the opponent, $\overline{q}_{-i}$, using (5.31).
4:     Compute the estimated frequency using $q_{-i} = C_{-i}^{-1}\overline{q}_{-i}$, $i = 1, 2$.
5:     Compute the best-response mixed strategy $\beta_i(q_{-i}(k))$ using (5.26).
6:     Randomly generate an action $a_i(k)$ according to the best-response mixed strategy $\beta_i(q_{-i}(k))$, such that the expectation $E[a_i(k)] = \beta_i(q_{-i}(k))$.
7: **end for**

Algorithm 5.8 provides the players with a way of compensating for the observation errors if the error probabilities are known. If, on the other hand, the players do not have any way of estimating these probabilities and ignore the errors, then the solution will deviate from the original one without errors. The next theorem formalizes this deviation.

Theorem 5.9 *Consider a two-player two-action fictitious play game with imperfect observations described by observation error matrices (5.36) and visualized in Figure 5.4. If the players ignore these errors while playing stochastic FP and the condition $(L^T G^1 C_2 L)(L^T G^2 C_1 L) \neq 0$ holds, then the solutions of the continuous-time FP with imperfect observations (5.33) satisfy*

$$
\begin{aligned}
\lim_{t \to \infty} p_1(t) &= \sigma\left(\frac{G^1 C_2 \lim_{t \to \infty} p_2(t)}{\tau_1}\right), \\
\lim_{t \to \infty} p_2(t) &= \sigma\left(\frac{G^2 C_1 \lim_{t \to \infty} p_1(t)}{\tau_2}\right),
\end{aligned}
\tag{5.40}
$$

where $\beta_i(p_{-i})$, $i = 1, 2$, are given in (5.26), $L := (1, -1)^T$, and $\sigma(.)$ is the soft-max function defined in (5.27).

Proof The proof basically follows the same procedure as in Section 5.2.1 to approximate the discrete-time FP by the continuous-time version. Consider a stochastic discrete-time FP where the recursive equations to update the empirical frequencies are (for $i = 1, 2$)

$$
q_i(k+1) = \frac{k}{k+1} q_i(k) + \frac{1}{k+1} \beta_i(C_{-i} q_{-i}(k)),
\tag{5.41}
$$

where C_i are the observation error matrices in (5.36). Approximating this discrete-time FP with its continuous-time version yields

$$
\begin{aligned}
\dot{q}_1(t) &= \beta_1(C_2 q_2(t)) - q_1(t), \\
\dot{q}_2(t) &= \beta_2(C_1 q_1(t)) - q_2(t).
\end{aligned}
$$

Now using the result in Theorem 5.5 with $\tilde{G}^1 = G^1 C_2$ and $\tilde{G}^2 = G^2 C_1$, one obtains

$$
\lim_{t \to \infty} \left(p_1(t) - \sigma\left(\frac{G^1 C_2 p_2(t)}{\tau_1}\right)\right) = 0,
\tag{5.42}
$$

$$\lim_{t \to \infty} \left(p_2(t) - \sigma \left(\frac{G^2 C_1 p_1(t)}{\tau_2} \right) \right) = 0, \tag{5.43}$$

which completes the proof. $\qquad\qquad\qquad\qquad\qquad\qquad\qquad\qquad\qquad\qquad\qquad\square$

Although the convergence of the continuous-time fictitious play with imperfect observations does not guarantee that of the discrete-time counterpart, Theorem 5.9 does provide the necessary limiting results for the discrete-time version.

Generalized formulation and inaccurate error

The effect of observation errors on the convergence of 2×2 fictitious play to the NE has been discussed in the previous section, where it has been shown that if each player has a correct estimate of observation error probabilities, they can reverse the effect of the "error channel" to obtain the NE of the original game. We now present a generalization of these results to two-player games with $m, n > 2$ actions and inaccurate error estimates.

Introduce a two-player fictitious play with imperfect observations where the observation errors are quantified in the following matrices

$$C_1 = \begin{pmatrix} \alpha_{11} & \alpha_{12} & \dots & \alpha_{1m} \\ \alpha_{21} & \alpha_{22} & \dots & \alpha_{2m} \\ & \dots & & \\ \alpha_{m1} & \alpha_{m2} & \dots & \alpha_{mm} \end{pmatrix}, \quad C_2 = \begin{pmatrix} \varepsilon_{11} & \varepsilon_{12} & \dots & \varepsilon_{1n} \\ \varepsilon_{21} & \varepsilon_{22} & \dots & \varepsilon_{2n} \\ & \dots & & \\ \varepsilon_{n1} & \varepsilon_{n2} & \dots & \varepsilon_{nn} \end{pmatrix}, \tag{5.44}$$

where α_{ij}, $i, j = 1, \dots, m$ is the probability that attacker's ($\mathcal{P}^1 = \mathcal{P}^A$) action $i \in \mathcal{A}^A$ is erroneously observed as action $j \in \mathcal{A}^A$, and ε_{ij}, $i, j = 1, \dots, n$ is the probability that defender's ($\mathcal{P}^2 = \mathcal{P}^D$) action i is erroneously observed as action j. Both matrices are stochastic, i.e. $\alpha_{ij} \geq 0$, $\sum_{j=1}^m \alpha_{ij} = 1$, $i = 1, \dots, m$ and $\varepsilon_{ij} \geq 0$, $\sum_{j=1}^n \varepsilon_{ij} = 1$, $i = 1, \dots, n$.

Unlike the earlier case, suppose now that the players have the following imperfect estimates of the observation error probabilities

$$\overline{C}_1 = \begin{pmatrix} \overline{\alpha}_{11} & \overline{\alpha}_{12} & \dots & \overline{\alpha}_{1m} \\ \overline{\alpha}_{21} & \overline{\alpha}_{22} & \dots & \overline{\alpha}_{2m} \\ & \dots & & \\ \overline{\alpha}_{m1} & \overline{\alpha}_{m2} & \dots & \overline{\alpha}_{mm} \end{pmatrix}, \quad \overline{C}_2 = \begin{pmatrix} \overline{\varepsilon}_{11} & \overline{\varepsilon}_{12} & \dots & \overline{\varepsilon}_{1n} \\ \overline{\varepsilon}_{21} & \overline{\varepsilon}_{22} & \dots & \overline{\varepsilon}_{2n} \\ & \dots & & \\ \overline{\varepsilon}_{n1} & \overline{\varepsilon}_{n2} & \dots & \overline{\varepsilon}_{nn} \end{pmatrix}, \tag{5.45}$$

where both matrices are stochastic, i.e. $\overline{\alpha}_{ij} \geq 0$, $\sum_{j=1}^m \overline{\alpha}_{ij} = 1$, $i = 1, \dots, m$, and $\overline{\varepsilon}_{ij} \geq 0$, $\sum_{j=1}^n \overline{\varepsilon}_{ij} = 1$, $i = 1, \dots, n$. Note that the players having imperfect error estimates means that

$$C_1 \neq \overline{C}_1, \quad C_2 \neq \overline{C}_2.$$

If both players have their estimates of the error probabilities (5.45), they can play the stochastic FP described in Algorithm 5.8 with $(\overline{C}_i)^{-1} \overline{q}_{-i}$ to compensate for observation errors. Again, using a procedure similar to those in the previous cases, the discrete-time FP

$$q_1(k+1) = \frac{k}{k+1}q_1(k) + \frac{1}{k+1}\sigma\left(\frac{G^1(\overline{C}_2)^{-1}C_2 q_2(k)}{\tau_1}\right),$$

$$q_2(k+1) = \frac{k}{k+1}q_2(k) + \frac{1}{k+1}\sigma\left(\frac{G^2(\overline{C}_1)^{-1}C_1 q_1(k)}{\tau_2}\right) \tag{5.46}$$

can be approximated by its continuous-time version in terms of best-response dynamics

$$\dot{p}_1(t) = \sigma\left(\frac{G^1(\overline{C}_2)^{-1}C_2 p_2(t)}{\tau_1}\right) - p_1(t),$$

$$\dot{p}_2(t) = \sigma\left(\frac{G^2(\overline{C}_1)^{-1}C_1 p_1(t)}{\tau_2}\right) - p_2(t). \tag{5.47}$$

Then, the pair of mixed strategies (p_1^*, p_2^*) that satisfies

$$p_1^*(t) = \sigma\left(\frac{G^1(\overline{C}_2)^{-1}C_2 p_2^*(t)}{\tau_1}\right),$$

$$p_2^*(t) = \sigma\left(\frac{G^2(\overline{C}_1)^{-1}C_1 p_1^*(t)}{\tau_2}\right),$$

is by definition the NE solution to the stochastic FP (5.47) where the players implement imperfect error compensation.

The convergence properties of stochastic FP with imperfect error compensation is investigated next in the continuous time. Linearizing the best-response dynamics (5.47) at the NE point (p_1^*, p_2^*) allows one to assess the local stability of this point by examining the eigenvalues of the appropriate Jacobian matrix. It is possible to write $p_i(t)$ as

$$p_i(t) = p_i^*(t) + \delta p_i(t).$$

Since both $p_i(t)$ and $p_i^*(t)$ are probability vectors and their elements add up to 1, the sum of the entries of $\delta p_i(t)$ must be zero for $i = 1, 2$. Therefore,

$$\delta p_1(t) = Q\tilde{p}_1(t), \quad \delta p_2(t) = R\tilde{p}_2(t), \tag{5.48}$$

for some $\tilde{p}_1(t)$ and $\tilde{p}_2(t)$. The matrix Q has dimensions $m \times (m-1)$ and matrix R has dimensions $n \times (n-1)$, and

$$\mathbf{1}^T Q = \mathbf{0}, \text{ and } Q^T Q = I,$$
$$\mathbf{1}^T R = \mathbf{0}, \text{ and } R^T R = I,$$

where I is the identity matrix and $\mathbf{1}$ and $\mathbf{0}$ are vectors of appropriate dimensions with all entries ones and zeros, respectively. Then, the linearized system dynamics are

$$\frac{d}{dt}\begin{pmatrix} \tilde{p}_1(t) \\ \tilde{p}_2(t) \end{pmatrix} = M \cdot \begin{pmatrix} \tilde{p}_1(t) \\ \tilde{p}_2(t) \end{pmatrix},$$

where the reduced-order Jacobian matrix M is

$$M := \begin{pmatrix} -I & Q^T \nabla\tilde{\beta}_1(p_2^*) R \\ R^T \nabla\tilde{\beta}_2(p_1^*) Q & -I \end{pmatrix}, \tag{5.49}$$

and

$$\tilde{\beta}_1(p_2) = \sigma \left(\frac{G^1(\overline{C}_2)^{-1} C_2 p_2}{\tau_1} \right),$$

$$\tilde{\beta}_2(p_1) = \sigma \left(\frac{G^2(\overline{C}_1)^{-1} C_1 p_1}{\tau_2} \right).$$

The following theorem now provides a necessary condition for the discrete-time FP process with observation errors to converge almost surely to an equilibrium point.

Theorem 5.10 *Let (p_1^*, p_2^*) be an NE of the FP system (5.47). If the Jacobian matrix M has a positive eigenvalue λ with $Re(\lambda) > 0$, then the equilibrium (p_1^*, p_2^*) is unstable under both the discrete- and continuous-time dynamics (5.46) and (5.47), in the sense that*

$$\text{Prob} \left\{ \lim_{k \to \infty} q_i(k) = p_i^* \right\} = 0, \ i = 1, 2. \tag{5.50}$$

When $m = n = 2$ as a special case, the equilibrium point (p_1^*, p_2^*) is globally stable for the continuous-time FP under some mild assumptions as summarized in the following theorem.

Theorem 5.11 *Consider a two-player two-action fictitious play with imperfect observations quantified by error matrices*

$$C_1 = \begin{pmatrix} 1 - \alpha & \gamma \\ \alpha & 1 - \gamma \end{pmatrix}, \ C_2 = \begin{pmatrix} 1 - \varepsilon & \mu \\ \varepsilon & 1 - \mu \end{pmatrix}, \tag{5.51}$$

and the players having the following estimates of the error probabilities

$$\overline{C}_1 = \begin{pmatrix} 1 - \overline{\alpha} & \overline{\gamma} \\ \overline{\alpha} & 1 - \overline{\gamma} \end{pmatrix}, \ \overline{C}_2 = \begin{pmatrix} 1 - \overline{\varepsilon} & \overline{\mu} \\ \overline{\varepsilon} & 1 - \overline{\mu} \end{pmatrix}. \tag{5.52}$$

In the stochastic FP, where the players compensate for observation errors using their estimates, if $(L^T M_1(\overline{C}_2)^{-1} C_2 L)(L^T M_2(\overline{C}_1)^{-1} C_1 L) \neq 0$, then the continuous-time FP with imperfect observations (5.47) converges to

$$\lim_{t \to \infty} p_1(t) = \sigma \left(\frac{G^1(\overline{C}_2)^{-1} C_2 \lim_{t \to \infty} p_2(t)}{\tau_1} \right),$$

$$\lim_{t \to \infty} p_2(t) = \sigma \left(\frac{G^2(\overline{C}_1)^{-1} C_1 \lim_{t \to \infty} p_1(t)}{\tau_2} \right). \tag{5.53}$$

Here, $\sigma(.)$ is the soft-max function defined in (5.27) and $L := (1, -1)^T$.

Remark 5.12 *The following remarks apply to Theorem 5.11:*

- *The result in this theorem can be extended to the case where only one player is restricted to two actions, and the other has more than two actions.*
- *When $\overline{C}_i = C_i = I$, $i = 1, 2$, the result reduces to convergence of stochastic FP in reference [164].*

- When $\overline{C}_i = I$, $i = 1,2$, the result reduces to convergence of stochastic FP with imperfect observations in Theorem 5.9.
- When $\overline{C}_i = C_i$, $i = 1,2$, the result reduces to convergence of stochastic FP with imperfect observations where players completely reverse the effect of erroneous channels.

5.2.3   Fictitious play with decision errors

The occurrence of observation errors is not the only issue that comes up in the application of fictitious play to security problems. A similar set of problems arises when players are not totally rational or the channels carrying commands are error prone. In many security scenarios the attackers or defenders may act irrationally by misjudging their preferences or due to emotional reasons. This situation is similar to the well-known *"trembling hand"* problem discussed in the game theory literature (e.g. see [31, Section 3.5.5]). Another case is when the action commands issued by attackers (e.g. using botnets) or defenders (e.g. IDPSs) may reach the actuators, i.e. actual systems implementing them, with some error. These cases and the resulting FP with decision errors can be analyzed similarly to that in Section 5.2.2. Since the methods are very similar to those used in the previous subsection, only the main results are presented here.

Consider a game where the attacker ($\mathcal{P}^1 = \mathcal{P}^A$) makes decision errors with probabilities α_{ij}'s, where α_{ij}, $i,j = 1,\ldots,m$, is the probability that $\mathcal{P}^1$ intends to play action $i \in \mathcal{A}^A$ but ends up playing action $j \in \mathcal{A}^A$, $\alpha_{ij} \geq 0$, $\sum_{j=1}^m \alpha_{ij} = 1$, $i = 1,\ldots,m$. Similarly, the decision error probabilities of the defender ($\mathcal{P}^2 = \mathcal{P}^D$) are given by ε_{ij}, $\varepsilon_{ij} \geq 0$, $\sum_{j=1}^m \varepsilon_{ij} = 1$, $i = 1,\ldots,n$. The corresponding decision error matrices are

$$D_1 = \begin{pmatrix} \alpha_{11} & \alpha_{12} & \cdots & \alpha_{1m} \\ \alpha_{21} & \alpha_{22} & \cdots & \alpha_{2m} \\ \cdots & & & \\ \alpha_{m1} & \alpha_{m2} & \cdots & \alpha_{mm} \end{pmatrix}, \quad D_2 = \begin{pmatrix} \varepsilon_{11} & \varepsilon_{12} & \cdots & \varepsilon_{1n} \\ \varepsilon_{21} & \varepsilon_{22} & \cdots & \varepsilon_{2n} \\ \cdots & & & \\ \varepsilon_{n1} & \varepsilon_{n2} & \cdots & \varepsilon_{nn} \end{pmatrix}. \tag{5.54}$$

In the special case of $m = n = 2$, the decision error matrices can be written as

$$D_1 = \begin{pmatrix} 1-\alpha & \gamma \\ \alpha & 1-\gamma \end{pmatrix}, \quad D_2 = \begin{pmatrix} 1-\varepsilon & \mu \\ \varepsilon & 1-\mu \end{pmatrix}. \tag{5.55}$$

As in Section 5.2.2, we first study the case where the players both have complete information about the decision error matrices D_i, $i = 1,2$. This formulation can be interpreted as a stochastic version of the "trembling hand" scenario. Specifically, suppose that the players want to randomize their empirical frequency $\overline{p}_i$ (instead of the frequency of their intended actions, or intended frequency, p_i) by including an entropy term in their utility function. Then, the corresponding utility functions are

$$U_i(p_i, p_{-i}) = p_i^T \tilde{G}^i p_{-i} + \tau_i H(D_i p_i), \quad i = 1,2, \tag{5.56}$$

where p_i's are the intended frequencies, $\tilde{G}^1 = D_1^T G^1 D_2$ and $\tilde{G}^2 = D_2^T G^2 D_1$. Using $\overline{p}_i := D_i p_i$, $i = 1,2$, the utility functions can now be written as

$$U_i(p_i, p_{-i}) = \overline{p}_i^T G^i \overline{p}_{-i} + \tau_i H(\overline{p}_i).$$

The game is thus reduced to the one without decision errors and the NE of the static game satisfies

$$\overline{p}_i^* = \beta_i \left(\overline{p}_{-i}^* \right), \tag{5.57}$$

or equivalently (with the assumption that D_i's are invertible)

$$p_i^* = (D_i)^{-1} \beta_i \left(D_{-i} p_{-i}^* \right), \tag{5.58}$$

where the players' best-response functions are

$$p_i = (D_i)^{-1} \beta_i (\overline{p}_{-i}) = (D_i)^{-1} \sigma \left(\frac{G^i \overline{p}_{-i}}{\tau_i} \right), \ i = 1, 2. \tag{5.59}$$

In the corresponding fictitious play (the *"trembling hand stochastic FP"*), the mean dynamics of the empirical frequencies are

$$\overline{q}_1(k+1) = \frac{k}{k+1} \overline{q}_1(k) + \frac{1}{k+1} \beta_1(\overline{q}_2(k)),$$
$$\overline{q}_2(k+1) = \frac{k}{k+1} \overline{q}_2(k) + \frac{1}{k+1} \beta_2(\overline{q}_1(k)). \tag{5.60}$$

The continuous-time approximation of these dynamics become

$$\dot{\overline{p}}_1(t) = \beta_1 \left(\overline{p}_2(t) \right) - \overline{p}_1(t),$$
$$\dot{\overline{p}}_2(t) = \beta_2 \left(\overline{p}_1(t) \right) - \overline{p}_2(t). \tag{5.61}$$

Each player $\mathcal{P}^i$ can observe here the opponent's empirical frequency $\overline{p}_{-i}$ directly, and hence does not need to know D_{-i} to compute the best-response. Thus, with knowledge of their own decision errors, the players can completely precompensate for these errors and the equilibrium empirical frequencies remain the same as those of the original game without decision errors. The following theorem provides a convergence result for the FP process with decision errors in the case of $m = n = 2$.

Proposition 5.13 *Consider a two-player two-action fictitious play where players make decision errors with invertible decision error matrices D_1 and D_2 in (5.55). Suppose that at each step, the players calculate their best-responses taking into account their own decision errors using (5.59). If $(L^T G^1 L)(L^T G^2 L) \neq 0$ holds, then the continuous-time FP with decision errors converges to*

$$\lim_{t \to \infty} p_1(t) = D_1^{-1} \sigma \left(\frac{G^1 D_2 \lim_{t \to \infty} p_2(t)}{\tau_1} \right),$$
$$\lim_{t \to \infty} p_2(t) = D_2^{-1} \sigma \left(\frac{G^2 D_1 \lim_{t \to \infty} p_1(t)}{\tau_2} \right), \tag{5.62}$$

where $\sigma(.)$ is the soft-max function defined in (5.27) and $L = (1, -1)^T$.

In the opposite case when the decision error probabilities are not known to the players and are ignored, the mean dynamics of the empirical frequencies become

$$\overline{q}_1(k+1) = \frac{k}{k+1}\overline{q}_1(k) + \frac{1}{k+1}D_1\beta_1(\overline{q}_2(k)),$$

$$\overline{q}_2(k+1) = \frac{k}{k+1}\overline{q}_2(k) + \frac{1}{k+1}D_2\beta_2(\overline{q}_1(k)). \tag{5.63}$$

The continuous-time approximations of these dynamics are

$$\dot{\overline{p}}_1(t) = D_1\beta_1(\overline{p}_2(t)) - \overline{p}_1(t),$$

$$\dot{\overline{p}}_2(t) = D_2\beta_2(\overline{p}_1(t)) - \overline{p}_2(t). \tag{5.64}$$

The equilibrium solution to (5.64) is naturally different from the original one.

The following two theorems are direct counterparts of Theorems 5.10 and 5.11 of Section 5.2.2.

Theorem 5.14 *Consider the two-player FP with decision errors where both players are unaware of the decision error probabilities. Let $(\overline{p}_1^*, \overline{p}_2^*)$ be an equilibrium point of system (5.64). If the Jacobian matrix*

$$M_D = \begin{pmatrix} -I & Q^T D_1 \nabla\beta_1(\overline{p}_2^*) R \\ R^T D_2 \nabla\beta_2(\overline{p}_1^*) Q & -I \end{pmatrix}$$

has an eigenvalue λ with $Re(\lambda) > 0$, then the equilibrium $(\overline{p}_1^*, \overline{p}_2^*)$ is unstable under the discrete- or continuous-time dynamics, (5.63) and (5.64), respectively, so that

$$Prob\left\{\lim_{k\to\infty}\overline{q}_i(k) = \overline{p}_i^*\right\} = 0, \ i = 1,2. \tag{5.65}$$

We note additionally that when $m = n = 2$, the point $(\overline{p}_1^*, \overline{p}_2^*)$ is globally stable for the continuous-time system under some mild assumptions.

Theorem 5.15 *Consider a two-player two-action fictitious play where players make decision errors according to D_1 and D_2 in (5.55). Suppose that both players are unaware of all the decision error probabilities and use the regular stochastic FP in Algorithm 5.4. If D_i, $i = 1,2$, are invertible and $(L^T M_1 D_2 L)(L^T M_2 D_1 L) \neq 0$, the solutions of continuous-time FP process with decision errors (5.61) satisfy*

$$\lim_{t\to\infty}\overline{p}_1(t) = D_1\sigma\left(\frac{G^1\lim_{t\to\infty}\overline{p}_2(t)}{\tau_1}\right),$$

$$\lim_{t\to\infty}\overline{p}_2(t) = D_2\sigma\left(\frac{G^2\lim_{t\to\infty}\overline{p}_1(t)}{\tau_2}\right), \tag{5.66}$$

where $\sigma(.)$ is the soft-max function defined in (5.27) and $L = (1, -1)^T$.

Note that the result in this theorem can be extended to the case where only one player is restricted to two actions, and the other has more than two actions [164]. Furthermore, the results in the subsection can easily be extended to address the general case where each player has estimates of the decision errors and includes them in the best-response to compensate for decision errors.

5.2.4 Time-invariant and adaptive fictitious play

It is possible to define a version of the fictitious play which uses time-invariant frequency updates by modifying the running or cumulative averaging of observations in the empirical frequency calculation

$$q_j(k+1) = \frac{1}{k+1} \sum_{l=0}^{k} v_{a_j}(l) = \frac{k}{k+1} q_j(k) + \frac{1}{k+1} v_{a_j}(k), \tag{5.67}$$

where $q_j = [q_{j1}, \ldots, q_{jN}]$ is the empirical frequency vector of player $\mathcal{P}^j$, k the time step, and $v_{a_j}(k)$ is a vector of ones and zeros with a single one at the index corresponding to action a_j of the player. The recursive empirical frequency computation in (5.67) is actually a low-pass filter. At the same time, it is a time-varying dynamical system that is more difficult to analyze compared to a time-invariant one. As an alternative, the observations can be filtered using an exponential moving average scheme with a (piecewise) fixed smoothing factor.

Consider the following alternative computation of time-invariant empirical frequency r by taking the exponential moving average

$$r_i(k+1) = (1-\eta)r_i(k) + \eta v_i(k), \quad r_i(1) = v_i(0), \tag{5.68}$$

where the parameter η is a positive smoothing constant. Then, the characteristics of the low-pass filter (e.g. bandwidth) can be adjusted by varying the parameter $0 < \eta < 1$. The time-invariant empirical frequencies in (5.67) can also be defined directly as estimates of mixed strategies,

$$r_i(k) = (1-\eta)^{k-1} v_i(0) + \cdots + (1-\eta)\eta v_i(k-2) + \eta v_i(k-1), \quad k \geq 2. \tag{5.69}$$

Note that the weight on older data points decreases exponentially in the computation, which places more emphasis on recent observations.

The *time-invariant fictitious play*[2] leads to the following time-invariant dynamic system for two players:

$$r_1(k+1) = (1-\eta)r_1(k) + \eta\beta_1(r_2(k)), \tag{5.70}$$
$$r_2(k+1) = (1-\eta)r_2(k) + \eta\beta_2(r_1(k)),$$

where β_i are the best-response functions of the players. We next establish the relationship between time-varying and time-invariant empirical frequencies, q and r.

Proposition 5.16 *The time-varying empirical frequencies q are related to the time-invariant ones r through*

$$q_i(k+1) = \frac{1}{k+1} \left(\frac{2\eta-1}{\eta} r_i(1) + r_i(2) + \cdots + r_i(k) + \frac{r_i(k+1)}{\eta} \right). \tag{5.71}$$

---

[2] With a slight abuse of terminology, we will henceforth be referring to "fictitious play with time-invariant frequency updates" as "time-invariant fictitious play."

Proof It immediately follows from the definition in (5.67) that

$$v_i(0) = r_i(1),$$

$$v_i(1) = \frac{1}{\eta}\left[r_i(2) - (1-\eta)r_i(1)\right],$$

$$\cdots$$

$$v_i(k) = \frac{1}{\eta}\left[r_i(k+1) - (1-\eta)r_i(k)\right].$$

$$(5.72)$$

Summing up the left- and right-hand sides of these equations yields

$$\sum_{j=0}^{k} v_i(j) = \frac{2\eta - 1}{\eta}r_i(1) + r_i(2) + \cdots + r_i(k) + \frac{r_i(k+1)}{\eta},$$

$$(5.73)$$

and hence the result in (5.71) follows.   □

Consider now the two-player two-action security game in Section 5.2.2 with the game matrices

$$G^1 = -G^A = \begin{matrix} (d) & (nd) \\ \begin{pmatrix} a & b \\ c & d \end{pmatrix} & \begin{matrix} (a) \\ (na) \end{matrix} \end{matrix} \quad , \quad G^2 = \left(-G^D\right)^T = \begin{matrix} (a) & (na) \\ \begin{pmatrix} e & g \\ f & h \end{pmatrix} & \begin{matrix} (d) \\ (nd) \end{matrix} \end{matrix} \quad , \quad (5.74)$$

as before, where all entries are positive. We make the following assumptions on these game parameters:

- $a < c$: when the defender ($\mathcal{P}^2$) defends, the benefit of an attack is less;
- $b > d$: when the defender does not defend an attack, the attacker has more to gain from it;
- $e > f$: when an attack is defended, the defender benefits more;
- $g < h$: when there is defensive action without attack (false alarm), the cost to the defender is higher.

These assumptions are consistent with previously discussed security games, e.g. the matrix intrusion detection game in (3.1).

For this game, it is possible to establish the local stability of the time-invariant fictitious play in (5.70), i.e. if the players start with sufficiently close (correct) estimates of each other's mixed strategies, then their empirical frequency estimates converge asymptotically to the real values, as shown below.

Theorem 5.17 *Consider the time-invariant fictitious play system (5.70) of the two-player two-action security game in Section 5.2.2. Assume that the positive game parameters satisfy $a < c$, $b > d$, $e > f$, and $g < h$ and the system admits an equilibrium solution $(\bar{r}_1, \bar{r}_2)$ such that*

$$\begin{matrix} \bar{r}_1 = \beta_1(\bar{r}_2) \\ \bar{r}_2 = \beta_2(\bar{r}_1) \end{matrix}.$$

Then, the empirical frequency estimates in (5.70) are asymptotically stable around the given equilibrium point if, and only if

$$\eta < \frac{2}{[(c-a)+(b-d)][(e-f)+(h-g)]\dfrac{\bar{r}_{11}\bar{r}_{12}\bar{r}_{21}\bar{r}_{22}}{\tau_1\tau_2}+1}. \tag{5.75}$$

Proof The time-invariant empirical frequency vectors for this game are

$$r_1 = [r_{11},\, 1-r_{11}]^T \text{ and } r_2 = [r_{21},\, 1-r_{21}]^T.$$

Thus, the system (5.70) can be reduced to two dimensions and written as

$$r(k+1) = F(r(k)), \text{ or } r_1(k+1) = F_1(r(k)) \text{ and } r_2(k+1) = F_2(r(k)).$$

Linearizing this system around the equilibrium point $(\bar{r}_1, \bar{r}_2)$ yields the Jacobian matrix

$$M := \frac{\partial F(r)}{\partial r} = \begin{pmatrix} \dfrac{\partial F_1(r)}{\partial r_{11}} & \dfrac{\partial F_1(r)}{\partial r_{21}} \\[3mm] \dfrac{\partial F_2(r)}{\partial r_{11}} & \dfrac{\partial F_2(r)}{\partial r_{21}} \end{pmatrix}. \tag{5.76}$$

These definitions directly yield

$$\frac{\partial F_1(r)}{\partial r_{11}} = \frac{\partial F_2(r)}{\partial r_{21}} = 1-\eta, \tag{5.77}$$

and

$$\frac{\partial F_1(r)}{\partial r_{21}} = \eta\frac{\partial \beta_1(r_2)}{\partial r_{21}}. \tag{5.78}$$

Likewise, from the definition of β,

$$\frac{G^1 r_2}{\tau_1} = \begin{pmatrix} \frac{1}{\tau_1}[ar_{21}+b(1-r_{21})] \\[2mm] \frac{1}{\tau_1}[cr_{21}+d(1-r_{21})] \end{pmatrix}, \tag{5.79}$$

$$\beta_{11}(r_2) = \frac{e^{\left(\frac{1}{\tau_1}[ar_{21}+b(1-r_{21})]\right)}}{e^{\left(\frac{1}{\tau_1}[ar_{21}+b(1-r_{21})]\right)} + e^{\left(\frac{1}{\tau_1}[cr_{21}+d(1-r_{21})]\right)}}, \tag{5.80}$$

and

$$\frac{d\beta_1(r_2)}{dr_{21}} = \frac{1}{\tau_1}[(a-c)+(d-b)]\beta_{11}(r_2)\beta_{12}(r_2) \tag{5.81}$$

follow. Therefore, the other entries of the Jacobian matrix are

$$\frac{\partial F_1(r)}{\partial r_{21}} = \frac{\eta}{\tau_1}[(a-c)+(d-b)]\beta_{11}(r_2)\beta_{12}(r_2), \tag{5.82}$$

and

$$\frac{\partial F_2(r)}{\partial r_{11}} = \frac{\eta}{\tau_2}[(e-f)+(h-g)]\beta_{21}(r_1)\beta_{22}(r_1), \tag{5.83}$$

which is computed similarly.

Thus, at the equilibrium point, the following hold:

$$\frac{\partial F_1(\bar{r})}{\partial r_{21}} = \frac{\eta}{\tau_1}[(a-c)+(d-b)]\bar{r}_{11}\bar{r}_{12}, \tag{5.84}$$

$$\frac{\partial F_2(\bar{r})}{\partial r_{11}} = \frac{\eta}{\tau_2}[(e-f)+(h-g)]\bar{r}_{21}\bar{r}_{22}. \tag{5.85}$$

Finally, one can show using straightforward algebraic manipulations that in order for the spectral radius of the Jacobian matrix (or absolute value of all its eigenvalues) to be less than one, the necessary and sufficient condition (5.75) has to be satisfied. □

Numerical example

The following numerical example illustrates the properties of stochastic and time-invariant discrete-time FP. The game (payoff) matrices of players are chosen as

$$M_1 = \begin{pmatrix} 1 & 5 \\ 3 & 2 \end{pmatrix}, \quad M_2 = \begin{pmatrix} 4 & 1 \\ 3 & 5 \end{pmatrix}. \tag{5.86}$$

Note that the players maximize their payoffs here, departing from the cost minimization convention. The static game with simultaneous moves admits the mixed strategy NE: $p_1^* = [0.4, 0.6]$ and $p_2^* = [0.6, 0.4]$. The entropy parameters are picked as $\tau_1 = \tau_2 = 0.2$.

The evolution of empirical frequency estimates for the time-varying stochastic FP in (5.67) and time-invariant FP in (5.68) with a step-size of $\eta = 0.01$ are depicted in Figures 5.6 and 5.7, respectively. In both variants of FP, there is ongoing fluctuation as

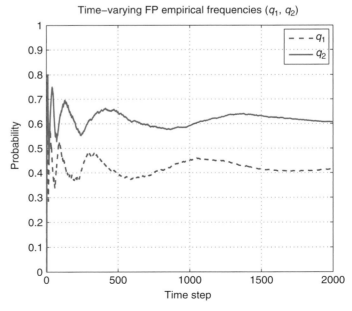

Figure 5.6 Evolution of observed empirical frequencies of players in time-varying stochastic fictitious play.

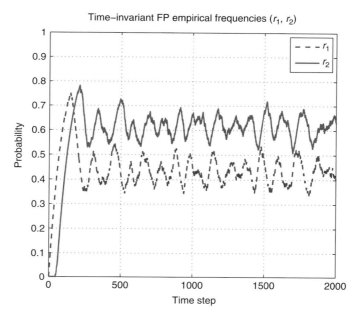

Figure 5.7 Evolution of observed empirical frequencies of players in time-invariant fictitious play with a step size of $\eta = 0.01$.

a result of randomization due to the entropy term in the player utilities. It is observed that time-invariant FP converges quickly to a random limit cycle with high variance since the step-size is fixed, while that of time-varying FP naturally decreases over time.

Adaptive time-invariant fictitious play

Choosing the step size of time-invariant FP to be large leads to high variance fluctuations, whereas keeping it small increases the convergence time. However, there is no reason to keep the step size constant and not vary it adaptively similarly to, for example, Newton's method in gradient descent. Adaptive time-invariant FP is based on this idea. One possible definition provided in Algorithm 5.18 differs from regular time-invariant FP through the dynamically updated step size, which is constant over each time period. In the specific example implementation, the step size is either kept fixed or halved, based on the variance of empirical frequency in the previous time window.

The adaptive time-invariant FP is studied using the numerical example above. Initial and minimum step sizes are chosen as $\eta_0 = 0.1$ and $\eta_{min} = 0.0005$, respectively. The time window for updating the step size is $T = 50$ steps. The evolution of the observed empirical frequencies are depicted in Figure 5.8, which shows that adaptive time-invariant FP converges much faster than the time-varying FP. The reason behind faster convergence is the adaptive update of step size, illustrated in Figure 5.9. Note that when compared to fixed $1/k$ decrease in the time-varying FP, the step sizes in the adaptive time-invariant FP are higher in the beginning and smaller afterwards, resulting in aggressive convergence first and less fluctuation in the stable phase.

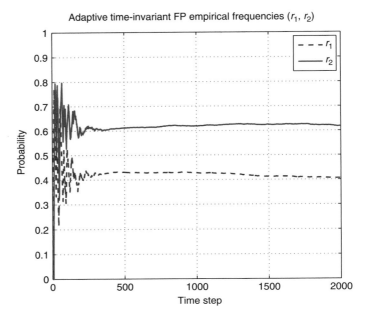

Figure 5.8 Evolution of observed empirical frequencies of players in adaptive time-invariant fictitious play described in Algorithm 5.18.

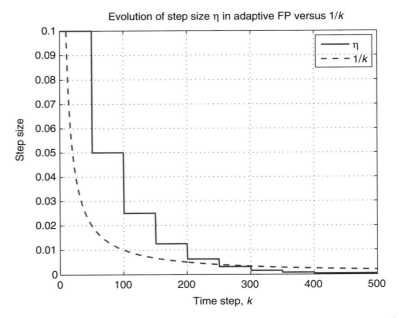

Figure 5.9 Evolution of step size in adaptive time-invariant fictitious play compared to $1/k$ decrease in the time-varying version.

Algorithm 5.18 Adaptive time-invariant fictitious play

1: Given game matrices G^1, G^2, initial step size η_0, minimum step size η_{min}, and window size T.
2: **for** $k \in \{0, 1, 2, \ldots\}$ **do**
3:     Update the observed empirical frequency of the opponent, q_{-i}, using (5.68).
4:     Compute the best-response mixed strategy $\beta_i(q_{-i}(k))$ using (5.26).
5:     Randomly play an action $a_i(k)$ according to the best-response mixed strategy $\beta_i(q_{-i}(k))$, such that the expectation $E[a_i(k)] = \beta_i(q_{-i}(k))$.
6:     **if** at the end of a time window,   mod $(k, T) = 0$, **then**
7:         Compute standard deviation (std) of observed empirical frequency in the time window $[q_{-i}(k), \ldots, q_{-i}(k-T)]$:

$$\text{std}(k) = \sqrt{\frac{1}{T} \sum_{k-T}^{k} \left(q_{-i}(k) - \left[\frac{1}{T+1} \sum_{k-T}^{k} q_{-i}(k) \right] \right)^2}$$

8:         **if** the computed std(k) has decreased compared to previous time window **then**
9:             Decrease step size: $\eta = 0.5\eta$ and $\eta = \max(\eta, \eta_{min})$.
10:         **else**
11:             Keep step size η constant.
12:         **end if**
13:     **end if**
14: **end for**

5.3   Discussion and further reading

Bayesian games were first introduced by Nobel Laureate John Harsanyi as a method for analysis of games of incomplete information. Considering the crucial role that information limitations play in security games, the Bayesian game approach and fictitious play discussed in this chapter should only be seen as a starting point.

Two illustrative Bayesian security games have been presented in Section 5.1: a Bayesian intrusion detection game based on reference [6] and a Bayesian game for wireless networks summarizing reference [153]. An earlier work applying Bayesian games to intrusion detection in wireless ad-hoc networks is reference [99].

Section 5.2 on security games with observation and decision errors builds upon the results reported in reference [128]. Properties of fictitious play without observation and decision errors have been analyzed earlier in references [34, 163, 164]. A more detailed discussion on time-invariant and adaptive fictitious play presented in Section 5.2.4 is in reference [130].

Part III

Decision making for network security

6 Security risk-management

Chapter overview

1. Quantitative risk-management
   - risk in networked systems and organizations
   - a probabilistic risk framework: Risk-Rank
   - dynamic risk mitigation and control
2. Security investment games
   - influence network and game model
   - equilibrium and convergence analysis
   - incentives and game design
3. Cooperative games for security investments
   - coalitional games and coalition formation

Chapter summary

Security risk assessment and response are posed as dynamic resource allocation problems. First, a quantitative risk-management framework based on the probabilistic evolution of risk and Markov decision processes is presented. Second, a noncooperative game model is analyzed for the long-term security investments of interdependent organizations. In addition, incentive mechanisms are investigated to achieve organization-wide objectives. Finally, a cooperative game is studied to develop a better understanding of coalition formation and operation between the divisions of large organizations.

6.1   Quantitative risk-management

6.1.1   Risk in networked systems and organizations

Networked systems have become an integral and indispensable part of daily business. Hence, system failures and security compromises have direct consequences for organizations in multiple dimensions. For a company in e-commerce, network downtime translates to millions of dollars lost per second. For a telecommunications company, stolen customer data may turn into a public relations nightmare. The municipality of an entire city can be held hostage by a single disgruntled former system administrator. In a simpler and more common scenario, a computer virus or worm infection may simply mean a free day for the entire office, resulting in significant productivity loss.

Such risks and security threats are the novel consequences of the IT revolution, which at the same time has brought immense productivity gains and new business opportunities. As organizations and enterprises are becoming increasingly aware of these risks, they have little option other than learning how to manage emerging IT and security risks. Risk-management in this specific area is, consequently, a young and vibrant field with substantial research challenges and opportunities.

Early IT and security risk-management research has been mostly empirical and qualitative in nature. This is partly due to the fact that IT risk-management has its roots more in social and management sciences. It can also be attributed to the young age of the field, for example when compared to financial risk-management. However, the research landscape dominated by qualitative principles, empirical calculations on spreadsheets, and white papers is increasingly enriched by quantitative models and approaches.

Quantitative and analytical frameworks not only formalize risk-management[1] processes but also provide a foundation for computer-assisted assessment and decision-making capabilities. Given the complexity of the problem and the underlying networked systems, a risk-management scheme that relies on too many manual processes cannot be expected to succeed. Even in personal computers, most of the security services such as antivirus software, firewalls, and patching of browsers rely on automated updates. Unsurprisingly, a variety and increasing number of software solutions exist to streamline and automate data collection and risk-management processes in large-scale organizations.

One of the objectives of quantitative risk models is to take this formalization trend one step further and create a solid analytical foundation for security and IT risk. Such a mathematical abstraction is useful to combine seemingly different problems under a single umbrella, facilitate future research, and develop computer-based scalable solutions that rely on rational principles and transparency. On the other hand, one should not forget that risk-management is not a purely technical problem and cannot be addressed by purely technical solutions. Organizational and human aspects play at least as important a part as do the technical ones.

---

[1] We will use the term risk exclusively in the context of security and IT in this chapter.

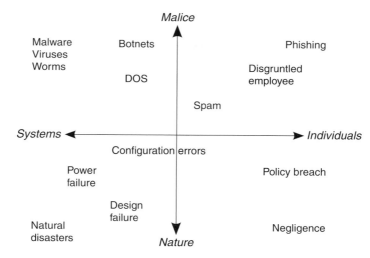

Figure 6.1 Properties of IT and security risks: nature versus malice and systems versus individuals.

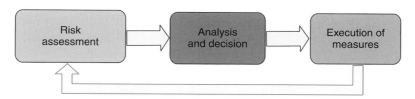

Figure 6.2 Three main components of the security risk-management.

When defining the **properties and sources of risks**, people and systems as well as nature and malice are among the fundamental factors. For example, system failures are often related to natural causes while botnets are systems organized by people with malicious intent. A cyber attack by a malicious person is different from an accidental configuration error introduced by a system administrator. Figure 6.1 depicts these factors by characterizing a few example risks.

While the sources, properties, and effects of security risks can be different, they share the same fundamental risk management **principles and steps**. The basic risk-management process can be divided into the following three steps:

- *risk assessment* involving identification of vulnerabilities and assessment of their potential effects;
- *risk analysis and decision making*, which includes creating a risk-management plan as well as deciding on what are the feasible countermeasures given organizational priorities and constraints; and
- *execution of measures* that may involve dynamic allocation of existing resources, organizational changes, and future investments.

These steps and the resulting risk-management cycle are visualized in Figure 6.2.

Risk assessment involves the identification of vulnerabilities against security attacks and IT failures, what effects these can have on business processes, their potential cost, and whether the countermeasures taken are satisfactory. Risk assessment is closely related to domain knowledge on and *observation* of the networked system in the broad sense including business processes. If the owners do not know about and observe what happens in their systems, then it is unrealistic to expect them to correctly analyze security threats and other risks. Therefore, risk assessment invariably includes building observation capabilities that enable better defenses against unwanted behavior and malicious security compromises to be built. There exists an increasing number of software solutions that focus on the problem of data collection.

Even with software support, implementing such a data collection scheme can be challenging and costly. One issue is to find the people in the organization who are knowledgeable in their domain and to motivate their involvement. Another problem is the fact that it is a repeating (sometimes even in real-time) process rather than a static one. This means that observation and risk assessment need to be well integrated to almost the entire organization. A final obstacle is the precision of the observations. Complexity of the underlying networked systems and human factors affect the reliability of the data provided even by the experts or carefully designed schemes.

The inherent uncertainty and imprecise nature of the input data makes **risk analysis** a challenging task. The analysis and subsequent decision making have to take into account information limitations as well as other constraints such as financial and organizational ones. Most of these decision-making processes can be formalized within the mathematical frameworks provided by decision and game theories. Such a formalization has multiple benefits ranging from efficiency, prioritization, and clear expression of tradeoffs in a quantitative manner to scalability and transparency. It also opens doors to software support for risk analysis, which is currently mostly lacking. Combined with telecommunications aspects, *computer-assisted decision support systems* may have profoundly positive consequences for the whole field of risk analysis and related decision making.

Risk analysis is closely related to **decision variables**, actors, and an available set of actions that may vary significantly in terms of scope and timescale. While some security threats such as fast, self-spreading malware require immediate and often automated action by an IDPS, others such as security investments require decision making at chief information officer level and have a time horizon of months. Three example scenarios with different actors, timescales, and decision variables are summarized in Table 6.1.

The third step of the risk-management cycle is the **execution of measures** or **implementation**. A number of issues have to be addressed for this important step to be successful. First, it requires significant commitment from management who has to overcome organizational resistance to change and transparency. Sometimes, psychological factors such as "nobody likes the bearer of bad news" also create nontrivial obstacles. If these issues are not properly addressed, then the previous two steps in the risk-management cycle may well degenerate into a futile bureaucratic exercise. The second issue is the limited number of realistic options available to the organization to counter the risks and security threats given budget and manpower limitations. In some cases,

Table 6.1 Decision variables in risk-management

Decision maker	Timescale	Actions
CIO	Months	Company-wide policies, major security investments
Dept. head	Days to hours	Department rules, allocation of manpower and other resources
IDPS	Seconds or less	Block ports, access control, packet inspection

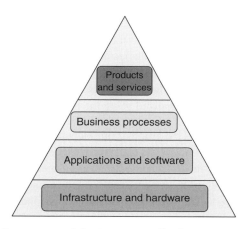

Figure 6.3 Business units encompass infrastructure, applications, processes, and services.

the best an organization can do is to be rationally aware of risks, accept them, and take the most reasonable course of action given the constraints. In general, having as much and as high quality information as possible about the systems and business processes, in other words transparency, is very useful to achieve a higher degree of control of risks. On the other hand, if the only tool available to a risk manager is a "hammer," then there is no need to buy a "magnifier" to collect detailed information on the networked system.

6.1.2   A probabilistic risk framework

As an example of an **analytical risk model**, a probabilistic risk-management approach is presented that provides a unified quantitative framework for investigation of the interdependence between various *business units*, the potential impact of various *vulnerabilities* or threats, and the risk implications of relationships between *people*. The term business unit encompasses infrastructure, software, business processes, and products as depicted in Figure 6.3. Vulnerabilities and security threats can be very diverse, ranging from generic malware to specifically targeted phishing attacks. Employees in an organization have diverse roles, responsibilities, and relationships with respect to each other and the business processes.

The business units, vulnerabilities, and people constitute the main three factors of the risk-management framework:

- *business units,* $\mathcal{N}_B = \{n_1^B, \ldots, n_{M_B}^B\}$, representing infrastructure elements (e.g. computing servers), applications (e.g. software), processes (e.g. billing, customer care), and products or services (e.g. DSL service, SMS service);
- *security threats* and vulnerabilities, $\mathcal{N}_S = \{n_1^S, \ldots, n_{M_S}^S\}$, targeting or adversely affecting the business units $\mathcal{N}_B$;
- the *people* $\mathcal{N}_P = \{n_1^P, \ldots, n_{M_P}^P\}$, e.g. managers and employees, who run the business units $\mathcal{N}_B$.

Here, the positive integers M_B, M_S, and M_P are the cardinalities of the respective sets. The chosen categorization and the formal definitions above enable a quantitative study of the complex interdependencies between these factors and how they contribute to the security risk profile of an organization.

Based on the definitions of the sets $\mathcal{N}_B$, $\mathcal{N}_S$, and $\mathcal{N}_P$, the intradependencies among their members are formalized using the following graphs:

1. $\mathcal{G}_B = (\mathcal{N}_B, \mathcal{E}_B)$ is the graph of intradependencies among business units in $\mathcal{N}_B$, where each edge $\varepsilon_{ij}^B = (i, j) \in \mathcal{E}_B$ represents the dependency or interaction between business units i and j in $\mathcal{N}_N$. Each edge is associated with a scalar weight w_{ij}^B denoting its "intensity" or propensity to transfer and cascade risk.

2. $\mathcal{G}_S = (\mathcal{N}_S, \mathcal{E}_S)$ is the graph of intradependencies among the security threats and vulnerabilities in $\mathcal{N}_S$, where each edge $\varepsilon_{ij}^S = (i, j) \in \mathcal{E}_S$ represents the dependency between threats i and j in $\mathcal{N}_S$ and is associated with a scalar weight w_{ij}^S. This weight represents the propensity to transfer and cascade risk such as the likelihood that exploiting one vulnerability i will lead to exploiting j subsequently.

3. $\mathcal{G}_P = (\mathcal{N}_P, \mathcal{E}_P)$ is the graph of intradependencies among people in the organization, where each edge $\varepsilon_{ij}^P = (i, j) \in \mathcal{E}_P$ represents the relationship between employees i and j in $\mathcal{N}_E$ and is associated with a scalar weight w_{ij}^P. It represents again the propensity to transfer and cascade risk such as compliance failure of employee i leading to compliance failure of employee j.

Consequently, the matrices $W^B = \{w_{ij}^B\}$, $W^S = \{w_{ij}^S\}$, and $W^P = \{w_{ij}^P\}$, which contain scalar intradependency weights, are associated with the respective graphs.

Bipartite graphs are utilized in order to model the *interdependencies* across the sets $\mathcal{N}_B$, $\mathcal{N}_V$, and $\mathcal{N}_P$. Consider first the bipartite graph $\mathcal{G}_{PB} = (\mathcal{N}_B, \mathcal{N}_P, \mathcal{E}_{PB})$, shown in Figure 6.4, that represents the cross-relationships between employees and business units. For example, it could reflect who is responsible for a process, or product, or for running an infrastructure system, such as a server. Here, $\mathcal{E}_{PB}$ denotes the edges between the sets $\mathcal{N}_P$ and $\mathcal{N}_B$. It is important to note that the edges of this graph are always between the members of the respective disjoint sets $\mathcal{N}_B$ and $\mathcal{N}_P$, but not within each set. The latter case is already modeled by the graphs $\mathcal{G}_B$ and $\mathcal{G}_P$. The cross-relationships across the other graphs can be modeled using similarly defined bipartite graphs.

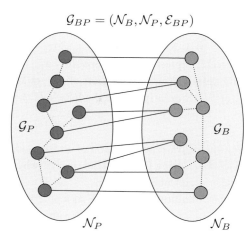

Figure 6.4 Regular and bipartite graphs can be used to represent the relationships between business units, people, and security threats.

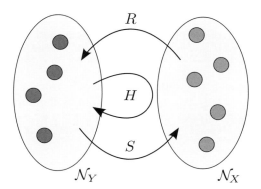

Figure 6.5 Risk diffusion on a bipartite graph and the "ping-pong" effect.

Cross-set risk-diffusion and equilibrium

The bipartite structure of the probabilistic risk framework is used to model how risk cascades and gradually spreads (diffuses) in an organization. In order to explore how business units, security vulnerabilities, and people affect and relate to each other with respect to risks, an approach based on "diffusion processes" over a graph [32, 112] is adopted. Diffusion processes, as means for computing similarities among the vertices of graphs and ranking them, have been investigated by several studies [2, 85, 193]. This methodology is a distinguishing feature of the approach here and is leveraged for security risk assessment. Thus, secondary and indirect relations between entities (nodes) are explored through computation of a risk-diffusion process over the respective bipartite graph [32] as shown in Figure 6.5.

Consider the risk-diffusion across a generic bipartite graph $\mathcal{G}_{XY} = (\mathcal{N}_X, \mathcal{N}_Y, \mathcal{E}_{XY})$ as in Figure 6.5 without any loss of generality. First, define the **relative risk probability vector (RRPV)**

$$v^X(t) = \left[v_1^X(t), \ldots, v_i^X(t), \ldots, v_{M_X}^X(t) \right] \tag{6.1}$$

at time slot $t \in \{0, 1, 2, \ldots\}$ such that $0 \le v_i^X(t) \le 1$, $\sum_i v_i^X(t) = 1$ $\forall t$. This normalized relative risk vector represents risk probability as a normalized measure of security compromises and failures over the members of the set $\mathcal{N}_X$. Note that this set can be any one of $\mathcal{N}_B$, $\mathcal{N}_V$, or $\mathcal{N}_P$. Similarly, define the RRPV of the set $\mathcal{N}_Y$ as

$$v^Y(t) = \left[v_1^Y(t), \ldots, v_i^Y(t), \ldots, v_{M_X}^Y(t) \right].$$

These definitions allow representation of an isolated risk factor associated with the i-th member of a set $\mathcal{N}_X$ using the RRPV $v^X = \mathbf{e}(i)$, where $\mathbf{e}(i) := [0, \ldots, 1, \ldots, 0]$ is defined as the i-th basis vector in $\mathbb{R}^{M_X}$.

Based on the definition of the RRPV as a probabilistic measure of relative risks faced by the elements of a set, the **relative risk value vector (RRVV)** is defined by the multiplication of business values of individual members of the set with the relative risk probabilities they individually face. In other words, given the business value vector of the set $\mathcal{N}_X$,

$$b^X = \left[b_1^X, \ldots, b_i^X, \ldots, b_{M_X}^X \right],$$

the RRVV at time t is the elementwise product of $v^X(t)$ and β^X:

$$r_i^X(t) := v_i^X(t) \cdot b_i^X \quad \forall i \in \{1, \ldots, M_X\}.$$

The one-step risk diffusion or cascade from a set $\mathcal{N}_X$ to $\mathcal{N}_Y$ over the bipartite graph $\mathcal{G}_{XY} = (\mathcal{N}_X, \mathcal{N}_Y, \mathcal{E}_{XY})$ is captured by the **normalized risk cascade matrix**, R. Consequently, one can write $v_i^Y(t+1) = \sum_{j=1}^{M_X} R_{ij} v_j^X(t)$ or in matrix form

$$v^Y(t+1) = R v^X(t),$$

where $R := R_{ij}$, $i \in \{1, \ldots M_Y\}, j \in \{1, \ldots, M_X\}\}$. Note that here any intranode risk transfer between $\mathcal{N}_Y$-nodes themselves are ignored for simplicity.

Similarly, define S_{ji} to be the normalized risk transfer from node n_i^Y of $\mathcal{N}_Y$ at t to node n_j^X of $\mathcal{N}_X$ at $t+1$. It follows that $v_j^X(t+1) = \sum_{i=1}^{M_Y} S_{ji} v_i^Y(t)$ or in matrix form

$$v^X(t+1) = S v^Y(t),$$

where $S = \{S_{ji}, i \in \{1, \ldots, M_Y\}, j \in \{1, \ldots, M_X\}\}$ is the normalized risk cascade matrix in the opposite direction of R with $\sum_{j=1}^{M_X} S_{ji} = 1$ for each $j \in \{1, \ldots, M_Y\}$. Hence, it provides the one-step risk-diffusion from Y to X within the bipartite graph $\mathcal{G}_{XY}$.

Combining normalized risk cascade matrices R and S, define the matrix

$$H := SR,$$

which captures the two-step risk-diffusion $Y \to X \to Y$ via a "ping-pong" risk cascade effect. As in S and R, the matrix H is also *column-normalized* (left-stochastic). This follows directly from

$$\sum_i H_{ij} = \sum_i \sum_k S_{ik} R_{kj} = \sum_k \left(\sum_i S_{ik} \right) = \sum_k R_{kj} = 1$$

for every $j \in \{1, ..., M_X\}$, since R and S are column-normalized (left-stochastic) matrices, i.e. $\sum_k R_{kj} = 1$ for all j and $\sum_i S_{ik} = 1$ for all k. It is worth remembering that the vectors $v^X(t)$ and $v^Y(t)$ are probability vectors. Hence, the matrix H is a Markov transition matrix on $v^Y(t)$.

Risk-Rank (RR) algorithm

Given an initial RRPV $v^Y(0)$ on $\mathcal{N}_Y$, it is possible to naively rank the nodes in $\mathcal{N}_Y$ based on their immediate relative risk probability, and declare the one with maximum immediate risk to be "in most danger." This naive ranking strategy, however, fails to recognize the diffusion or cascade of risk from $\mathcal{N}_Y$ to $\mathcal{N}_X$ and back, which influences the initial risk vector through the risk "ping-pong" effect. On the other hand, if the diffusion effect is taken to the limit, then it leads eventually to the stationary RRPV $v^{Y*} = \lim_{t \to 0} v^Y(t)$ via the iteration $v^Y(t+1) = H v^Y(t)$ as $t \to \infty$. In this case, the resulting RRPV v^{Y*} does not retain any memory of the initial one $v^Y(0)$; the memory fades away and risk diffuses through the system.

It is desirable to devise a general algorithm that is flexible enough to take into account both the immediate risk and diffusion effects. The RR algorithm, which aims to satisfy this goal, is defined as the following dynamic stochastic process

$$v^Y(t+1) = \alpha H v^Y(t) + \beta v^Y(0), \tag{6.2}$$

where $\alpha, \beta \in [0, 1]$ are the relative weights of immediate versus cascaded risks and $\alpha + \beta = 1$. The algorithm basically describes the cascading effect of an initial event $v^Y(0)$ or a member of the set $\mathcal{N}_Y$ on other members through evolution of risk probabilities v^Y.

In the special case of $\alpha = 1$ in (6.2), the RR iteration

$$v^Y(t+1) = H v^Y(t)$$

converges to a unique stationary risk probability vector (distribution) v^{Y*} as $t \to \infty$, if the sets N_X and N_Y are finite and the stochastic matrix H is irreducible. In general, when $0 \leq \alpha < 1$, the Markov process (6.2) converges to

$$v^{Y*} = \lim_{t \to \infty} v^Y(t) = \beta(I - \alpha H)^{-1} v^Y(0), \tag{6.3}$$

where I denotes the identity matrix, under the assumption of $[I - \alpha H]$ being invertible. The RR vector v^{Y*} can then be used to rank the nodes with respect to their immediate as well as cascaded risk probabilities, balancing between current and future risk as desired using parameters α and β.

The mapping $\beta(I - \alpha H)^{-1}$ in (6.3) constitutes a diffusion kernel. At the same time, the RR algorithm in (6.2) roughly corresponds to the iterative version of the adjusted page rank procedure used by the Google search engine [93]. However, a distinguishing feature of the RR algorithm is that it operates over a bipartite graph rather than an adjacency one.

Remark 6.1 *The RR approach can also be used to describe the cascading effect of an initial event $v(0)^X$ (or member of the set $\mathcal{N}_X$) on $\mathcal{N}_Y$ through evolution of risk probabilities,*

$$v^Y(t+1) = \alpha H v^Y(t) + \beta R v^X(0), \tag{6.4}$$

which converges to

$$v^{Y*} = \lim_{t \to \infty} v^Y(t) = \beta(I - \alpha H)^{-1} R v^X(0),$$

again under the assumption that $I - \alpha H$ is invertible.

Given, for example, an adjacency matrix W^Y as a representation of the intra-dependencies of the graph $G_Y = (\mathcal{N}_Y, \mathcal{E}_Y)$, it is possible to extend the RR algorithm to:

$$v^Y(t+1) = \alpha H v^Y(t) + \gamma W^Y v^Y(t) + \beta v^Y(0), \tag{6.5}$$

where $\alpha + \beta + \gamma = 1$ and $0 \leq \gamma \leq 1$. If the matrix $\alpha H + \gamma W^Y$ is aperiodic and irreducible, then the process again converges to the unique solution [148]

$$v^{Y*} = \beta(I - \alpha H - \gamma W^Y)^{-1} v^Y(0),$$

which can be used by risk managers to rank relative risks.

The parameters α, β, and γ provide a way to adjust the weights of the following three distinct factors when computing the solution. The first factor (α) emphasizes the risk-diffusion process which results in a stationary risk distribution irrespective of the initial starting point. The second one (β) is the counterpart of the first and represents the immediate risks. The third one (γ) provides a way of taking into account the effects of dependencies between members of the same set on risk probabilities.

Remark 6.2 *It is possible to capture mathematically the relationships between the nodes in a single generic graph and study risk diffusion on it. However, this assumes full knowledge of all internal and external relationships, which may not be the case in many scenarios. The RR approach imposes a special structure on the set of nodes based on the natural partitioning of business units, security threats, and people. Subsequently, cross-relationships are studied using bipartite graphs. This model is arguably better suited to the problem domain, for it captures inherent properties and enables the study of cases where only partial information is available.*

The application of the RR algorithm to operational situations can be illustrated with the following example. Let $\mathcal{N}_Y$ be the set of business units $\mathcal{N}_B$ in an enterprise, and $\mathcal{N}_X$ the set of people $\mathcal{N}_P$ running the business units. If an employee fails to perform a server maintenance function such as applying a security patch, this may result in a compromise of the server by a malicious attacker. The attacker may then disable some check functions which may lead to potential failures of other employees to perform further maintenance functions. Thus, the risks cascade to other business units and processes. Several such risk-diffusion steps via "ping-pong" iterations may in the end result in catastrophic failures that affect the whole organization.

Scenarios similar to the one above are not at all unusual in complex systems. Well-known accidents with catastrophic consequences from Titanic to space shuttle Discovery are results of similar cascading processes. Other examples, such as the NASA metric conversion fiasco or more mundane events such as failure of wide area networks (Telecom) can be, fortunately, less damaging but still significantly costly. If the risks in such complex systems can be analyzed in a principled way and prioritized (ranked) properly, then it may be feasible to address them within the given resources of the respective projects and organizations. The next section addresses this risk mitigation problem from an optimization and control perspective.

6.1.3 Dynamic risk mitigation and control

The previous section focused on risk assessment as a first step in risk-management (Figure 6.2) and specifically on the RR algorithm, which can be used to prioritize risks based on the collected data. This section presents a quantitative framework for taking appropriate actions to perform *risk mitigation and control* as the next step once the assessment phase is completed. Hence, it is shown how the RR approach can also be used to evaluate risk mitigation strategies. More specifically, the question of how to control the risk-diffusion process defined in the previous section over time is addressed in order to achieve a more favorable risk distribution across the assets of an organization.

The control actions available to the decision maker (e.g. risk manager) range from policies and rules to the allocation of security resources or updating system configuration. These actions change the dependencies in the organization, and hence directly affect the evolution of the RRPV $\mathbf{v}^Y(t)$ (6.1). For example, patching a server decreases the weight between a virus threat node and the server node in the organization graph.

The objective of risk control is to achieve a more favorable risk distribution. Suppose that every element in $\mathcal{N}_Y$ has a value $z_y \in \mathbb{R}^+$ for the organization such that

$$z^Y := \left[z_1^Y, z_2^Y, \ldots, z_{M_Y}^Y \right].$$

For example, if $\mathcal{N}_Y$ is a set of business units, then each unit would have an associated value that represents the cost incurred if that unit is compromised. Then, the *risk cost* incurred by the elements of $\mathcal{N}_Y$ during time period t can be quantified according to

$$c(t) = v^Y(t) \cdot z^Y,$$

where " $\cdot$ " denotes the dot product between two vectors. This definition supports the intuition that the risk manager's expected cost increases as the higher-value nodes gain a larger proportion of the total risk. Therefore, the goal of the risk mitigation actions should be to drive risk away from high-value nodes.

Due to the normalization inherent to the RR approach, controlling the risk-diffusion process over $\mathcal{N}_Y$ will result in a *redistribution* of risk over this set of nodes. However, the framework as it is does not model the *absolute* amount of risk. In order to allow the absolute amount of risk to decrease as a result of possible control actions, it is convenient to define a *risk sink* as an artificial node with zero value and add it to the set $\mathcal{N}_Y$ such that its cardinality is now $M_Y + 1$. Any proportion of risk which accumulates on this node can be thought of as having left the network. Note that this requires appropriate modifications of H, $v^Y(t)$, and z^Y, but that once these are made, the mechanics of the risk-diffusion process are basically the same. A related assumption is: if no control actions are taken, no risk will diffuse into the risk sink.

The control actions can now be applied to modify the evolution of relative risk (6.2) and reduce the risk cost accumulated over time. The theory of MDPs [148] provides a suitable framework for analyzing this stochastic optimization problem. MDPs have also been discussed in the context of Markov security games in Section 4.1 and the underlying dynamic programming principles are presented in Appendix A.3. An MDP is defined by four key elements: a state space, a set of possible actions, a set of transition probabilities between states given each action, and a cost or reward function. Each of these elements is described next in the context of security risk-management.

It is possible to use the RRPV, $\mathbf{v}^Y(t)$, as the risk state of the system at time t. However, the relative risk probability simplex

$$\mathcal{R}^X := \left\{ p \in \mathbb{R}^X : p_i \in [0,1] \;\; \forall i \text{ and } \sum_{i=1}^X p_i = 1 \right\}$$

is not suitable as a state space since in actual deployments of the proposed risk analysis techniques the state space is more likely to be quantized to a finite number of levels. This is a consequence of limited system *observability*. As the relative risk evolves, it is very difficult for an organization to track it with a high level of precision, and it is much more likely that only a rough estimate of the relative risk probability can be obtained. In addition, continuous state spaces are typically infeasible for evaluating MDPs numerically.

Therefore, the finite state space

$$S = \{s_1, s_2, \ldots, s_k, \ldots, s_K\}$$

is defined as an alternative after quantizing the relative risk probability simplex $\mathcal{R}$ into K *risk regions* (such as those in reference [111]). Figure 6.6 shows a visualization of an example partitioning of the probability simplex $\mathcal{R}^3$ into risk regions. Each risk region or state k is associated with an average risk vector $\bar{v}_k^Y$, which is taken to be the arithmetic mean of risk vectors which belong to partition k. One way that this average vector can be determined is through Monte Carlo simulation, i.e. by generating random vectors according to a uniform distribution over the probability simplex. Note that the definitions of risk regions can be very application-specific, due to the diversity of risk metrics employed in industry today. An example partitioning will be discussed as part of a numerical example later in this section.

To reach a desirable risk state or region, the risk manager takes an action a_l out of a set of available actions

$$\mathcal{A} = \{a_1, a_2, \ldots, a_l, \ldots, a_L\}.$$

Without loss of any generality, it is assumed here that only one action can be taken in each time step. Each action changes the weights of one or more edges in the bipartite graph model of Section 6.1.2. In terms of risk-diffusion dynamics (6.2), these actions correspond to modified versions of the matrix H, resulting in the set

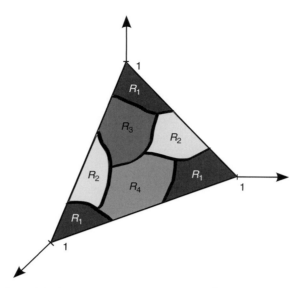

Figure 6.6 Example partitioning of the probability simplex $\mathcal{R}^3$ into four risk regions.

$$\mathcal{H} = \{H_1, H_2, \ldots, H_l, \ldots, H_L\},$$

of the same cardinality as the actions.

The probability of transitioning from the current state $s(t) = s_k$ to another state $s(t+1) = s_{k'}$, given that action $a(t) = a_l$ is taken, is

$$Prob\,(s(t+1) = s_{k'}|s(t) = s_k, a(t) = a_l).$$

For each action a_l, the $K \times K$ probability transition matrix $P(a_l)$ is constructed, where the entries are

$$P_{ij}(a_l) = Prob\,(s(t+1) = j|s(t) = i, a(t) = a_l).$$

The transition probabilities which make up $P(a_l)$ can be estimated, for example using Monte Carlo simulation. It is important to distinguish here between H, which describes the evolution of relative risk probabilities, and $P(a_l)$, which denotes the transition probabilities from one state (risk region) to another.

Given the transition probabilities, the evolution of states over time is described next. Let the vector $x^Y(t) \in \mathbb{R}^K$ represent the probabilities of being in each state at time t. Then, taking action a_l leads to

$$x^Y(t+1) = P(a_l)x^Y(t).$$

As the last step of MDP formulation, define an additive cost function that sums the risk cost accumulated by the system over a finite number of time steps $1, \ldots, T$. At time t, the current state contributes a cost of $c_s(s(t))$, while the action taken contributes a cost $c_a(a(t))$. The cost c_a is usually application-specific and captures the cost of performing some action, for example, hardening a node against attack. The cost of the risk state c_s is based on the definition of risk cost above. In particular, it is given by

Table 6.2 Example vulnerability classes

Class 1	Windows vulnerabilities
Class 2	Microsoft SQL server vulnerabilities
Class 3	IIS server vulnerabilities
Class 4	Microsoft Office vulnerabilities
Class 5	Phishing-related vulnerabilities

$$c_s(s_i) = \bar{v}_i^Y \cdot z^Y,$$

where $\bar{v}_i^Y$ is the representative (mean) probability vector of state (region) i. Therefore, the total cost $C(T)$ accumulated up to time T is

$$C(T) = \gamma^T c_s(s(T)) + \sum_{t=0}^{T-1} \gamma^t \left[c_s(s(t)) + c_a(a(t)) \right], \qquad (6.6)$$

where $\gamma \in (0,1]$ is a discount factor which captures the typical scenario that risk reductions closer to the present are of more value than those in the future. The goal of the risk manager is then to minimize $C(T)$, where T is the time span of interest. Notice that it is also possible to define infinite horizon versions of the problem as in the MDP literature.

Given the MDP formulation above, standard methods such as value or policy iteration [148] can be used to solve for the optimal risk cost-minimizing strategy. The risk control approach described in this section is particularly applicable to scenarios in which a fixed resource must be redeployed over a given time period in response to a dynamic, rapidly evolving risk situation. Such a scenario is discussed next as an illustrative numerical example.

Numerical example

The dynamic risk control and mitigation framework presented is illustrated with a numerical example scenario. Consider an IT manager responsible for patching five corporate subnets that consist of multiple computers and servers. On each subnet, there is a number of high-priority vulnerabilities that need to be patched. Each vulnerability belongs to one of five common *vulnerability classes* in a Windows environment, listed in Table 6.2.

A set of costs is associated with the subnets, signifying the productivity losses incurred if a subnet is compromised by the exploitation of a vulnerability. Assume that, since patching is a time-intensive process [112], the IT manager can work only on one subnet at a given time period. Hence, it is important to determine the proper order of patching, which can have a significant financial impact.

A simplistic approach would be to start with the subnet which contains the highest number of vulnerabilities, patch them, and then proceed to the next subnet. However, this strategy is far from optimal, for several reasons. One of these is due to the case when a vulnerability on one of the subnets with a low number of vulnerabilities is exploited in an attack, all other subnets are at a much higher risk. Another reason is the fact that attacks exploiting different vulnerabilities may vary significantly in terms of their financial and productivity impact. Therefore, it is expected that an optimized patching strategy that relies on the presented framework would lead to much better results.

Let $\mathcal{N}_S$ be the set of vulnerability classes, indexed by $i = 1, \ldots, 5$, and $\mathcal{N}_B$ be the set of subnets, indexed by $j = 1, \ldots, 5$. The productivity losses incurred by the failures of subnets make up the vector z^Y. Time is discrete and indexed by $t = 0, 1, \ldots, T$. The number of vulnerabilities on each subnet at time $t = 0$ is q_j, and the number of class i vulnerabilities on each subnet at time $t = 0$ is q^i_j.

Given the number of vulnerabilities in each class and consequently for each subnet, the IT manager can determine the relative risk posed to each subnet by each vulnerability class using the simple equation $\theta_{ji} := q^i_j / q_j$, so that θ_{ji} denotes the proportion of the vulnerabilities on subnet j which belong to class i. These values can be interpreted as the relative probabilities that each subnet will serve as an entry point into the overall corporate network for a vulnerability of the associated class.

At the same time, the IT manager also knows how many class i vulnerabilities exist across the entire network, as well as the distribution of these vulnerabilities over the various subnets. Suppose that there are b_i total class i vulnerabilities in the system. Then, the proportion of these vulnerabilities which lies on each subnet is $\phi_{ij} := q^i_j / b_i$, which can be thought of as the relative probability that if a class i vulnerability is exploited on the network, it would occur on subnet j.

Adopting the notation in Section 6.1.2, let the matrix S represent the cross-set risk-diffusion from $\mathcal{N}_B$ to $\mathcal{N}_S$, and R the cross-set risk-diffusion back from $\mathcal{N}_S$ to $\mathcal{N}_B$. Then, the entries of S and R are defined respectively as $S_{ji} = \theta_{ji}$ and $R_{ij} = \phi_{ij} \; \forall i, j$. The diffusion matrix H from $\mathcal{N}_B$ to $\mathcal{N}_B$ is consequently $H = S \cdot R$, which is given in this example by (prior to the inclusion of the risk sink):

$$
H = \begin{pmatrix}
0.61 & 0.05 & 0.03 & 0.25 & 0.06 \\
0.05 & 0.69 & 0.04 & 0.18 & 0.04 \\
0.04 & 0.05 & 0.56 & 0.29 & 0.06 \\
0.06 & 0.04 & 0.06 & 0.78 & 0.06 \\
0.07 & 0.04 & 0.05 & 0.26 & 0.58
\end{pmatrix} .
$$

Suppose that the IT manager uses the proportion of the total number of vulnerabilities on each subnet as an estimate of the initial risk vector $v^Y(0)$. Choosing $\alpha = 0.95$ and $\beta = 0.05$ in the RR algorithm (6.2), which suggests that the risk evolution is highly dynamic, results in a very different risk estimate than the initial one, due to the cascade of risks, as shown in Figure 6.7. For example, subnet number 4 is much more at risk in the long run than estimated initially due to interdependencies in the system.

The risk assessment obtained through the RR algorithm is now used as a basis for developing an optimal patching strategy for risk mitigation. First, a risk sink is added to the set $\mathcal{N}_B$, which is called node n^B_6. This results in the following change to the matrix **H**: the risk sink is incorporated in a way such that no risk can diffuse into the risk sink, but a small amount of risk can diffuse out of it. It can be interpreted as a small amount

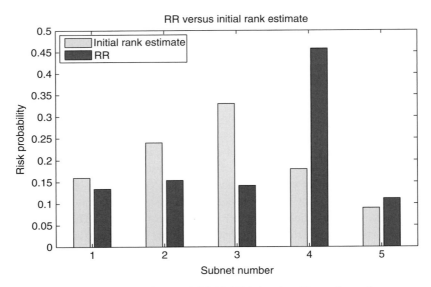

Figure 6.7 Initial versus the RR estimate of risk highlighting the effects of complex interdependencies.

of additional risk entering the network over time, perhaps due to the announcement of new vulnerabilities.

Next, the resulting probability simplex $\mathcal{R}^6$ is partitioned into risk regions by quantizing each entry $v_j^Y(t)$ of $v^Y(t)$ into five levels through the function $\mathcal{L}: \mathcal{R}^6 \longmapsto Q = \{1,2,3,4,5\}$. These five levels represent {very low, low, medium, high, very high} risk levels.

The set of actions available to the IT manager for risk mitigation can be, for example, assigning employees to patch subnets, purchasing automatic software upgrades, etc. For simplicity, it is assumed that in each time period t, the IT manager takes one of these actions on a particular subnet or does nothing. The patching actions are labeled as $a_1, \ldots, a_5$ for the respective subnets, and the action corresponding to doing nothing is denoted by a_6.

When an action is taken, the effect manifests itself as a modification of the matrix H as discussed earlier. Furthermore, each action is associated with a cost reflecting the effort required to take that action. Here, this cost is assumed to be small compared to the productivity losses incurred if a subnet is compromised. In addition, the following assumptions are made: if no action is taken the matrix remains the same, $H_6 = H$. If the subnet n_j^B is patched, then the corresponding row of H is modified such that

$$H_l(j,j') = \begin{cases} (1-\delta)H(j,j') & \text{, if } j' = 1,\ldots,5 \\ \delta & \text{, if } j' = 6 \end{cases}$$

where $\delta \in [0,1]$ represents the proportion of risk that is taken out of the network by taking patching action on a subnet. This change to H represents the fact that patching a subnet will reduce the risk it transfers to other subnets.

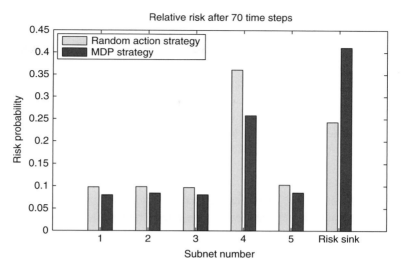

Figure 6.8 Relative risk under random action and MDP risk control strategies.

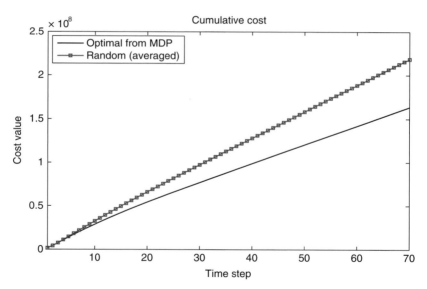

Figure 6.9 Aggregate cost of the optimal MDP-based risk mitigation strategy versus the average cost of random strategies.

Thus, this example is an instance of MDP and solved numerically for the given matrix H with

$$z^Y = [500, 400, 30, 8000, 100] \cdot 10^3,$$

$\delta = 0.1$, $\gamma = 1$, $\alpha = 0.95$, and $\beta = 0.05$. The time horizon is chosen to be $T = 70$. The results in Figures 6.8 and 6.9 clearly show that, in comparison to choosing actions at random, the dynamic risk mitigation method can provide considerable risk reduction.

6.2   Security investment games

The increasing interaction and collaboration between various organizations and companies create **complex interdependencies** on a global scale. Consequently, these organizations often share sophisticated information and communication infrastructures providing data access, business applications, and services. In such a complex environment, the security of one organization depends on not only its own defensive actions but also the preventive measures taken by others. Likewise, the vulnerability of one organization may lead to cascading failures and compromises for others. In addition, a single large-scale organization, such as a multinational company, often consists of autonomous yet interdependent units with different incentives and agendas.

Ensuring that interdependent organizations and units collaborate with each other in a mutually beneficial security prevention framework is a challenge for security officers defending their networked systems. Even in cases where there is willingness for collaboration between enterprises or units within a company to improve their security, it is not immediately clear how individual systems, services, and business processes affect and interact with each other. Hence, making informed decisions on how to take preventive measures, allocate limited resources, and address vulnerabilities across the units and organizations remains a significant challenge.

To address the questions posed above, a linear influence network-based **security investment game framework** is presented in this section for evaluating risks, benefits, incentives, and investments by independent organizations or autonomous units with interdependent operations.[2] The framework builds a connection between economic (game-theoretic) modeling techniques and security problems. Based on game-theoretic techniques, it provides a quantitative evaluation of possible investments, given sufficiently accurate quantitative estimates of the governing parameters. In addition, reasonable qualitative insight can be obtained using rough estimates of relevant input quantities. The framework also investigates the interdependent nature of security investments by modeling the relationships between players in a flexible way.

The **influence network**, which models the interdependence between a security investment or vulnerability at one organization and the resulting security benefit or cost at another is assumed to be linear. While nonlinear relationships may exist in some situations, it is reasonable to use a linear approximation. Furthermore, in many situations of interest, obtaining meaningful estimates of numerical parameters for more complex nonlinear models is not possible. Hence, a linear approximation is consistent with the accuracy of available numerical input to the model.

A **security investment game** played among the organizations is defined based on the linear influence networks described. In this noncooperative game, each organization (player) decides on its own individual security investment level, which is a positive quantity. The security investments or preventive measures need not have a symmetric

---

[2] In the remainder of the chapter, the two scenarios of independent organizations collaborating with each other and autonomous units of a single organization will be used interchangeably.

effect between organizations. A decision by one player can be either beneficial or detrimental to neighboring players resulting in positive or negative interactions. An investment in security by one player may benefit others because the investment reduces a risk shared by both players. A negative interaction may result from the fact that given two potential victims, an attacker will likely choose the path of least resistance. In other words, an investment in security by one player may increase the likelihood that the others within the network are attacked instead.

Each player is associated with a **cost function** that is defined as the difference between a linear pricing function denoting the price of security investments and a utility function, which represents the nonlinear relationship between improved security through preventive investments and the total value of all deployed security mechanisms to the organization. The use of a nonlinear utility function is an important feature of the game-theoretic model. This feature distinguishes the investment game from purely probabilistic fault tolerance and failure ones that only aim to provide probability estimates of break-ins or failures. The players choose a non-negative investment level to minimize their own cost functions in a selfish manner. Thus, the game is defined as a noncooperative one that models the misalignment of incentives between players. The players are assumed to observe others' actions rather accurately, either because they are independent but collaborating organizations, or because they are divisions of the same umbrella company.

In its pure form, the investment game differs from the security games in Chapter 3 as it is played among agents (organizations) who are selfish but not hostile to each other. The interaction between agents is based mainly on how their security investment decisions affect each other through their respective utility functions. However, the model is flexible enough to **incorporate malicious attackers** as players in the game by reinterpreting their respective decision variables and cost functions. In this case, the decision variable of the attacker is the attack intensity and the negative weights in the linear influence model indicate how much other agents are affected by the attacks. The cost function of the attacker can be, for example, the difference between the price of launching an attack and gain from it, quantified by the utility function.

6.2.1 Influence network and game model

The linear influence network modeling the interdependency between players of the investment game is similar to those in Section 4.3.2. It is a weighted directed graph $\mathcal{G}(\mathcal{N}, \mathcal{E})$, where $\mathcal{N} = \{n_1, \ldots, n_N\}$ is the set of vertices or nodes and $\mathcal{E}$ is the set of edges or links. Each node is associated with a unique player (e.g. an enterprise or autonomous unit) resulting in an N-player game. The link $e_{ij} \in E$ between players i and j captures the effect of player i's actions on player j. The real value or weight of an individual link, $\psi_{ij} \in \mathbb{R}$, denotes the degree of influence of node i on node j. The weighted influence graph can be represented by the $N \times N$ matrix $W = [w_{ij}]_{N \times N}$ whose entries are defined as

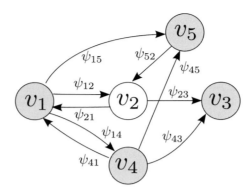

Figure 6.10 Example linear influence network with five nodes.

$$w_{ij} := \begin{cases} 1, \text{ if } i = j \\ \psi_{ij}, \text{ if } e_{ij} \in E \\ 0, \text{ else} \end{cases} \quad . \tag{6.7}$$

Notice that the matrix W is not necessarily symmetric. An example influence network is shown in Figure 6.10 with the associated matrix

$$W = \begin{bmatrix} 1 & \psi_{12} & 0 & \psi_{14} & \psi_{15} \\ \psi_{21} & 1 & \psi_{23} & 0 & 0 \\ 0 & 0 & 1 & 0 & 0 \\ \psi_{41} & 0 & \psi_{43} & 1 & \psi_{45} \\ 0 & \psi_{52} & 0 & 0 & 1 \end{bmatrix}.$$

In the security investment game, each player i chooses a security investment level $x_i \in [0, \infty)$. Then, the investments of all players are given by the vector $x := (x_1, \dots, x_N) \in \mathbb{R}^N$. Based on the influence network defined through matrix W, the scalar value $(W^T x)_i$ (the i-th element of the vector $W^T x$) represents the *total effective security investment* of player i, taking into account the effect of all other players in the network. The quantity

$$y_i := (W^T x)_i$$

at the same time constitutes the argument of the utility function of this player. One can think of this utility function as a function that translates the effective investment in security into the total "benefit" experienced by that player.

The i-th player's cost function is defined as

$$J_i(x) := c_i x_i - U_i\left((W^T x)_i\right), \tag{6.8}$$

where c_i is a user-specific pricing parameter. The term $c_i x_i$ is the linear price due to the level of effort or investment. The utility function $U_i(\cdot)$ represents the benefit received from the effective investment made by itself and all of its neighbors in security. Since a single unit of investment does not necessarily translate into the same unit amount in benefit, there is a nonlinear relationship between the investment and the benefit. We now make a number of assumptions on the utility function.

Assumption 6.3

1. The utility function $U_i(\cdot)$ is continuous, twice differentiable, and strictly concave on $[0, \infty)$, $\forall i$.

2. $U_i(0) = 0$, $\left. \dfrac{dU_i(y_i)}{dy_i} \right|_{y_i=0} > c_i$, and $\lim_{y_i \to \infty} \dfrac{dU_i(y_i)}{dy_i} < c_i$, for $c_i > 0$ in (6.8), $\forall i$.

The first assumption ensures mathematical tractability and diminishing returns for each additional unit of security investment following the marginalist principle in economics. The second assumption suggests that if no effective investment is made, there is no benefit; investment prices are low enough that some amount of investment is feasible for each user; and the optimal investment level for each user is finite.

The resulting N-player noncooperative game will now be solved to obtain the NE in pure strategies; that is, we are looking for an $x^* := (x_1^*, \ldots, x_N^*)$ satisfying

$$J_i\left(x_i^*, x_{-i}^*\right) \leq J_i\left(x_i, x_{-i}^*\right), \quad \forall x_i \in [0, \infty), \forall i, \tag{6.9}$$

where x_{-i}^* denotes the NE strategy of all players except the i-th one. Alternatively, as we have seen earlier, NE can be defined in terms of "best-response" functions. Let

$$r_i(x_i, x_{-i}) := \arg\min_{x_i \geq 0} J_i(x_i, x_{-i}) \tag{6.10}$$

be the best-response function of player i. Then, x^* is an NE solution of the security investment game if, and only if,

$$x_i^* = r_i\left(x_i^*, x_{-i}^*\right) \quad \forall i. \tag{6.11}$$

In other words, x^* is a fixed point of the best-response functions, r_i $\forall i$.

6.2.2 Equilibrium and convergence analysis

Under Assumption 6.3, the strictly convex cost function of player i in (6.8) admits a unique positive minimum solution. Let b_i be defined as the single positive value such that

$$\left. \frac{dU_i(y_i)}{dy_i} \right|_{y_i=b_i}$$

from (6.8). The value b_i represents the optimal level of investment made by player i independent of network effects or externalities. It then follows from the first-order optimality conditions that any equilibrium must satisfy

$$\begin{aligned} (W^T x)_i &= b_i \text{ if } x_i > 0 \\ (W^T x)_i &\geq b_i \text{ if } x_i = 0, \end{aligned} \tag{6.12}$$

which are also sufficient by the concavity assumptions made.

Equivalently, the optimality conditions for the equilibrium solution can be expressed in terms of vectors x and y such that

$$y = W^T x - b$$
$$y^T x = 0 \qquad\qquad (6.13)$$
$$x \geq 0, \ y \geq 0.$$

Any solution (x^*, y^*) denotes both the NE investment levels and the "slacks" on those players who invest nothing. The conditions thus take the form of the classic, extensively studied *linear complementarity problem* (LCP). Since these optimality conditions are both necessary and sufficient, it follows that finding an NE of the security investment game is equivalent to solving the associated LCP for x^*. By leveraging results from the latter, strong existence, uniqueness, and convergence results can be obtained.

The existence and uniqueness of NE are closely related to the influence matrix W being strictly diagonally dominant. In general, a matrix $W \in \mathbb{R}^{N \times N}$ is said to be *strictly diagonally dominant* in the row sense if $\sum_{j \neq i} |w_{ij}| < |w_{ii}| = 1 \ \forall i$. If W is strictly diagonally dominant, then this has the following interpretation here: the fixed amount of investment by all players other than i produces less value for player i than an individual investment of the same fixed amount. On the other hand, due to differences in utility functions, an agent's investment in itself may still end up being more valuable to another player. In other words, a fixed amount of investment by one company may translate into a smaller benefit for itself when compared to the benefit of another company which has a higher utility for that investment.

The following uniqueness theorem establishes the relationship between the diagonal dominance of W and the existence of NE.

Theorem 6.4 *If Assumption 6.3 holds and W is strictly diagonally dominant, then the security investment game admits a unique NE.*

Proof First, it is established that if W is diagonally dominant, then it is a *P-matrix*. A P-matrix is a matrix in which every real eigenvalue of each principal submatrix is positive. If W is a diagonally dominant matrix with diagonal elements of ones, then it has only positive real eigenvalue (from the Gershgorin circle theorem). Since each submatrix of a diagonally dominant matrix with positive diagonal elements is also diagonally dominant with positive diagonal elements, it must be a P-matrix. Therefore, if W is diagonally dominant, then it is a P-matrix, and thus the associated LCP of (6.13) has a unique solution for any b. Therefore, the game admits a unique NE solution. □

Although the NE can be obtained in the case of complete information on all players and centrally solving the LCP of (6.13), in reality the organizations repeatedly play this game using "best-response" strategy updates and "solve" it in a distributed manner without knowing the utility functions of other players. Suppose that players update their investment levels in discrete time instances and in parallel but without any coordination among themselves. In order to formalize the game dynamics, let the time be slotted and indexed as $t = 0, 1, 2, \ldots$ Each agent updates its investment level x_i asynchronously at times t_i according to its "best-response" function

$$r_i(x) = \max(0, [(I - W^T)x]_i + b_i),$$ (6.14)

where I denotes the identity matrix. Then, the formal asynchronous update algorithm is described in Algorithm 6.5.

Algorithm 6.5 Asynchronous update algorithm

1: Given $x(0) \geq 0$
2: Set $t \leftarrow 0$
3: **repeat**
4:   **for** $i = 1 \ldots N$ **do**
5:     **if** $i \varepsilon$ {set of updating players} **then**
6:       $x_i(t+1) = r_i(x(t))$
7:     **else**
8:       $x_i(t+1) = x_i(t)$
9:     **end if**
10:    **end for**
11:    $t \leftarrow t + 1$
12: **until** converged

Assume that the set of update times is infinite for each player, i.e. each player updates its strategy infinitely often even though these updates need not be regular or periodic. Then, the global asymptotic convergence of this algorithm is established in the following theorem.

Theorem 6.6 *Let Assumption 6.3 hold. If W is strictly diagonally dominant, then Algorithm 6.5 globally asymptotically converges to the unique NE point of the game. In other words, starting from any feasible point $x(0) \geq 0$, the system converges, $x(t) \to x^*$, as $t \to \infty$ under Algorithm 6.5.*

Proof Let $M := |I - W^T|$, which is a non-negative matrix with all zero diagonal elements and a maximum row (or column) sum strictly less than one by definition of W. From the Gershgorin circle theorem [78] it necessarily follows that the spectral radius of matrix M, $\rho(M) < 1$. Hence, there exists an N-component positive vector, $w > 0$, such that the weighted infinity matrix norm is less than one, $\|M\|_\infty^w < 1$. The weighted infinity matrix norm is defined here as

$$\|M\|_\infty^w := \max_{x \neq 0} \frac{|Mx|_\infty^w}{|x|_\infty^w},$$

where the weighted infinity vector norm is defined as

$$|x|_\infty^w := \max_i \frac{|x_i|}{w_i}.$$

In reference [3], it is proven that the synchronous algorithm satisfies:

$$|x(t+1) - x^*| \leq M |x(t) - x^*|.$$ (6.15)

Taking the weighted infinity norm of both sides yields

$$
\begin{aligned}
|x(t+1)-x^*|_\infty^w &\leq |M|x(t)-x^*||_\infty^w \\
&\leq ||M||_\infty^w |x(t)-x^*|_\infty^w \\
&= \beta |x(t)-x^*|_\infty^w
\end{aligned}
\tag{6.16}
$$

for some constant $0 < \beta < 1$. Thus, the synchronous algorithm represents a *contraction* with respect to the weighted infinity norm.

Using the notation from reference [38], define the sets

$$
X(k) = \{x \in \mathbb{R}_+^N : |x-x^*|_\infty^w \leq \beta^k |x(0)-x^*|_\infty^w\}.
\tag{6.17}
$$

Then, it follows that

- $\ldots \subset X(k+1) \subset X(k) \subset \ldots \subset X(0)$;
- $x(k+1) \in X(k+1) \ \forall k$ and $x \in X(k)$;
- for any sequence $x^k \in X(k) \ \forall k$, $\lim_{k\to\infty} = x^*$;
- for each k, one can write $X(k) = X_1(k) \times X_2(k) \times \ldots \times X_n(k)$ for sets $X_i(k) \subset \mathbb{R}_+$.

Subsequently, it follows from the *asynchronous convergence theorem* in reference [38, p. 464] that the asynchronous update algorithm globally asymptotically converges to the unique equilibrium solution. $\qquad\square$

Although the result in Theorem 6.6 ensures global convergence of Algorithm 6.5, it is asymptotic in nature. Therefore, usually a stopping criterion is used in practice to decide when to stop the updates, i.e. to decide on the convergence in step 12 of the algorithm. One possible stopping criterion is $\|x-r(x)\|_\infty < \varepsilon$ for a chosen sufficiently small $\varepsilon > 0$.

6.2.3   Incentives and game design

One of the important features of the game-theoretic framework presented is the property that investments by one enterprise can produce externalities on its neighbors. When a player benefits from positive externalities produced by a neighbor, the affected player ultimately invests less than it would in isolation. This is commonly referred to as "**free riding**" in economics. For example, while a company in isolation would make a security investment of b by optimizing its own cost (6.8), with positive externalities from its neighbors the same company no longer needs to bear the full burden of investing in security for itself and makes an investment in an amount less than b. On the other hand, when an enterprise receives negative externalities from its neighbors, it is forced to invest more than it would in isolation (b). These relationships can be formalized by defining a metric called "free riding ratio."

The *free riding ratio*, γ_i, for a user i is defined as

$$
\gamma_i = \frac{(W^T x)_i - x_i}{b_i},
\tag{6.18}
$$

for given security investment game parameters and functions W, $U_i(\cdot)$, and $c_i \ \forall i$. In the vector form it is defined for all users as $\gamma = [\gamma_1, \ldots, \gamma_N]$. Verbally, the free riding ratio of a player, $\gamma_i \in \mathbb{R}$, is the ratio of the externalities produced by i's neighbors over the amount it would invest in isolation. Since $b_i > 0$ and $|x| < \infty$, this ratio is always finite and well defined. If γ_i is negative, then player i is forced to overinvest, since the neighbors' contributions are a net negative. If $\gamma_i = 0$, then the player invests as if it is isolated. If $0 < \gamma_i < 1$, then the player enjoys limited free riding, yet it still makes a positive investment, i.e. $b_i > 0$. If $\gamma_i \geq 1$, however, the player i is completely "free riding" and contributes nothing, i.e. $b_i = 0$.

The incentive mechanisms for security investments are complicated even in single large-scale enterprises with autonomous yet interdependent units with their own agendas and incentives. Security divisions in such enterprises have the difficult task of ensuring that these autonomous units collaborate with each other within a mutually beneficial security prevention framework in order to reach the security objectives of the entire company. The influence model discussed provides a quantitative scheme for capturing the interaction between individual systems, services, and processes of these business units. In addition, the decision-making mechanisms are quantified within the game-theoretic framework.

The equilibrium outcome of a security investment game played between autonomous business units of an enterprise may be suboptimal with respect to the goals of the entire organization. In this case, the results of the game can be influenced by the *security division* of the enterprise by deploying explicit incentive mechanisms such as subsidies to other units. Thus, the NE of the game can be optimized or "designed" [18] or optimized using a *subsidy mechanism* according to some enterprise-wide security objectives, such as enforcing a certain security investment policy.

The subsidy mechanism used by the security division of the organization is formalized by subtracting a *subsidy term* $\alpha_i x_i$ from the cost function of player i, which is an autonomous unit within the company,

$$J_i(x) := c_i x_i - U_i\left((W^T x)_i\right) - \alpha_i x_i, \tag{6.19}$$

where α_i is a player-specific subsidy constant. In this case, the matrix W represents the effects of a security investment decision of units on others within the company.

Let x^* be the NE solution of the N-player security investment game with the cost functions (6.19) and for a given vector of subsidies $\alpha := [\alpha_1, \ldots, \alpha_N]$. Then, a game mapping $\mathcal{T}$ can be defined that maps the subsidy α to the unique NE point x^* such that $x^* = \mathcal{T}(\alpha)$. Similarly, the inverse mapping $\hat{\mathcal{T}}$ is defined as $\alpha = \hat{\mathcal{T}}(x^*)$. Notice that the mappings $\mathcal{T}$ and $\hat{\mathcal{T}}$ are highly nonlinear, often not explicitly expressible except from special cases.

The security policy of the company can be expressed in terms of a desired region Ω of investments, $x \in \Omega$, or a function $F(x)$ which is maximized using the subsidy α. Consider the special case of the function $F(x)$ being twice continuously differentiable, strictly concave, and unimodal such that it admits a unique maximum $\hat{x} := \arg\max_{x \geq 0} F(x)$ without any loss of generality.[3] Then, there is a unique subsidy

---

[3] The methods discussed are also applicable to the more general cases of a desired region Ω and multiple maxima of $F(x)$.

$\hat{a} = \hat{\mathcal{T}}(\hat{x})$ which ensures that the NE of the game coincides with the desired security investment profile $x^* = \hat{x}$ [18]. This result is summarized in the following theorem.

Theorem 6.7 *For the security investment game with the cost structure given in (6.19) and assuming that the policy enforcer has full access to the player investment levels, there is a unique subsidy vector, α, which locates the unique NE point of the game to any desirable feasible point, $\hat{x} \in \Omega$ ($\hat{x} := \arg\max_{x \geq 0} F(x)$), where Ω is a desired investment region ($F(x)$ an investment objective function) of the organization.*

Proof The proof immediately follows from the first-order necessary optimality conditions of player cost optimization problems due to the convexity of the cost structure and uniqueness of NE,

$$c_i - \alpha_i - \frac{\partial U_i(W^T \hat{x})}{\partial x_i} = 0 \Rightarrow \hat{\alpha}_i = c_i - \frac{\partial U_i(W^T \hat{x})}{\partial x_i} \ \forall i,$$

and for any feasible $\hat{x}$.   □

Theorem 6.7 establishes that if there is an additional incentive mechanism, then the security investment decisions of the business units can be influenced such that a company-wide security objective is achieved. A similar system can be devised for independent yet collaborating organizations through a neutral overseeing body which enforces industry-wide security investment objectives such as United States Computer Emergency Readiness Team (US-CERT).

6.3   Cooperative games for security risk-management

Risk-management and related decision making by autonomous yet interdependent organizations and divisions have been studied in Section 6.2 within a noncooperative game framework. This section presents an alternative model for **cooperation** among a number of divisions in an organization. At the same time, various risk-management factors such as interdependencies, vulnerabilities, security resources, and organizational frictions are taken into account. The model is based on coalitional game theory, where divisions can form a cooperative group or coalition after evaluating the potential benefits and costs from this cooperation. In addition, various coalitional structures as well as the conditions needed for cooperation are presented.

6.3.1   Coalitional game model

Consider an organization with N divisions denoted by the set $\mathcal{N}$. Each division affects its neighbors both positively and negatively. The positive influence is due to the effect of security resources such as the budget, investments, and expertise. At the same time each division is under certain threats and has vulnerabilities that can affect neighboring divisions negatively. The divisions being "neighbors" often refers to their interdependencies due to their roles in the organization rather than geographical location.

The security relationships between the divisions are represented by linear influence graphs as in Section 6.2.1. The graph $G_p(\mathcal{N}, \mathcal{E}_p)$ models the positive influence of each division's resources on others and $G_n(\mathcal{N}, \mathcal{E}_n)$ captures the negative influences. The graph G_p is represented by the $N \times N$ matrix $W^p = [W_{ij}]$ with elements

$$W_{ij}^p = \begin{cases} 1, & \text{if } i = j, \\ \psi_{ij}, & \text{if } e_{ij} \in \mathcal{E}_p, \quad i, j = 1, \ldots, N, \\ 0, & \text{otherwise}, \end{cases}$$

where $0 < \psi_{ij} \leq 1$ is a real number that quantifies the degree of positive influence from the resources of division i on division j, and e_{ij} is the edge between those divisions. Similarly, the negative influence graph G_n is captured by an $N \times N$ influence matrix $W^n = \left[W_{ij}^n \right]$ where each element is defined as

$$W_{ij}^n = \begin{cases} 1, & \text{if } i = j, \\ \eta_{ij}, & \text{if } e_{ij} \in \mathcal{E}_n, \quad i, j = 1, \ldots, N. \\ 0, & \text{otherwise}. \end{cases}$$

Here, $0 < \eta_{ij} \leq 1$ quantifies how much the vulnerabilities of division i influence or threaten division j. Note that own vulnerabilities affect the division fully corresponding to unit weight.

Let

$$x := [x_1, x_2, \ldots, x_N]$$

be the vector of *security resources* of all divisions that can be used to defend against security risks, and

$$v := [v_1, v_2, \ldots, v_N]$$

be the vector of *threats* or *vulnerabilities* of respective divisions. The *total effective security resources* of a division $i \in \mathcal{N}$ is $(W^{pT} \cdot x)_i$ and the *total effective threats* of division $i \in \mathcal{N}$ is $(W^{nT} \cdot v)_i$. Define the utility function of division i, which it tries to maximize, as

$$U_i(x, v) = b((W^{pT} \cdot x)_i) - c((W^{nT} \cdot v)_i). \tag{6.20}$$

The functions $b(\cdot)$ and $c(\cdot)$ are the actual resource benefit and threat cost of division i, which take its total effective resources and vulnerabilities respectively as arguments.

Cooperative model

Unlike the security investment games of Section 6.2, the players here have fixed resources x and vulnerabilities v, which they cannot change for a certain time period. Nevertheless, they can cooperate by forming coalitions (cooperative groups) as an alternative way of improving their effective security resources and reducing their effective threats. By forming a coalition, $S_1 \subseteq \mathcal{N}$, the divisions can strengthen the positive effect on each other and improve the weights $W_{ij}^p \; \forall i, j \in S_1$, for example by working together more effectively. It is further assumed that even if two divisions have no influence noncooperatively, i.e. $W_{ij} = 0$, they might still be able to create a positive influence by cooperating. At the same time the divisions can reduce the negative effect of threats on each other and decrease the weights $W_{ij}^n \ \forall i, j \in S_1$ by sharing information and collaboratively addressing vulnerabilities.

Define a coalition structure $S = \{S_1, \ldots S_M\}$, $S_i \subset \mathcal{N} \ \forall i$ such that $S_i \cap S_j = \emptyset$, if $i \neq j$ and $\bigcup_{i=1}^M S_i = \mathcal{N}$. Its positive and negative effects are captured by modified matrices $\overline{W}^p(S)$ and $\overline{W}^n(S)$, respectively. For notational convenience, the arguments are dropped to obtain $\overline{W}^p$ and $\overline{W}^n$, when the coalition structure is clear from the context. Given any two divisions $i, j \in \mathcal{N}$, and a coalition $S_1 \subseteq \mathcal{N}$, the elements of the matrix $\overline{W}^p$ are defined as

$$\overline{W}_{ij}^p := \begin{cases} 1, & \text{if } i = j, \\ W_{ij}^p, & \text{if } i \notin S_1 \text{ or } j \notin S_1, \quad i, j = 1, \ldots, N, \\ f\left(W_{ij}^p\right), & \text{if } i, j \in S_1, \end{cases} \tag{6.21}$$

where $f\left(W_{ij}^p\right) \geq W_{ij}^p$ is the cooperative improvement in the positive influence W_{ij}^p when the two divisions i and j belong to the same coalition. Similarly, the matrix $\overline{W}^n$ with elements

$$\overline{W}_{ij}^n := \begin{cases} 1, & \text{if } i = j, \\ W_{ij}^n, & \text{if } i \notin S_1 \text{ or } j \notin S_1, \quad i, j = 1, \ldots, N, \\ g\left(W_{ij}^n\right), & \text{if } i, j \in S_1, \end{cases} \tag{6.22}$$

represents the reduction of the negative influence whenever the two divisions i and j join the same coalition. The underlying assumption here is: when two divisions are in the same coalition, they are able to operate cooperatively and reduce their threats on each other.

Coalitions often entail, in addition to benefits, certain costs due to cultural, economical, or social reasons. For example, it may be quite costly for a well-organized division to cooperate with a badly organized one. There are usually natural frictions between divisions due to business culture or social differences that need to be overcome to establish a coalition. Furthermore, as the number of employees in a coalition increases, various challenges emerge, e.g. coordination or scaling of existing structures. These frictions and size effects can lead to non-negligible impediments to the potential cooperation between divisions as well as the whole organization.

The frictions between various divisions can be modeled using a friction graph $Q(\mathcal{N}, \mathcal{E}_Q)$ defined over the set of divisions $\mathcal{N}$. The effects of the friction graph Q are captured by the friction matrix Q where each element is given by

$$Q_{ij} := \begin{cases} \chi_{ij}, & \text{if } e_{ij} \in \mathcal{E}_Q \text{ and } i \neq j, \\ 0, & \text{otherwise}, \end{cases} \tag{6.23}$$

and $\chi_{ij} > 0$ is a positive real number indicating the degree of friction between divisions i and j. In addition to the friction cost, for any coalition $S \subseteq \mathcal{N}$ an increase in its size

$|S|$, defined as the cardinality of the set, yields an additional cost pertaining to the extra effort that the employees in S need to invest for coordination within their coalition.

Taking into account the friction matrix Q and the size of coalition S, define a cost function $p(Q, S)$. It is natural to assume that the cost is an increasing function of the total friction induced by the graph $Q(\mathcal{N}, \mathcal{E}_Q)$ and of the coalition size $|S|$, which reflects the cost of coordination within large divisions.

The **value of a coalition** $S \subseteq \mathcal{N}$ is given by

$$V(S) = b((\overline{W}^{pT} \cdot x)_S) - c((\overline{W}^{nT} \cdot v)_S) - p(Q, S), \tag{6.24}$$

where Q is the associated friction graph,

$$(\overline{W}^{pT} \cdot x)_S := \sum_{i \in S} (\overline{W}^{pT} \cdot x)_i$$

is the total cooperative effective *security resources* and

$$(\overline{W}^{nT} \cdot x)_S := \sum_{i \in S} (\overline{W}^{nT} \cdot v)_i$$

is the total cooperative effective *threats and vulnerabilities* for coalition S. For the simplicity of the analysis, it will be assumed here that both the "resource benefit" function $b(\cdot)$ and "threat cost" function $c(\cdot)$ are linear, such that

$$V(S) = (\overline{W}^{pT} \cdot x)_S - (\overline{W}^{nT} \cdot v)_S - p(Q, S). \tag{6.25}$$

The model described can be analyzed as a $(\mathcal{N}, v)$ coalitional game [123, 151] with the players being the divisions and the value function given by (6.25). Due to the presence of a cost for cooperation as per (6.25), traditional solution concepts for coalitional games such as the core [123] may not be applicable. In fact, in order for the core to exist as a solution concept, a coalitional game must ensure that the grand coalition, i.e. the coalition of all players, will form. However, in the formulated game, the grand coalition may not always form due to the friction between the divisions. Instead, independent and disjoint coalitions may emerge in the organization. Thus, the game is classified as a *coalition formation game* [151] where the objective is to characterize the coalitional structure that will possibly form between the players.

6.3.2 Coalition formation under ideal cooperation

The functions f and g in (6.21) and (6.22) depend strictly on the cooperative protocol of the divisions within a single coalition. The cooperative model of Section 6.3.1 is quite generic and accommodates any kind of cooperative protocol. It is, nonetheless, useful to present an example to illustrate the properties of the described model.

As a basic example, consider an *ideal cooperation protocol*. Under this protocol, a coalition S of divisions can maximize the positive effects of cooperation and totally eliminate the negative effect of threats on each other. Thus, this cooperation protocol is specified by

$$f\left(W_{ij}^p\right) = 1, \quad \forall i, j \in S \tag{6.26}$$

and

$$g\left(W_{ij}^n\right) = 0, \quad \forall i, j \in S \text{ and } i \neq j. \tag{6.27}$$

The protocol describes an ideal case where, by sharing expertise, skills, and resources, a group of cooperating divisions can effectively eliminate the vulnerabilities and have a perfect synergy among themselves. Although in practice this assumption may not hold, it still provides valuable insights on possible coalitions.

As a first step of analyzing cooperation possibilities among the divisions, the merger of two coalitions is discussed next. In fact, cooperation between two coalitions constitutes a building block for the organization-wide cooperation. The following theorem states the necessary and sufficient condition for the merger of two coalitions.

Theorem 6.8 *Consider two disjoint coalitions $S_1 \subseteq \mathcal{N}, S_2 \subseteq \mathcal{N}, S_1 \cap S_2 = \emptyset$ in the defined ideal cooperation environment and with value functions (6.25). The value of a merger between these two coalitions for the organization is larger than their aggregate value when being separate, i.e.*

$$V(S_1 \cup S_2) \geq V(S_1) + V(S_2),$$

if, and only if, the following condition on the cost functions holds

$$p(Q, S_1 \cup S_2) - (p(Q, S_1) + p(Q, S_2)) \leq \gamma, \tag{6.28}$$

where

$$\gamma := \sum_{j \in S_2} x_j \left(|S_1| - \sum_{i \in S_1} W_{ji}^p\right) + \sum_{j \in S_1} x_j \left(|S_2| - \sum_{i \in S_2} W_{ji}^p\right)$$

$$+ \left(\sum_{i \in S_2} \sum_{j \in S_1} W_{ji}^n v_j + \sum_{i \in S_1} \sum_{j \in S_2} W_{ji}^n v_j\right)$$

is the total effective benefit of this merger for the organization.

Proof The value of coalition S_1 is from (6.25),

$$V(S_1) = \sum_{i \in S_1} (\overline{W}^{pT} \cdot x)_i - \sum_{i \in S_1} (\overline{W}^{nT} \cdot v)_i - p(Q, S_1),$$

$$\Rightarrow V(S_1) = \sum_{i \in S_1} \sum_{j \in \mathcal{N}} \overline{W}_{ij}^{pT} \cdot x_j - \sum_{i \in S_1} \sum_{j \in \mathcal{N}} \overline{W}_{ij}^{nT} \cdot v_j - p(Q, S_1).$$

Using the definitions of f and g in (6.26) and (6.27), respectively, this value becomes

$$V(S_1) = |S_1| \sum_{i \in S_1} x_i + \sum_{i \in S_1} \sum_{j \in \mathcal{N} \backslash S_1} W_{ji}^p x_j - \sum_{i \in S_1} \sum_{j \in \mathcal{N} \backslash S_1} W_{ji}^n v_j$$

$$- \sum_{i \in S_1} v_i - p(Q, S_1).$$

The value of coalition S_2,

$$V(S_2) = |S_2| \sum_{i \in S_2} x_i + \sum_{i \in S_2} \sum_{j \in \mathcal{N} \setminus S_2} W_{ji}^p x_j - \sum_{i \in S_2} \sum_{j \in \mathcal{N} \setminus S_2} W_{ji}^n v_j$$

$$- \sum_{i \in S_2} v_i - p(Q, S_2),$$

and the one of $S_1 \cup S_2$,

$$V(S_1 \cup S_2) = |S_1 \cup S_2| \sum_{i \in S_1 \cup S_2} x_i + \sum_{i \in S_1 \cup S_2} \sum_{j \in \mathcal{N} \setminus (S_1 \cup S_2)} W_{ji}^p x_j$$

$$- \sum_{i \in S_1 \cup S_2} \sum_{j \in \mathcal{N} \setminus (S_1 \cup S_2)} W_{ji}^n v_j - \sum_{i \in S_1 \cup S_2} v_i - p(Q, S_1 \cup S_2),$$

are obtained similarly.

Simply substituting these expressions into $V(S_1 \cup S_2) \geq V(S_1) + V(S_2)$ yields then the necessary and sufficient condition in the theorem. □

Theorem 6.8 provides a quantitative criterion in the form of a condition on the cost functions, which allows an organization (or divisions within) to assess when a merger of two potential coalitions is beneficial. This result can be further investigated by considering a special case with a concrete cost function. In many practical scenarios, the cost p of a coalition S can be defined as a linear function of the total friction and the coalition size,

$$p(Q, S) = \begin{cases} \alpha \cdot \sum_{i \in S} \sum_{j \in S} Q_{ij} + \beta \cdot |S|, & \text{if } |S| > 1, \\ 0, & \text{otherwise,} \end{cases} \tag{6.29}$$

where Q is the friction matrix. Note that this matrix does not have to be symmetric. The parameters $\alpha > 0$ and $\beta > 0$ quantify the price of forming a coalition with $|S| > 1$ per unit friction and per unit size unit, respectively. The following result is a special case of Theorem 6.8 for the cost function defined.

Theorem 6.9 *Consider two disjoint coalitions $S_1 \subseteq \mathcal{N}$, $S_2 \subseteq \mathcal{N}$, $S_1 \cap S_2 = \emptyset$ in the defined ideal cooperation environment with the value function (6.25), where the cost is defined in (6.29). If both coalitions have more than one division, $|S_1| > 1$ and $|S_2| > 1$, they cooperate to form the coalition $S_1 \cup S_2$ for the benefit of the organization if, and only if*

$$\alpha \leq \frac{\gamma}{T(S_1 \cup S_2)},$$

where

$$T(S_1 \cup S_2) := \sum_{i \in S_1} \sum_{j \in S_2} (Q_{ij} + Q_{ji}) \tag{6.30}$$

is the total friction between the members of S_1 and the members of S_2.

Proof The cost of the coalition $S_1 \cup S_2$ is

$$p(Q, S_1 \cup S_2) = \alpha \cdot \sum_{i \in S_1 \cup S_2} \sum_{j \in S_1 \cup S_2} Q_{ij} + \beta |S_1 \cup S_2|$$

$$\Rightarrow p(Q, S_1 \cup S_2) = \alpha \sum_{i \in S_1} \sum_{j \in S_1} Q_{ij} + \alpha \sum_{i \in S_2} \sum_{j \in S_2} Q_{ij}$$

$$+ \alpha \sum_{i \in S_1} \sum_{j \in S_2} (Q_{ij} + Q_{ji}) + \beta |S_1| + \beta |S_2|.$$

Applying the definitions of cost p in (6.29) and total friction T in (6.30), to the cost function in (6.31) yields

$$p(Q, S_1 \cup S_2) = p(Q, S_1) + p(Q, S_2) + \alpha T(S_1 \cup S_2).$$

Then, the result follows immediately from condition (6.28) in Theorem 6.8.   □

Theorem 6.9 supports the intuition that the benefit of cooperation among two coalitions of divisions in a company mainly depends on the ratio between the total effective benefit γ to the total friction T between the members of the two coalitions. Furthermore, the friction between the members of a single coalition does not affect whether or not this coalition will cooperate with another one. Finally, the theorem provides an upper-bound on the price per unit of friction above which no cooperation is possible between any two coalitions of size larger than one.

The next example illustrates the results of Theorems 6.8 and 6.9 in the particular case of cooperation among two single divisions. On the other hand, there are many generalizations and extensions to the presented results. One direction is to investigate nonlinear "resource benefit" $b(\cdot)$ and "threat cost" $c(\cdot)$ functions in (6.25). Another natural extension is to study the case of a generic nonideal cooperation protocol where the improvement in the positive and negative weights is partial, i.e. f and g are different from those in (6.26) and (6.27), respectively.

Example: merger of two divisions
As an illustrative example, consider the particular case of cooperation among two single divisions. Applying the results in Theorems 6.8 and 6.9, the condition for the two divisions to merge is

$$\alpha(Q_{12} + Q_{21}) + 2\beta \leq x_1 \left(1 - W_{12}^p\right) + x_2 \left(1 - W_{21}^p\right) + W_{12}^n v_1 + W_{21}^n v_2. \tag{6.31}$$

Applying (6.31) to the following particular case yields interesting results. When the two divisions have no threats on each other, i.e. $v_1 = v_2 = 0$, β is very small, $\beta \approx 0$, and $q := Q_{12} = Q_{21}$, the condition in (6.31) reduces to

$$\left(1 - W_{12}^p\right) x_1 + \left(1 - W_{21}^p\right) x_2 > 2\alpha q. \tag{6.32}$$

This condition has several interesting properties and interpretations:

1. If there is no friction, $q = 0$, then cooperation is always beneficial.
2. If there is already strong positive influence between divisions, $W_{12}^p, W_{21}^p \to 1$, then they have almost no incentive to cooperate.

3. If one division already benefits from strong positive influence, e.g. $W_{12}^p \to 1$, then the second division needs to have the sufficient resources x_2 to overcome the friction and cooperate.

4. In the case of no prior positive influence, $W_{12}^p = W_{21}^p = 0$, it is sufficient that the total resources of both divisions are greater than the friction, $x_1 + x_2 > 2\alpha q$.

6.4 Discussion and further reading

There are very few quantitative studies in the area of risk-management for decision making in the context of information technology and network security [70, 111, 112]. Section 6.1 presents a novel probabilistic risk analysis and dynamic control framework summarizing results in references [9] and [121], respectively. A popular book that provides valuable insights to cascading risks, effects of corporate culture, and social factors in risk-management is reference [183].

The security investment games in Section 6.2 are based on references [113, 114]. The game-design approach to incentive mechanisms in security is an application of the framework introduced in references [18, 19]. The cooperative game-based analysis presented in Section 6.3 is based on reference [150] and utilizes the coalitional game models of reference [151].

7 Resource allocation for security

Chapter overview

1. An optimization approach to malware filtering
   - traffic centrality measures
   - centrality-based problem formulations
   - constrained problems
2. Robust control framework for security response
   - network traffic filtering model
   - derivation of optimal controller and state estimator
   - discussion on system implementation
3. Optimal and robust epidemic response
   - epidemic models and feedback response
   - multiple network case
   - malware removal using optimal and robust control methods

Chapter summary

The decision and control-theoretic approach quantifies implicit costs and formalizes decision-making processes for resource allocation in network security. The optimal and robust control methods utilized are illustrated with three example scenarios: placement of malware filters, dynamic filtering of suspicious packets, and robust response to malware epidemics.

7.1   An optimization approach to malware filtering

Placement of malware filters in the presence of an intelligent attacker has been modeled within a stochastic security game framework and investigated in Section 4.4. This section studies deployment and configuration strategies for next-generation network traffic filters within an optimization framework. In contrast to the high-level and long-term security decision-making models earlier in Section 6.2, the objective here is to analyze decision-making schemes at a technical level and closer to real time, using the filter deployment and configuration problem as an illustrative setting.

Network traffic filtering aims to stop malware packets while they traverse network links by sampling packets or sessions and either comparing their contents to known malware signatures or looking for anomalies likely to be malware. Filtering capabilities are increasingly integrated into routers themselves to reduce hardware deployment costs and to allow for more adaptive security. Traffic filters are expected to be configurable, networked, and even autonomous. The objective here is to investigate the deployment and configuration of such devices within an optimization framework.

In this section, a network of configurable, networked routers is considered with traffic filtering capabilities which can be dynamically and remotely set by a (centralized) server. Some subsets of these routers act as source routers, another subset as destination, and the remaining as core routers. The model is independent of the effectiveness of malware filters. It is **assumed** that packets are marked only once. In addition, the network administrator has full knowledge of the network traffic, possibly with some delay. Due to the small percentage of dropped malware packets within the total packet volume, the effect of filtering on traffic flow volumes is ignored. Alternatively, a flag can be inserted to the headers of suspicious packets instead of discarding them. It is worth noting that, although the focus here is on packet filtering, all of the theoretical results also apply to the filtering of sessions.

One of the main aspects of the framework discussed is the expression of the value or importance of a single router or link in a network with a given traffic pattern. Using modified versions of well-known centrality measures from graph theory, it is shown that when source–destination pairs are weighted based on traffic magnitude in centrality algorithms, the resulting centrality measure for each router is equivalent to the traffic seen at each router under certain assumptions on routing algorithms. This intuitive but not immediately obvious result allows the utilization of centrality measures in cost functions, and facilitates capturing common network security objectives within a quantitative model.

The quantitative optimization framework introduced is used as a basis for the derivation of optimal *filtering strategies*, which indicate at what rate to filter packets or sessions at a specific configurable filtering router or device at a given time and traffic pattern. In order to demonstrate the applicability of the approach to a variety of scenarios, multiple optimization problems are formulated and solved to determine the optimal filtering strategy for different objectives involving security level, centrality, and costs.

Furthermore, various hardware- and security-level constraints are taken into account in these problems.

7.1.1 Traffic centrality measures

Traffic centrality measures capture the importance of routers (or links) on a network, and hence are helpful when defining objective functions for malware filtering problems [45]. Two specific centrality measures applicable to communication networks are studied for use within the context of the filter placement problem. Traditional centrality measures used in social networking usually involve source–destination pairs, but each pair is weighted identically. Moreover, they consider every node to be a potential source or destination. Modifying these assumptions, a more relevant *traffic centrality measure* is obtained. First, the source–destination pairs are weighted according to the magnitude of traffic that travels between them. Second, only nodes that are in fact sources and/or destinations on the network (i.e. no core routers) are used in computations.

Traffic betweenness centrality (TBC) is a betweenness centrality metric that differs from traditional ones due to the two changes mentioned above. Let the undirected connected graph (V, E) represent a network, where $V = \{v_1, \ldots, v_N\}$ is the set of network nodes and E is the set of links weighted by the amount of traffic passing through them. Let $S \subset V$ be the set of all the sources, $D \subset V$ be the set of all destinations, and L be the set of all source–destination pairs (s, d). The number of shortest paths between $s \in S$ and $d \in D$ is denoted by $\sigma_{sd} \geq 1$, which immediately follows from the connectivity assumption. The number of these shortest paths that pass through some router $r \in V$ is $\sigma_{sd}(r)$. Moreover, the amount of traffic between s and d is u_{sd}, which is assumed to be positive. It is assumed that if there are multiple shortest paths between a source and a destination, then they are equally likely to be used, which corresponds to a load-balancing scheme. Then, the TBC of a router r is defined as

$$C_B(r) := \sum_{(s,d) \in L} u_{sd} \frac{\sigma_{sd}(r)}{\sigma_{sd}}. \tag{7.1}$$

In other words, it is the fraction of shortest paths of all source–destination pairs that pass through a particular router, with each source–destination pair being weighted by its traffic magnitude. If at least one shortest path for any source–destination pair passes through a given router r, then $C_B(r) > 0$. Note that a traditional betweenness centrality measure would not restrict the sum to only those pairs of vertices that are in the set of actual source–destination pairs, L.

The *traffic stress centrality (TSC)* is defined as a special case of TBC, where only one path is active at a given time between a source–destination pair. In other words, there is no load-balancing on the network. In this case, if a node is on this active shortest route, then $\sigma_{sd} = 1$, otherwise it is zero. The TSC of a router r is defined as

$$C_S(r) := \sum_{(s,d) \in L} u_{sd} \sigma_{sd}(r). \tag{7.2}$$

The following theorem establishes an intuitive but not immediately obvious relationship between the traffic centrality measures defined here and the actual traffic that passes through a router.

Theorem 7.1 *Let the TBC and TSC of a node r on a given network be defined by (7.1) and (7.2), respectively.*

1. *If the traffic on the network is routed using a load-balancing shortest path routing scheme, where all the packets from a source to a destination are equally likely to be delivered through multiple shortest paths between them, then the amount of traffic on a router is equal to its TBC measure.*
2. *If the network uses a simplest shortest path routing scheme, where all packets between a source and a destination follow a single shortest route, then the amount of traffic on a router is equal to its TSC measure.*

Proof In the first part of the theorem, if there are multiple shortest paths between a source–destination node pair, then they are equally likely to be used. Consider a single source–destination pair $i \in L$ with a traffic flow u_i and σ_i aggregate number of shortest paths. Let $\sigma_i(r)$ of these paths pass through router r. Consequently, the amount of traffic for this source–destination pair that passes through router r is simply

$$u_i(r) = u_i \frac{\sigma_i(r)}{\sigma_i}. \tag{7.3}$$

Then, the total traffic at router r is obtained by summing (7.3) over all source–destination pairs L resulting in

$$u_r = \sum_{i \in L} u_i(r) = \sum_{i \in L} u_i \frac{\sigma_i(r)}{\sigma_i} = C_B(r), \tag{7.4}$$

which establishes the first part.

In the second part, it is assumed that there is only one shortest path available or in use. Then, the result follows directly as a special case where σ_i is always one and $\sigma_i(r)$ is one if router r is on the used shortest path and zero otherwise. Thus,

$$u_r = \sum_{j \in L} u_j \sigma_j(r) = C_S(r), \tag{7.5}$$

which completes the proof. $\qquad\square$

7.1.2 Filtering problem formulations

In this section, multiple optimization problems are formulated for network-filtering scenarios and solved to determine the optimal filtering strategy for different objectives involving security level, centrality, and costs. Each formulation utilizes centrality measures and takes into account various hardware- and security-level constraints. A brief overview of the relevant background on optimization is available in Appendix A.1.3.

Centrality-based problem

The first filtering problem formulation is based on the assumption that the damage caused by malware within the network is quadratically increasing.[1] Let p_r be the traffic-filtering rate at a router. Then, the cost function is defined as

$$\sum_{r \in V} M_r(p_r), \text{where } M_r(p_r) = C_S(r)^2 (1 - p_r)^2, \tag{7.6}$$

which is a strictly convex function. This cost function admits an interpretation in terms of malware traffic. Let m_r be the number of malignant packets to traverse a router $r \in V$ in a time interval. Then, $m_r(t)(1 - p_r)$ such packets will successfully pass by the router undetected. From Theorem 7.1, if the amount of malware at a router is proportional to the amount of traffic at the router, then it is also proportional to the centrality of the router. Thus, the traffic centrality measure $C_S(r)$ in (7.6) can be substituted with $m_r(t)$ and would still lead to the same cost expression.

The centrality-based malware filtering problem is defined as

$$\min_p \sum_{r \in V} M_r(p_r), \text{ such that } \sum_{r \in V} u_r(t) p_r \leq \theta, \tag{7.7}$$

where $M_r(p_r)$ is defined in (7.6). The positive constant θ denotes here an upper bound on the total packet-filtering capacity, which follows from technological restrictions on filtering devices. This problem is a convex optimization problem with a strictly convex cost function, and hence admits a unique solution. Furthermore, depending on the routing properties of the network, C_B can be equivalently used instead of C_S in this formulation.

Although the filtering problem (7.7) has additional constraints such as the natural one on the filtering rate $p_r \in [0,1] \; \forall r$, an analytical solution is discussed first in its current form to gain further insight and for illustrative purposes. Toward this end, the problem (7.7) is rewritten as

$$\min_p \quad \frac{1}{2} p^T B p + c^T p$$
$$\text{subject to: } u^T p \leq \theta, \tag{7.8}$$

where

$$c^T = [-C_{TS}(1)^2 \; -C_{TS}(2)^2 \; \cdots \; -C_{TS}(N)^2],$$

$$B = \text{diag}(-c), \quad u^T = [u_1 \; u_2 \; \cdots \; u_N],$$

and N is the number of routers. Note that $\text{diag}(\cdot)$ denotes here a diagonal matrix with elements from its (vector) argument, and hence, B is positive definite.

---

[1] This analysis can be extended under different assumptions on the relationship between malware traffic and cost to the network (linear, logarithmic, etc.).

The corresponding dual problem is the scalar concave maximization problem:

$$\max_{\mu} \quad -\frac{1}{2}\mu^2 Q - t\mu - \frac{1}{2}c^T B^{-1} c$$
$$\text{subject to: } \mu \geq 0 \tag{7.9}$$

where both $Q > 0$ and t are scalars:

$$Q = u^T B^{-1} u \text{ and } t = \theta + u^T B^{-1} c.$$

This maximization problem admits the unique solution

$$\mu^* = \max\left(0, \frac{-\theta - u^T B^{-1} c}{u^T B^{-1} u}\right). \tag{7.10}$$

By substituting it back into the relation $p = -B^{-1}(c + u\mu)$, which is the unique solution to (7.8), the optimal p_r values are obtained:

$$p_r^* = 1 - \frac{\mu^* u_r}{C_{\text{TS}}(r)^2}. \tag{7.11}$$

This expression is simplified significantly using the equality between $C_{\text{TS}}(r)$ and u_r to obtain

$$p_r^* = 1 - \left[\frac{\bar{u} - \bar{\theta}}{u_r}\right], \tag{7.12}$$

under the assumption of $\theta + u^T B^{-1} c < 0$. Here, $\bar{u}$ refers to the average traffic at all routers and $\bar{\theta} = \theta/N$.

In this case, the optimal filtering rate p_r^* can be found with localized information, aside from the average traffic on all links. Hence, it can be computed in a distributed manner. As the filtering capacity per router ($\bar{\theta}$) approaches the average traffic per router ($\bar{u}$), the filtering probabilities increase. Similarly, as the traffic at a router (u_r) increases, the filtering probability at that router increases. An inherent assumption here is that $p_r \in (0, 1)$, i.e. an inner solution, which occurs when $\bar{\theta} < \bar{u}$ and $\bar{u} - \bar{\theta} > u_r$.

Constrained centrality-based problem

The problem (7.7) in the previous section is redefined as the *constrained centrality-based problem* when additional constraints are enforced, First, the filtering rate p_r cannot be negative or more than 100 percent. Second, if there is a marking mechanism for detected packets that prevents them from being processed again, certain restrictions can be imposed upon the *effective sampling rate* for source–destination pairs.

Let the index j refer to a source–destination pair in the set of source–destination pairs L and $\xi_j(r)$ denote the fraction of shortest paths that a router r lies on for a particular source–destination pair j. If there is only one shortest path or only one that is used for source–destination pair j, then $\xi_j(r)$ is equal to one for all routers on that path and zero for all other routers. Finally, let y_r be a discrete variable indicating whether or not filtering is enabled on the router r. Thus, considering that packets are not sampled twice because of packet marking, the effective sampling rate for pair j becomes

$$\rho_j = \sum_{r \in V} \xi_j(r) p_r y_r. \tag{7.13}$$

Here, p_r still refers to the proportion of the *total* amount of traffic at a router as a simplifying assumption.

It is possible to define a *minimum effective sampling rate* $0 < \alpha < 1$ on ρ. These constraints are then summarized as

$$p_r \in [0,1], \ r \in R$$
$$\rho_j \in [\alpha, 1], \ j \in P \tag{7.14}$$

Due to the strict convexity of the cost function and convexity of the constraint set, the constrained centrality-based problem has a unique optimal solution. However, this problem usually does not have a simple closed-form solution as in the unconstrained version. In this case, a centralized server calculates optimal filtering for the current time step based on traffic measurements from the last time step by solving the quadratic programming problem (7.7) with the constraints (7.14). This problem can be solved using a gradient projection method [35].

Centrality and sampling cost-based problem
As a variation of the formulations in Section 7.1.2, the strict upper bound on the filtering capacity (number of packets filtered networkwide) can be replaced by a cost on packet sampling, which is included within the centrality-based metric in the objective to be minimized. Hence, the objective now is to minimize the sum of the $M_r(p_r)$ and a cost on the sampling rate weighted by $\gamma \bar{u}$. The weight parameter $\bar{u}$ allows for more accurate comparisons of γ values across networks with different traffic loads. Thus, the problem is formulated as

$$\min_p \sum_{r \in V} [M_r(p_r) + \gamma \bar{u} u_r p_r c_r], \tag{7.15}$$

where c_r is a per-packet sampling cost on link r. The constraints on the effective sampling rate remain as in (7.14):

$$p_r \in [0,1], \ r \in R$$
$$\rho_j \in [\alpha, 1], \ j \in P \tag{7.16}$$

This is again a convex optimization problem over a compact set and admits a globally optimal solution. Therefore, the same methods used for solving the previous problem also apply here. Notice that if $\gamma = 0$ and the constraint on the total quantity of filtering θ is added, the centrality-based problem of Section 7.1.2 is obtained as a special case.

Effective sampling rate and filtering cost-based problem
In this variant of the filtering problem, the objective function combines the cost of enabling traffic filters at the routers, a per-packet sampling cost, and a utility term that captures the benefit of higher effective sampling rates. A constraint-effective sampling is imposed such that each source–destination pair has some minimum effective sampling rate. Assume that f_r is the cost of implementing filtering at a particular router

$r \in V$ per time step. This cost will depend on the particular hardware being used as well as the potential effects of filtering on traffic flow. The discrete variable $y_r \in \{0, 1\}$ indicates whether or not filtering is enabled at a particular router. In addition, the utility of sampling a source–destination pair traffic is taken into account explicitly. In certain cases it may be beneficial to filter more than the minimum effective sampling rates. For illustrative purposes, define the linear utility function

$$U(\rho_j, j) = \rho_j s_j d_j, \qquad (7.17)$$

as one possible version. Here, the effective sampling rate ρ is weighted by the value of the destination d_j and the suspicion level of the source s_j, each represented by a positive real number. The problem is then formulated as a mixed-integer nonlinear program (MINLP) as

$$\min_{p, y} \sum_{r \in V} [f_r y_r + y_r c_r p_r u_r] - \gamma \bar{u} \sum_{j \in L} U(\rho_j, j). \qquad (7.18)$$

An interesting special case of this problem is when $\gamma = 0$. Then, the objective is simply to achieve the constraints on p and the effective sampling rates ρ_j while minimizing the filtering costs.

The constraints on the effective sampling rate are similar to those in the previous problems. In addition, let $p_{max}(r)$ denote the maximum filtering rate available at a router $r \in V$. If the filter can only filter packets at a certain rate z, then $p_{max}(r)$ could be set equal to z/u_r or in the worst case, z/υ_r, where υ_r is the capacity of router r. The resulting set of constraints is

$$\rho_j \in [\alpha, 1] \; \forall j$$
$$p_r \in [0, p_{max}(r)], \; r \in R$$
$$y_r \in \{0, 1\}, \; r \in R \qquad (7.19)$$

This formulated problem admits a global optimal solution which may not be unique. For each given filter activation pattern, i.e. a set of enabled filters on the network, the problem turns into a linear optimization problem on a compact set, which will have an optimal (potentially nonunique) solution. The global optimum can only be found by solving 2^R such problems, one for each possible filter configuration.

7.2   A robust control framework for security response

Section 7.1 has focused on how to decide on the sampling rates of adaptive malware filters based on the measurements of dynamic traffic patterns. In this section, a robust dynamic filtering framework is introduced and analyzed that is not a function of overall traffic measurements, but one of suspected malware traffic. Instead of deciding what fraction of traffic on the network should be sampled as a precautionary measure, filtering rules are derived here to drop packets or sessions given their suspicion levels as a security response.

Distinguishing malware from legitimate network traffic is a very challenging task. A variety of detection schemes, e.g. machine learning or statistical methods, have been proposed in the literature to differentiate malicious traffic from regular traffic. Given that it is not possible to know with certainty whether or not a packet or session contains malware, when should a packet or session be dropped?

The objective here is to formalize this decision process within a quantitative framework under the assumption of imperfect detection or classification. The introduced framework captures the tradeoff between false alarms (e.g. dropping clean packets) and not filtering suspected malware. Specifically, H^∞-optimal control, a robust optimal control-theoretical tool, is used to analyze this problem (see Appendix A.3.4 for a brief overview). A feedback filtering rate controller is derived that explicitly considers the cost of malware infection as well as the cost of filtering. The resulting dynamic malware filtering algorithms are aggressive in the face of malware attacks and thus likely to prevent or at least delay their spread.

The H^∞-**optimal control framework** allows for dynamically changing filtering rules or parameters in order to ensure a certain performance level. In H^∞-optimal control, disturbances (malware) to the network can be interpreted as an intelligent maximizing opponent in a dynamic zero-sum game who plays with knowledge of the minimizer's control action. Hence, one evaluates the network under the worst possible conditions. This approach applies naturally to the problem of malware response as the traffic deviation resulting from a malware attack is not merely random noise, but represents the efforts of an intelligent attacker. Moreover, the objectives of the defense and attackers are diametrically opposed, so the zero-sum assumption is accurate. Thus, H^∞-optimal control provides a worst-case framework that ensures a performance guarantee in the form of a minimum security level.

7.2.1   Network traffic filtering model

In this section, a linear system model for malware traffic is presented and used as a basis for studying the problem of malware filtering. The optimal filtering strategies derived here can also be applied to other types of filtering problem in addition to network traffic filtering, such as spam filtering, distributed denial of service attacks, etc.

One common way to provide network security is to implement network traffic filters (firewalls) that eliminate suspicious packets. For example, packets corresponding to a certain port number or from a subclass of IP addresses might be filtered because these characteristics indicate that the packets are probably not legitimate. These filters are dynamic if they can easily be reconfigured to respond to changing network circumstances. The objective here is to investigate the criteria that should be used for selecting which packets to filter; in other words, to answer the question: how many packets are suspicious enough to be filtered? Given the difficulty in differentiating legitimate packets from those containing malware and the inherent costs associated with false decisions, this is not a question with an easy answer.

Consider a computer network under the control of a single administrative unit, such as a corporate network. Assume the network is divided into subnetworks for administrative and security purposes. While the model is described within the context of this scenario, the corresponding control framework can be applied to other contexts as well by redefining the entities in question. For example, a large telecommunications company may apply this model and control schemes to much larger units, such as the clients it serves.

Let $x(t)$ represent the number of malware packets that traverse a link on their way to the destination subnetwork at time t originating from infected sources outside the subnetwork. This **malware flow** to the subnetwork is modeled using a linear differential equation with control and disturbance terms – a framework commonly used in control theory:

$$\dot{x}(t) = ax(t) + bu(t) + w^a(t), \tag{7.20}$$

where $u(t)$ represents the number of packets that are filtered at a particular time. Usually, only some proportion, b, of the packets filtered are actually malware related. Thus, $(1 - b)$ is the proportion of filtered packets that are not malware related (false alarms).

The term $w^a(t)$ represents the number of malware packets added to the link at time t intentionally by malicious sources or unintentionally by hidden software running on hosts, both located outside the subnetwork considered. Alternatively, it represents a **worm attack**, expressed in terms of the number of malware packets sent from a subnetwork to other subnetworks at each time instant. Note again that no assumption is made on the form of the attack.

Consequently, $u(t)$ and $w^a(t)$ represent the packet-filtering rate and malware infiltration rate for this specific subnetwork, respectively. The negative value a represents the instantaneous proportion of malware packets on the link that are actually delivered to the subnetwork. Figure 7.1 visualizes the model considered within the context of malware traffic filtering.

Increasing the dimension of the model in (7.20) leads to a set of linear differential equations:

$$\dot{x}(t) = Ax(t) + Bu(t) + Dw^a(t), \tag{7.21}$$

which captures the **multiple networks case**. In this case both A and B are obtained simply by multiplying the identity matrix by a and b, respectively. The D matrix imposes a propagation model on the attack and quantifies how malware is routed and distributed

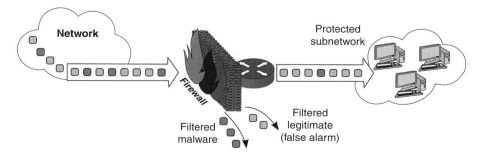

Figure 7.1 Graphical description of the malware traffic filtering model.

on this network. Here, we take it to have zeros for its diagonal terms (intra-subnetwork malware traffic does not leave the subnetwork), and each column must sum to one to ensure conservation of packets.

Overall, this model simplifies actual network dynamics by assuming a linear system and using a fluid approximation of traffic flow.

Denote by $y(t)$ the number of measured inbound malicious packets prior to filtering. Note that the separation between detection ($y(t)$) and response ($u(t)$) is only at the conceptual level. In the implementation, both may occur on the same device. Inaccuracies in $y(t)$ are inevitable due to the challenging problem of distinguishing malicious packets from legitimate ones [24]. To capture this uncertainty formally, define $y(t)$ as

$$y(t) := Cx(t) + Ew^n(t), \tag{7.22}$$

where $w^n(t)$ is measurement noise of any form. To simplify notation, the measurement noise and attack disturbance are denoted both as respective parts of the vector $w := [w^a \; w^n]^T$. In addition, define $N := EE^T$ and assume that it is positive definite, meaning that the measurement noise impacts each dimension of the measured output. The C matrix models the fact that $y(t)$ is higher than and proportional to $x(t)$, the measurement prior to filtering. When implemented, this constant could be measured from an analysis of packet filtering and the calculations required to determine that the optimal controller could be rerun periodically.

Note that no assumption is made on how $y(t)$ is obtained. It could be the result of some statistical analysis comparing the expected traffic to the measured traffic or be based on a set of rules where packets with certain characteristics are assumed to be malicious.

The model at hand contains several simplifications and assumptions to ensure analytical tractability:

- The components of the B matrix are set to be constants, although in reality the values of these components are variable and dependent on the specific attack type.
- The assumption of a constant value for the C matrix is an approximation, as in reality the number of malware packets prior to filtering will probably not be linearly dependent upon the number after filtering.
- Network dynamics are modeled using a fluid approximation of traffic flow.

7.2.2 Derivation of optimal controller and state estimator

In this subsection, an algorithm or controller for traffic filtering is designed under a given imperfect detection (measure) of inbound malicious packets.

As part of the H^∞-optimal control analysis and design, first the *controlled output* is defined

$$z(t) := Hx(t) + Gu(t), \tag{7.23}$$

where $G^T G$ is assumed to be positive definite, and no cost is placed on the product of control actions and states: $H^T G = 0$. The matrix H represents a cost on malicious packets arriving at a subnetwork. The conditions for the H^∞-optimal control theory to apply are that (A, B) and (A, D) be stabilizable, and (A, H) and (A, C) be detectable.

If x becomes negative, then this means that legitimate packets are filtered from the link. A balanced policy is adopted by penalizing underfiltering malicious traffic equally to overfiltering, which prevents legitimate network traffic from traversing the link. The cost on filtering legitimate traffic can be defined in further detail. If a cost of f_l is assigned to filtering legitimate packets, a $(1 - b)$ portion of total traffic, and a cost of f_a to the filtering action itself, then the components g of G can be specified as $g = f_l(1 - b) + f_a$.

The overall cost for the purpose of H^∞ analysis is subsequently defined by

$$F(x, u, w) = \frac{\|z\|}{\|w\|}, \tag{7.24}$$

where $\|z\| := \left(\int_{-\infty}^{\infty} |z(t)|^2 dt\right)^{1/2}$, the L^2-norm of the function $z(\cdot)$, and a similar definition applies to $\|w\|$. Although it is a cost ratio, it will be referred to as the cost for simplicity. It captures the proportional changes in z due to changes in w. More intuitively, it is the ratio of the cost incurred by the system to the corresponding attacker and measurement noise "effort." Notice that the focus here is on the malware traffic rather than the effects of it (e.g. disabling hosts or servers), which usually depend on the malware type.

H^∞-optimal control theory not only applies very directly and appropriately to the problem of worm response, but also guarantees that a performance factor (the H^∞ norm) will be met. This norm can be thought of as the worst possible value for the cost L and bounded above by

$$\gamma^* := \inf_u \sup_w F(u, w), \tag{7.25}$$

which can also be viewed as the optimal performance level in this H^∞ context.

In order to actually solve for the optimal controller $\mu(y)$, the number of packets to filter as a function of the inaccurately measured number of inbound malicious packets, a corresponding differential game is defined between the attackers and the malware filtering system, which is parameterized by γ, where $\gamma < \gamma^*$

$$L_\gamma(u, w) = \|z\|^2 - \gamma^2 \|w\|^2. \tag{7.26}$$

The malicious attackers try to maximize this cost function in the worst case by varying w while the malware filtering algorithm minimizes it via the controller u. This can be seen as a security game similar to those discussed in Chapter 3.

The optimal filtering strategy $u = \mu_\gamma(\hat{x})$ is derived from this differential game formulation for any $\gamma > \gamma^*$ as[2]

$$\mu_\gamma(\hat{x}) = -(G^T G)^{-1} B^T \bar{Z}_\gamma \hat{x}, \tag{7.27}$$

---

[2] The theory behind this and the derivation of this solution can be found in reference [29].

where $\bar{Z}_\gamma$ is solved from the so-called generalized algebraic Riccati equations (GARE)

$$A^T Z + ZA - Z(B(G^T G)^{-1} B^T - \gamma^{-2} DD^T)Z + H^T H = 0, \qquad (7.28)$$

as its unique minimal positive definite solution, and $\hat{x}$ is generated by

$$\dot{\hat{x}} = \left[A - (B(G^T G)^{-1} B^T - \gamma^{-2} DD^T)\bar{Z}_\gamma \right] \hat{x}$$
$$+ \left[I - \gamma^{-2} \bar{\Sigma}_\gamma \bar{Z}_\gamma \right]^{-1} \bar{\Sigma}_\gamma C^T N^{-1}(y - C\hat{x}), \qquad (7.29)$$

where $\bar{\Sigma}_\gamma$ is the unique minimal positive definite solution of the GARE

$$A\Sigma + \Sigma A^T - \Sigma(C^T N^{-1} C - \gamma^{-2} H^T H)\Sigma + DD^T = 0. \qquad (7.30)$$

Here, $\hat{x}$ is an estimate for x, corresponding to the worst possible values of w under the criterion (7.26). This is a linear feedback controller operating on a state estimate $\hat{x}$, and γ^* is the smallest γ such that the spectral radius condition $\rho\left(\bar{\Sigma}_\gamma \bar{Z}_\gamma\right) < \gamma^2$ holds.

The H^∞-optimal controller derived can be calculated offline using only the linear quadratic system model. The online calculation is simply a multiplication by the estimate of the system state. Also note that this controller requires a networkwide knowledge of the system state estimate and thus, this is a centralized control solution.

There are a few assumptions implicit in this specific controller formation. The various filters will have to send control packets to each other, indicating their y values. Moreover, it is assumed that these filters are able to convert a number of packets to filter per time step ($u(t)$) into a filtering rule that will implement that filtering rate. The packets that are most likely to be malicious should be filtered first. Exactly how this is done depends on the system implementation. For example, a rule-based filter could implement more rules (block more ports or IP addresses) or the sensitivity of an anomaly-based detector could be increased when $u(t)$ increases.

Remark 7.2 *The H^∞-optimal controller derived, (7.27), is a centralized control solution due to the D matrix, which imposes a specific malware propagation model. However, the same framework can be applied to each subnetwork separately by using (7.20) for each. This leads to a decentralized solution consisting of multiple independent scalar H^∞-optimal controllers.*

Discussion on system implementation

The numerical analysis and simulation results in reference [43] based on the analytical framework presented are promising and may be extended to an experimental analysis. It is possible to implement the robust control framework in a realistic environment as part of a dynamic malware filtering system using a procedural description of the H^∞ controller given in Algorithm 7.3. Such an experimental evaluation is on the one hand specific to the system setup and implementation. On the other hand, it can provide deeper insights into the practical aspects of the problem and illustrate the applicability of the developed algorithms.

The robust and dynamic malware filtering framework presented allows security engineers to make malware filtering decisions (e.g. setting filter thresholds) in a principled

---

Algorithm 7.3 H^∞-Optimal robust malware filtering

1: Choose the preference parameters H and G in (7.23) and a sufficiently large γ.
2: Estimate the parameters A, B, C, D, E in (7.21) and (7.22).
3: **while** Filtering system is operational **do**
4:      **repeat**
5:          Decrease γ.
6:          Compute $\bar{Z}_\gamma$ using (7.28).
7:          Compute $\bar{\Sigma}_\gamma$ using (7.30).
8:      **until** spectral radius $\rho(\bar{\Sigma}_\gamma \bar{Z}_\gamma) = \gamma^2 - \varepsilon$, for some small $\varepsilon > 0$.
9:      **for** t in the next fixed time period
10:         Compute $\hat{x}(t)$ using (7.29).
11:         Compute the optimal filtering strategy $u(t) = \mu_\gamma(\hat{x}(t))$ from (7.27).
12:         Apply the filter to the system using computed u
13:     **end for**
14:     Observe the system to re-estimate A, B, C, D, E.
15: **end while**

---

way. The costs and benefits in the system are modeled quantitatively, which enables the administrators to explicitly express their preferences. The quantitative approach brings significant advantages over heuristic decision making. For example, the quantitative framework is more scalable as it translates preferences to decisions in a semi-automated manner rather than configuring each threshold manually. In practice, this enables a smaller expert team to oversee a larger system.

Another significant advantage of the robust control framework is its relative insensitivity to parameter variations. Some of the parameters of the model have to be determined through an additional observation and estimation system that operates on a longer timescale than actual filtering dynamics. However, this process can be imperfect or parameters may shift in the intervals between estimation instances. The robust control framework is perfectly suited to such situations, increasing its potential practical applicability.

7.3   Optimal and robust epidemic response

Despite various preventive measures, computer networks continue to be infected with malware. Self-spreading attacks such as *worm epidemics* are costly not only due to the damage they cause but also due to the challenge of preventing and removing them. These attacks often exploit the inherent difficulty of differentiating legitimate from illegitimate network use, network security resource constraints, and other vulnerabilities. When such malware epidemics cannot be prevented and start to infect a large number of hosts, timely and efficient response becomes very important. This section utilizes a decision and control approach to provide security officers and system administrators with a quantitative and optimal response framework for malware removal.

The objective is optimization of patching response strategies to a worm epidemic within a **quantitative cost-benefit framework** incorporating epidemic models. While the hosts infected with a worm are costly for a network, patching rates are constrained and patching itself has a nonzero cost. For example, production systems must be tested extensively to ensure that they will function well after the patch has been applied. Furthermore, patching itself takes system administrators time and effort. These costs are expressed explicitly and balanced against each other within an optimal and robust control framework based on the epidemic models.

Classical **epidemic models** have been successfully applied to model the spread of computer worm epidemics. However, we show that the widely accepted proportional patching response rate assumption in such epidemic models is suboptimal. Optimal control theory allows one to explicitly specify the costs of infected hosts and the effort required to patch them. The resulting cost function is used in conjunction with the dynamic epidemic model differential equations to derive the optimal feedback patching strategies.

Single and multiple network versions of the classical epidemic model are considered to analyze and **derive the optimal response** in each case. Determining these expressions in the single network case involves the use of the Hamilton–Jacobi–Bellman (HJB) equation to solve for a value function. By considering the response in the case of several networks, we obtain a multi-dimensional model. In this case we linearize the system and derive controllers using pole placement, linear quadratic regulator (LQR) optimal control theory, and H^∞-optimal control theory.

The advantage of H^∞-**optimal control** theory in particular is that it accounts for (as we have seen in the previous section) worst-case system and measurement noise, which captures model inaccuracies and noisy measurements that have a non-negligible impact on performance. In addition, the challenging nature of detecting malware, i.e. expected inaccuracies in detection, justifies the need for a robust response solution.

A brief introduction to optimal and robust control can be found in Appendix A.3. Reference [29] contains further information on H$^\infty$-optimal control theory.

7.3.1   Epidemic models

The quantitative framework studied is based on the **classical epidemic model** [75], which uses a differential equation to model the spread of a virus or worm (also in a computer network). For a single network, this classical model is described by

$$\dot{x}(t) = \beta \left[N - x(t) \right] x(t) - u(t), \tag{7.31}$$

where $u(t)$ is the number of patches applied at a given time, $x(t)$ is the number of infected hosts, N is the number of hosts in the system, and β is a parameter that captures the rate of spread of the epidemic and is referred to as the *pairwise rate of infection*.

This model can be readily extended to the **multiple networks case**. Given M networks, let $x_i(t)$ denote the number of infected hosts in network i, where $i = 1, 2, \ldots, M$.

Likewise, let $u_i(t)$ be the malware removal rate for network i. Let α be the *cross-network pairwise rate of infection*. Note that the more security measures are used between various networks, the smaller is α relative to β. Further, let N_i denote the number of hosts on a particular network i. In general, because computers on a network are more likely to communicate with each other than those on different networks, and because individual networks typically have independent security measures, malware will be assumed to spread more rapidly within a network than between networks. Therefore, $\beta > \alpha$. Overall, one arrives at the model

$$\dot{x}_i(t) = \beta[N_i - x_i(t)]x_i(t) + \sum_{j=1, j \neq i}^{M} \alpha[N_i - x_i(t)]x_j(t) - u_i(t),$$

for $i = 1, \ldots, M$.

Another epidemic model considers the case where hosts that have had malware removed are no longer susceptible to malware infection. This model is referred to as the **epidemic model with removals**. When only one network is considered, this model becomes

$$\dot{x}_1(t) = \beta[N - x_1(t) - x_2(t)]x_1(t) - u(t)$$
$$\dot{x}_2(t) = u(t). \tag{7.32}$$

Note that for each network, there are two state variables in this case. The first is the number of infected hosts in the network. Its dynamics are very similar to those of the regular epidemic model. The second keeps track of the number of hosts that have been patched and thus are no longer vulnerable to attack.

The epidemic model with removals can also be extended to the case where there are multiple networks, as described above. This leads to the set of $2M$ coupled differential equations

$$\dot{x}_i(t) = \beta[N_i - x_i(t) - x_{M+i}(t)]x_i(t)$$

$$+ \sum_{j=1, j \neq i}^{M} \alpha[N_i - x_i(t) - x_{M+i}(t)]x_j(t) - u_i(t) \tag{7.33}$$

$$\dot{x}_{M+i}(t) = u_i(t),$$

for $i = 1, \ldots, M$. Here, $x_1, \ldots, x_M$ are the number of infected hosts in networks $1, \ldots, M$, and $x_{M+1}, \ldots, x_{2M}$ are the number of patched hosts in networks 1 through M, respectively.

Traditionally, when patching infected hosts, it is assumed that a particular proportion of them are patched at each time instance, referred to as a **proportional patching controller**,

$$u_i(t) = \kappa_i x_i(t),$$

for some $\kappa_i \in (0, 1)$, and for all $i = 1, \ldots, M$. The coefficient κ is known as the *removal rate* of infectious hosts.

In order to properly define an optimal control strategy, first a **cost function** must be chosen as the criterion of optimality. Traditionally, quadratic costs are implemented on both the state (number of infected hosts) and control (patching rate). This structure is theoretically reasonable and mathematically tractable. Consider the cost function

$$J(x(\cdot), u(\cdot)) = \int_0^\infty \left[x^T(t)Qx(t) + u^T(t)Ru(t) \right] dt, \tag{7.34}$$

where x and u are vectors of the state and control variables. In the classical epidemic model the Q and R matrices are chosen as diagonal matrices, with the (i,i) entry designating the cost of an infected host in network i (for Q) and a particular patching response rate in network i (for R). In the epidemic model with removal, the Q matrix is similarly structured but with no cost placed on states x_{M+1} to x_{2M}, as these merely keep track of the number of patched hosts. The R matrix is unchanged in this case.

7.3.2   Feedback response for malware removal

Optimal malware removal feedback controllers of the form

$$u = \mu(x)$$

are derived in this subsection using standard optimal control methods for the single network epidemic models. This derivation utilizes continuous-time dynamic programming and the ensuing HJB equation. The critical quantity (function) of interest here is the value function, $V(x,t)$, which is the minimum value of (7.34) subject to the state dynamics, when the integration interval is $[t, \infty)$, and the state $x(\cdot)$ at time t is $x(t) = x$. The value function, V, satisfies the HJB equation, written as

$$-V_t(x) = \min_{u \in U} \left[g(x,u) + V_x(x)^T f(x,u) \right], \tag{7.35}$$

where we have suppressed the dependence of V on t; $g(x,u)$ refers to the quantity inside the integral in the cost equation (7.34) and $f(x,u)$ is the right-hand side of the system dynamic equation (7.31). $V_t(x)$ and $V_x(x)$ refer to the partial derivative of the value function with respect to t and x, respectively.

A couple of observations are in order here toward **derivation of the feedback controller**. First, in the infinite time horizon case $V_t(x) \equiv 0$; second, the minimum control in (7.35) is obtained by differentiating the HJB equation with respect to u, setting the resulting expression equal to zero, and solving for $u^*(t)$, yielding

$$u^*(t) = \frac{V_x(x)}{2r}. \tag{7.36}$$

Substituting this into the HJB equation leads to a quadratic equation in V_x, which can be solved for V_x:

$$V_x(x) = 2ra(x)x + 2r\sqrt{a^2(x)x^2 + \frac{q}{r}x^2}, \tag{7.37}$$

where $a(x) = \beta(N - x)$. Using this in (7.36) yields the **optimal feedback controller**

$$\mu^*(x(t)) = \left(a(x) + \sqrt{a^2(x) + \frac{q}{r}} \right) x. \tag{7.38}$$

The derivation of the optimal controller for the single network epidemic model with removal (7.32) is similar. In this case the HJB equation becomes

$$0 = \min_u \left\{ qx_1^2 + ru^2 + V_{x_1} \left[\beta \left[N - x_2 - x_1 \right] x_1 - u \right] + V_{x_2} u \right\}. \tag{7.39}$$

The minimizing $u^*(t)$ becomes

$$u^*(t) = \frac{V_{x_1} - V_{x_2}}{2r}. \tag{7.40}$$

Substituting this back into (7.39) yields

$$0 = qx_1^2(t) - \frac{1}{4r}(V_{x_1} - V_{x_2})^2 + V_{x_1} \left[\beta (N - x_1(t) - x_2(t)) x_1(t) \right]. \tag{7.41}$$

Solving for V_{x_1} and V_{x_2} explicitly proves difficult in this case. Instead, two approximate solutions are investigated for the optimal controller with removal, $\mu_r(x(t))$.

The first approximation is obtained by starting with a quadratic structure for V:

$$V = k_0 + k_1 x_1 + k_2 x_2 + k_3 x_1^2 + k_4 x_1 x_2 + k_5 x_2^2. \tag{7.42}$$

The partial derivatives of this assumed form with respect to x_1 and x_2 are substituted back into (7.41). When the coefficients of the various terms in this equation (x_1, x_2, x_1^2, etc.) are set to zero, a system of nine equations in five variables is obtained. A study of these equations yields two possible solutions. Numerical analysis of these two solutions confirms that **one approximate optimal feedback controller** is

$$\mu_{r1}(x) \approx \left(\beta N + \sqrt{\beta^2 N^2 + \frac{q}{r}} \right) x_1. \tag{7.43}$$

The **second controller** can be inferred by investigating and slightly altering the optimal solution derived for the classic epidemic model (7.38):

$$\mu_{r2}(x) \approx \left(a(x_1 + x_2) + \sqrt{a^2(x_1 + x_2) + \frac{q}{r}} \right) x_1, \tag{7.44}$$

where the term x in $a(x)$ is simply replaced with $(x_1 + x_2)$.

Each of the derived strategies (7.38), (7.43), and (7.44) describe the (approximately) optimal patching or malware removal rate in the form of a feedback controller for a given set of cost parameters. Notice that all of these strategies differ significantly in form from a proportional patching controller.

7.3.3 Multiple networks

A generalization of the results presented in the previous subsection to the multiple networks case is not straightforward. It may not be possible to derive closed-form analytical expressions for feedback controllers of the nonlinear system models (7.32) and (7.33). Therefore, we next focus instead on derivation of stabilizing feedback controllers. Subsequently, LQR and H^∞-optimal malware feedback response strategies will be derived.

Stabilizing response

The nonlinear models (7.32) and (7.33) have particular properties which can be exploited to derive **suboptimal** but reasonable **stabilizing feedback controllers** or malware response strategies. One crucial observation is that $x_i(t)$ has to be non-negative for all i. This leads to the insight that all of the cross-terms and squared terms in the models (7.32) and (7.33) decrease the magnitude of the infection rates $(\dot{x}_i(t))$. Therefore, if these helpful squared and cross-terms are disregarded, then one would be working with systems of equations that are actually more difficult to stabilize than the original models. Moreover, when these terms are disregarded, the models reduce to the same **linear model**

$$\dot{x}(t) = Ax(t) + Bu(t), \tag{7.45}$$

where

$$A = \begin{bmatrix} \beta N_1 & \alpha N_1 & \alpha N_1 & \cdots & \alpha N_1 \\ \alpha N_2 & \beta N_2 & \alpha N_2 & \cdots & \alpha N_2 \\ \vdots & & \ddots & & \vdots \\ \vdots & & & \ddots & \vdots \\ \alpha N_M & \alpha N_M & \cdots & \alpha N_M & \beta N_M \end{bmatrix} \tag{7.46}$$

and B is simply the negative identity matrix of dimension $M \times M$.

Notice that the epidemic models have the inherent physical constraints

$$0 \le x_i \le N_i, \quad i = 1, \ldots, M.$$

However, if $x_i = N_i$ for any i, then $\dot{x}_i < 0$ for the original nonlinear system (7.32) under the condition $u < 0$, which is discussed in detail later in this section. In other words, the trajectory leaves the boundary $[N_1, \ldots, N_M]$ immediately. A similar argument can also be made for (7.33). Therefore, the upper bounds are disregarded in the simplified linear system (7.45) and the analysis focuses on the lower bounds (positivity)

$$x_i \ge 0, \quad i = 1, \ldots, M. \tag{7.47}$$

Under (7.47), the **constrained linear model** becomes

$$\dot{x}_i = \begin{cases} [Ax + Bu]_i, & \text{if } \begin{cases} x_i > 0 \text{ or} \\ [Ax + Bu]_i \ge 0 \text{ and } x_i = 0 \end{cases} \\ 0, & \text{else} \end{cases} \tag{7.48}$$

for all $i = 1, \ldots, M$.

While it is known that a linear feedback controller can stabilize the linear model (7.45), whether such a controller also stabilizes the nonlinear models (7.32), (7.33), and (7.48) is a question which is investigated next.

Stability analysis

The stability of the system (7.48) is analyzed and established under the set of boundary constraints (7.47) when controlled by the linear feedback controller

$$u_s = -Kx, \qquad (7.49)$$

where K is the feedback matrix. Obviously, the origin constitutes the unique equilibrium for this system.

A **sufficient condition for stability** can be found by considering the special structure of this problem. Since it is known that the components of x will never become negative, the closed-loop system matrix does not even need to be Hurwitz, i.e. all of its eigenvalues have a strictly negative real part. A sufficient condition for stability is that the diagonal elements are negative and that the non-diagonal elements are nonpositive. This condition is easy to verify upon inspection of the closed-loop matrix.

Theorem 7.4 *The nonlinear system of the form in (7.48) is stable under feedback control $u = -Kx$, if the closed-loop matrix $(A - BK)$ has negative diagonal entries and nonpositive off-diagonal entries.*

Proof In the context of the system (7.48), components of x can only be positive or zero. In this case the diagonal entries of the closed-loop matrix $(A - BK)$ are assumed to be negative and the off-diagonal entries are nonpositive. Clearly, this implies that each component of $\dot{x}$ is nonpositive.

However, this is not enough to guarantee stability. It must also be shown that any positive component of x will decrease to zero. This is ensured because the diagonal elements of the closed-loop system matrix are assumed to be negative. Therefore, all positive components of x will decrease to zero at a rate faster than or equal to that specified by the corresponding diagonal entry in $(A - BK)$. $\qquad \square$

The feedback controllers to be derived later, using LQR and H^∞-optimal control, may not have this property. Nevertheless, in many situations controllers derived with optimal control theory will meet this condition and therefore can be used to stabilize the system under consideration. In cases where such controllers violate the sufficient conditions, they can be appropriately modified to obtain suboptimal but stabilizing counterparts without having a significant effect on the performance.

It is next argued that **the stabilizing controller** (7.49) which meets the conditions of Theorem 7.4 also **stabilizes the actual nonlinear systems** (7.32) and (7.33). Given that $x \geq 0$, the nonlinear terms in these equations will only decrease the magnitude of the components of $\dot{x}$. If x could become negative, this may destabilize the system. However, in this case it only adds additional negative drift, leading to faster stabilization. In conclusion, the stabilizing condition in Theorem 7.4 ensures stability over the entire state space, even in the nonlinear case.

Linear quadratic regulator optimal response

Determining the optimal malware removal strategy relative to the cost (7.34) for the epidemic models (7.32) and (7.33) is nontrivial. When multiple networks are considered, even if only two networks are studied and a very simple form for the value function is assumed, the approach presented in Section 7.3.2 leads to an overdetermined set of nonlinear equations, which is not tractable. Therefore, a more tractable but suboptimal approach to this problem is investigated next, based on the linear model (7.45).

The linear model (7.45) and the quadratic cost function (7.34) lead to the well-studied LQR optimal control problem, whose optimal solution can readily be obtained:

$$u_0(t) = -R^{-1}B^T Px(t), \tag{7.50}$$

where P is the unique positive definite solution to the algebraic Riccati equation (ARE)

$$A^T P + PA - PBR^{-1}B^T P + Q = 0. \tag{7.51}$$

Here, the fact that (A, B) is controllable is used, which follows because B is the negative identity matrix. Also, note that because $Q > 0$, the observability condition holds, and consequently the solution to (7.51), $P > 0$, exists and is unique, and the closed-loop matrix $A - BR^{-1}B^T P$ is Hurwitz.

H^∞-optimal response

While the model (7.45) in Section 7.3.3 is useful, it entails several assumptions. First, the nonlinear terms in the more precise models (7.32) and (7.33) have been ignored. Second, availability of perfect measurement of the number of infected hosts in each network has been assumed. And third, the original epidemic models (7.32) and (7.33) themselves only approximate malware propagation.

To capture these approximations and imperfections, the **linear model** (7.48) **is altered to include a noise term**. Then, the problem is formulated within an H^∞-optimal control framework. Let

$$\delta_i = [Ax + Bu + Dw_a]_i.$$

Then

$$\dot{x}_i = \begin{cases} \delta_i, & \text{if } \begin{cases} x_i > 0 \text{ or} \\ \delta_i \geq 0 \text{ and } x_i = 0 \end{cases} \\ 0, & \text{else.} \end{cases} \tag{7.52}$$

Here, $w_a(t)$ is a noise term that accounts for model assumptions and approximations. It is important to note that this disturbance term differs from that in Section 7.2 in its interpretation. While it represents actions of malicious attackers in the model of Section 7.2, it merely captures modeling errors here.

The D matrix describes how this noise term impacts the dynamics of $x(t)$ and will be set to the identity matrix. In addition, a measurement error can be introduced: if $y(t)$ is the **measured number of infected hosts**, then

$$y(t) = x(t) + w_n(t), \tag{7.53}$$

which says that the noise vector $w_n(t)$ impacts the measurement of the number of infected hosts on each network (each element of $y(t)$).

In order to develop the H^∞-optimal controller, several definitions and assumptions have to be made. First, a **controlled output**, z, is defined as

$$z(t) = Hx(t) + Gu(t). \tag{7.54}$$

It is assumed here that $G^T G$ and $H^T H$ are positive definite and that $H^T G = 0$. This says that there is no cost placed on the product of patching response and infected hosts, although each of those quantities individually contributes to the cost. To compare the cost $\|z\|^2$ (defined below) with the cost (7.34) for the LQR controller, let $Q = H^T H$ and $R = G^T G$. A few other conditions that must be met for this H^∞-optimal control theory to apply are that (A, B) and (A, D) be stabilizable, and (A, H) and (A, I) be detectable, which, however, all hold given that $B = -I$, $D = I$, and $Q > 0$. Define $w := \begin{bmatrix} w_a^T & w_n^T \end{bmatrix}^T$ as the total disturbance to the system. Let the **cost ratio** used in the H^∞ analysis be

$$F(x, u, w) = \frac{\|z\|}{\|w\|}, \tag{7.55}$$

where $\|z\|^2 := \int_{-\infty}^{\infty} |z(t)|^2 dt$ and a similar definition applies to $\|w\|^2$. This captures the proportional changes in z due to changes in w, as discussed earlier.

H^∞-optimal control theory also produces a **performance bound** (the H^∞ norm) that one can guarantee will be met. This norm can be thought of as the worst possible value, γ, for the cost L. The lowest possible value of γ is

$$\gamma^* := \inf_u \sup_w F(u, w), \tag{7.56}$$

which can also be viewed as the optimal performance level in the H^∞-control context.

In order to actually solve for the optimal controller $\mu(y)$, a corresponding differential game is introduced, which is parameterized by γ. The **optimal worst-case controller** $u_w = \mu_\gamma(y)$ can be determined from this differential game for any $\gamma > \gamma^*$, following an analysis similar to that in Section 7.2.2. Note that here, γ^* is the smallest γ such that $\bar{\Sigma}_\gamma$ and $\bar{Z}_\gamma$ exist, and the spectral radius bound $\rho(\bar{\Sigma}_\gamma \bar{Z}_\gamma) < \gamma^2$ holds.

The linear H^∞-optimal feedback controller provides a robust malware response or epidemic removal strategy based on the estimate of the number of infected hosts. It can be calculated offline using only the linear quadratic system model. The numerical results in reference [44] indicate that the H^∞-optimal feedback controller leads to an aggressive malware removal strategy which may be suboptimal in some cases, especially when compared to the LQR controller, but is still the only approach that ensures a minimum performance bound and exhibits robustness under noise and modeling errors.

7.4 Discussion and further reading

This chapter summarizes results from the MS thesis of M. Bloem and related publications [42–45]. Section 7.1 is based on reference [45], which includes extensive analyses of and simulations on optimization formulations for malware filtering. Earlier work on this subject using a similar approach are references [52, 169]. The malware filter placement problem has also been analyzed within a stochastic game framework in Section 4.4. In a related study [88], another (deterministic) game formulation for the monitor placement problem has been presented.

Although Section 7.2 focuses on malware traffic filtering, the framework is quite general in nature and can be applied to different contexts such as email (spam). The approach was introduced first in reference [42] which contains additional simulation results and discussion. Reference [29] contains all the background knowledge on H^∞-optimal control theory needed to follow the derivations in this chapter and more.

The well-known classical epidemic model, on which the analysis in Section 7.3 is based, has been extensively used to study the propagation of and response to worm epidemics in computer networks [57, 116, 146, 196]. Quarantine strategies in order to contain worm epidemics by partitioning networks into subnetworks have been proposed in reference [57]. Another study adopting an optimal control approach to prevent worm epidemics is reference [107]. An earlier version of the material in Section 7.3 has appeared in reference [42] and later in reference [44], which additionally contains a numerical example and analysis.

8   Usability, trust, and privacy

Chapter overview

1. Security and usability
   - complex relationship between security and usability
   - a system for security alert dissemination
   - effective administrator response scheme
2. Digital trust in online communities
   - community trust game
   - equilibrium, dynamics, convergence
   - numerical analysis
3. Location privacy: a game-theoretic analysis
   - a location privacy model
   - location privacy games

Chapter summary

Security has a strong social dimension, connected to human factors and areas such as usability, trust, and privacy. This chapter presents example applications of the decision and game-theoretic approach to social aspects of security using various models. First, the complex relationship between security and usability is discussed and two example schemes, one for improving the usability of security alert dissemination and another one for effective administrator response, are investigated. Next, the community effects in the evolution of trust to digital identities in online environments are studied using a specific noncooperative trust game. Finally, a game-theoretic analysis of location privacy in wireless networks is presented.

8.1   Security and usability

The relationship between security and usability is a complex one. Some in the research community claim that they go hand in hand whereas others argue that security and usability are inversely proportional to each other. *Security* is not a purely engineering issue but has also social, economic, and psychological dimensions. Likewise, usability is a field that goes beyond pure engineering and intersects with security in these aspects. While the question of "how security and usability relate to each other" does not have a plain and simple answer, it is relevant and deserves an exploration from multiple angles.

Usability, similar to security, is a rich concept that is difficult to capture within a widely agreed formal definition. One possible description of usability, provided by the US Department of Health and Human Services, is given below. In addition, the word "usability" may refer to methods for improving ease of use during the design process [131].

Definition 8.1 *Usability is the measure of the quality of a user's experience when interacting with a product or system – whether a website, a software application, mobile technology, or any user-operated device. Usability is a combination of factors that affect the user's experience with the product or system, including ease of learning, efficiency of use, memorability, error frequency and severity, and subjective satisfaction.*

Usability is relevant to security simply because the users of networked systems can easily render carefully planned security mechanisms useless. A confusing system that is hard to configure and use increases the probability of its users making mistakes, leading to various vulnerabilities. It is a known fact that a significant portion of vulnerabilities stems from configuration errors. Furthermore, there is a whole class of attacks using psychological and social engineering factors as discussed in Section 2.1.2. The dictum "Only amateurs attack machines; professionals target people"[1] summarizes this fact elegantly and can be supported by numerous examples from real life.

Although the human factor is an important aspect of security in practice, it is still not sufficiently emphasized in security systems and research. This is partly due to lack of incentive mechanisms (e.g. pushing responsibility conveniently to users) and partly due to limited collaboration between these research fields.

Making a system usable is not trivial, just as a good design is hard to achieve. Many companies (e.g. Apple) that distinguish themselves from competition mainly on usability and elegant design can be seen as anecdotal evidence for this. When the requirements of securing a system are added to those of usability, these two create additional challenges for system developers. Even if the claim that usability and security go hand in hand is accepted, this extra cost at the design and implementation phase cannot be ignored.

---

[1] Attributed to Bruce Schneier [23].

The economic costs of properly integrating usability and security into systems play a significant role in incentive mechanisms of the systems' stakeholders. The misalignment of incentives among users, security officers, and management has been discussed in Section 2.2. Adding usability considerations as a factor naturally does not simplify these relationships. Although the customers (users) desire usability and security as features, they do not necessarily want to pay for them. Hence, a difficult situation emerges from the business perspective as making a system secure and usable requires additional development and testing effort, e.g. implementation of additional functionality and conducting user studies. Such multi-dimensional incentive mechanisms lie at the center of the complex relationship between usability and security, which can be explained only if the underlying incentives are properly understood.

The objective of the decision and game-theoretic approach is to make these hidden costs and incentives explicit by quantifying them. At the same time, the approach aims to formalize the security decision processes as a first step to disentangle the complexities described. Once objectives and the way to achieve them are formalized and handled in a principled way, a methodology for alignment of conflicting incentives emerges. An insecure system may seem usable but this illusion breaks down once attacks render it useless! A difficult-to-use security system pushed to end users may seemingly relieve a security officer of responsibility but once users start to circumvent it and even bigger security problems emerge later, it proves to be inefficient. A networked system can be optimized with respect to usability, security, and design criteria, if priorities are expressed explicitly over the chosen time horizon. A rich set of decision and game-theoretic tools already exists to address such problems as demonstrated below.

8.1.1   A system for security alert dissemination

The number of **security vulnerabilities** on networked systems has increased significantly in recent years. The trend is expected to continue as networked systems become more complex and ubiquitous. The statistics issued by CERT at Carnegie Mellon University, which are shown in Figure 8.1, demonstrate the magnitude of the problem faced by the security community. Each of these vulnerabilities can be exploited by malicious attackers to compromise computer systems, and has potential financial and productivity related consequences for enterprises and organizations.

There are multiple sources which inform and warn system administrators about recently discovered security vulnerabilities. The "@RISK: The Consensus Security Vulnerability Alert" newsletter issued weekly by the SysAdmin, Audit, Network, Security (SANS) Institute is one example. It summarizes the warnings that matter most, tells what damage they do and how to protect from them. However, each warning is not necessarily relevant to all systems as vulnerabilities are almost always platform- and application-dependent. For example, in a company with Linux servers, only the Windows server-related security warnings are irrelevant. Furthermore, even if a warning is relevant, it is time-consuming to find and test all solutions. The security officers must filter the incoming data and prioritize their defensive actions according to the priorities of their own systems.

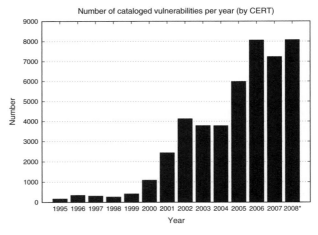

Figure 8.1 Number of cataloged vulnerabilities by CERT during 1995–2008. (*2008 estimate*)

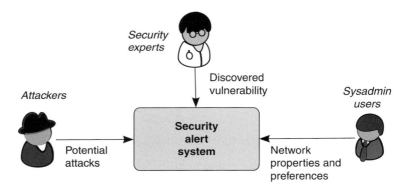

Figure 8.2 Overview of the game-theoretic recommendation system for efficient security alert dissemination.

This section presents a recommendation system (Figure 8.2) that aims to improve the efficiency of security alert dissemination process and the evaluation of security alerts. The objective is to design a mechanism which brings together two communities: security experts, who issue warnings, and system administrators (*sysadmins*), who evaluate warnings and implement respective measures on their own networks. The system computes a strategy recommendation for each of its sysadmin users based on the local network description of and security alerts issued by experts. This strategy filters out irrelevant reports and focuses on local systems which may be potentially targeted as a result of discovered vulnerabilities. It suggests how to allocate limited defense resources through prioritization. At the same time, it partly automates security-related decision making. Thus, the recommendation system helps to increase productivity of security personnel. It acts as a security service which builds a bridge between security researchers and sysadmins on the field.

A **security game** modeling attacker behavior provides the engine of the recommendation system. Each discovered vulnerability is a potential threat and paves the way for malicious attacks. While security vulnerabilities are rapidly increasing, the resources allocated to counter them remain limited. Each new vulnerability may require a specific configuration change or patching of a system which costs time and effort. This creates a widening gap between the resources needed to prevent security attacks and the resources available. Consequently, the system administrators and security personnel have to allocate their limited resources efficiently in order not to become overwhelmed. Prioritizing individual devices on their network is one way of increasing efficiency. These factors are brought together in a security game, similar to those in Chapter 3, which provides a quantitative method for formal decision making. The equilibrium solution of the game constitutes the recommended strategy to sysadmins.

Recommendation game formulation

The **recommendation game** models the interaction between attackers and defenders (sysadmins) by taking into account the threat posed by discovered vulnerabilities and the priorities of the sysadmins. It is based on the same security game framework introduced in Chapter 3. The defender, $\mathcal{P}^D$, patches and removes vulnerabilities from the devices on the network, $\mathcal{A}^D = \{d_1, \ldots, d_N\}$. On the other hand, the action set of the attacker, $\mathcal{P}^A$, consists of potential attacks exploiting the discovered vulnerabilities $\mathcal{A}^A = \{a_1, \ldots, a_N\}$. An inherent assumption here is that a vulnerability can be used by attackers sooner or later as a basis for an attack on the defended network.

The **game matrix**, P, maps player actions (attacks exploiting vulnerabilities and defensive/preventive actions such as patching or reconfiguration) to outcomes, i.e. payoff and cost for the attacker and defender, respectively. The diagonal entries of the matrix are zero based on the assumption that if a sysadmin (defender) properly patches a vulnerability or reconfigures a device, the attacker cannot harm that device anymore. The remaining entries of the game matrix are functions of the importance of each device and the vulnerability of the respective device category. Here, the functions are chosen to be bilinear, i.e. the vulnerability and preference values of the device are multiplied with each other to determine the outcome. Accordingly, the game matrix, P, is defined as

$$P = [P(i,j)] := \begin{cases} v_i r_i, & \text{if } i \neq j \\ 0, & \text{if } i = j, \ \forall i,j \in \mathcal{A}^D \end{cases} \tag{8.1}$$

where v denotes the vulnerability level of a device and r the importance attached to the device by the sysadmin. Both vulnerability and importance values are represented by five levels {very low, low, medium, high, very high} and quantified by $v, r \in \{1, 2, 3, 4, 5\}$. The vulnerability level of a device is chosen to be the maximum of the categories it belongs to.

The mixed-strategy solution of this two-player zero-sum matrix game, which is known to exist, can be obtained using standard methods [31]. The equilibrium strategy of the attacker can be interpreted as expected attack probabilities which follow from the vulnerability level and value of the devices on the network. The computed defense

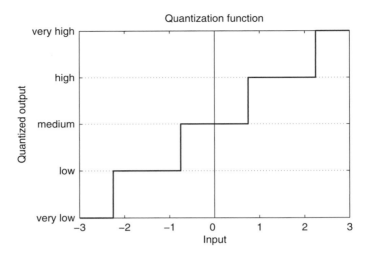

Figure 8.3 Possible quantization function for defense strategy.

strategy is quantized by the function shown in Figure 8.3 as one possibility. The result is then used as a guideline to decide where to allocate limited defense resources.

Demonstrative prototype

A proof-of-concept prototype is discussed to illustrate the core functionality of the introduced system. The system allows for two types of participant: (security) experts and (sysadmin) users. Given a categorization or taxonomy of platforms and applications, the security experts enter the vulnerability levels for each device category based on the most recent information. Each sysadmin user, on the other hand, enters the system description along with own preferences. Subsequently, the system computes the recommended strategy for each user separately using the security game formulation in Section 8.1.1 and displays the results.

The prototype is implemented as a **web application** to improve its flexibility and designed for access from mobile devices. As the programming language, *Python* is chosen for its efficiency and library support. The web application server is realized using the *TurboGears* web framework, which allows for rapid application development. The *cvxopt* library is used in solving linear programs associated with the zero-sum games. A screenshot of the system for an example network is shown in Figure 8.4.

The current implementation adopts a *security service* approach possibly as a paid service, where users register to a remote system and receive recommendations from experts based on recently discovered vulnerabilities. The privacy of the users are protected as the experts do not see details of individual users. Instead, a device categorization maps vulnerabilities discovered by experts to individual user networks. This can be taken a step further and an alternative open community approach can be adopted. Then, the system can be deployed locally in a single user scheme and receive vulnerability information from a central repository which is updated by voluntary security experts in the community.

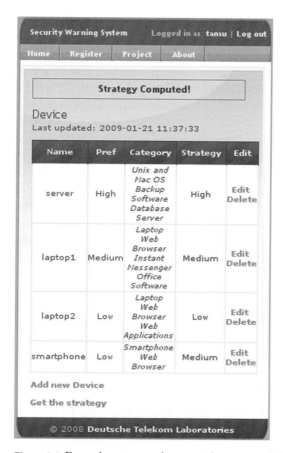

Figure 8.4 Example system and computed recommended strategy.

8.1.2 Effective administrator response

One of the often overlooked constraints of security systems is the limited manpower available for handling security incidents and addressing security vulnerabilities. System administrators and security officers usually operate under severe time limitations while each incident separately requires time and attention. This shortcoming will continue to exist until the development of advanced "artificial intelligence" systems. Taking also into account the attacks conducted by human opponents, contemporary security systems have to rely on human intervention for the foreseeable future.

Given the mentioned time constraints, it is not realistic to expect that system administrators handle every reported anomaly or suspected attack with the same level of attention. Therefore, a computer support system, however imperfect, would be beneficial to increase efficiency. The algorithm considered in this section describes one possible approach for deciding on which attacks are forwarded to the administrator and which are not.

The approach presented next relies on machine-learning methods such as clustering and classification to separate most commonly encountered and very rare security incidents from others. These are arguably the most "interesting" security events observed in the networked system and are natural candidates for handling by a system administrator rather than being processed and responded to automatically. There exists a variety of advanced machine-learning methods to address this attack classification problem as long as a certain probability of error is taken into account (see, e.g., Chapter 9). Then, the decision to be made is what percentage of the incidents to forward to the administrator and what percentage to handle automatically. This resource allocation problem can be formalized within a standard optimization (linear programming) framework.

A security incident can be represented as a finite feature vector in most cases. This vector may include features such as the resource (device) under attack, IP address of the attacker, and time of attack. Then, the task of the classification algorithm is to sort out the set of incidents observed in a time interval into three categories: frequent, normal, and rare events. Thus, the problem is turned into a static (vector) classification problem, which can be solved by a variety of existing methods. For example, Kohonen self-organizing maps (SOMs) can be used as a well-known clustering algorithm to detect frequent features. SOMs can be trained using a standard *Kohonen learning rule* with existing dataset to specify "neurons" corresponding to representative abstract incidents. Subsequently, a distance measure (e.g. Euclidean distance) can be defined to compute frequent events (close to neurons) and rare ones (distant to neurons). In order to maximize the applicability of the discussed algorithm to diverse and future systems, the formulation here is kept abstract rather than focusing on a specific system or a specific machine learning method.

Automatic or administrator response algorithm

The decision on whether to alert the system administrator or respond automatically to the outlying security incident is formulated as a resource allocation problem. As a quantitative decision criterion, relative costs are assigned to these two options. A parameter $c_1 > 0$ represents the cost of automated response to incidents, which also includes the risk of making mistakes due to imperfections in the automated process. Another parameter $c_2 > 0$ quantifies the cost of time and effort spent by the system administrator per incident. In many cases c_1 is less than c_2 unless the automated response system performs unacceptably poorly. An analogous case is, for example, voice recognition for customer service systems, where the exceptional incidents that cannot be handled by the automated recognition are forwarded to human operators.

Let x_1 represent the proportion of incidents that are sent to the event classification mechanism for further analysis and response. The value $x_2 = 1 - x_1$ is then the proportion of incidents that are randomly forwarded to the system administrator. Of those sent to the event classification mechanism, let λ be the proportion that are chosen to be forwarded to the administrator. Finally, let N be the average number of incidents per time slot and L be the number of cases the system administrator can handle per given time slot. A "time slot" refers to a period of time between the recalibrations of these parameters.

The overall cost function, which reflects the tradeoffs mentioned in the previous section, is defined for a given time slot as

$$J(x_1, x_2) := Nx_1[(1 - \lambda)c_1 + \lambda c_2] + Nx_2 c_2. \tag{8.2}$$

Thus, the constrained optimization problem which constitutes the basis of the automatic or administrator response (AOAR) algorithm is

$$\min_{x_1, x_2} J(x_1, x_2) = Nx_1[(1 - \lambda)c_1 + \lambda c_2] + Nx_2 c_2$$
$$\text{such that } N(\lambda x_1 + x_2) \leq L$$
$$x_1 + x_2 = 1$$
$$x_1, x_2 \geq 0. \tag{8.3}$$

The individual steps of the AOAR are given in Algorithm 8.2, and Figure 8.5 graphically depicts its operation. We refer to Appendix A.1.3 for an overview on how to solve such optimization problems.

Although the variable λ is treated as a parameter in the problem formulation (8.3), it is an implicit decision variable that depends on the specific event classification algorithm used. For example, in the case of the SOM clustering method, the value of λ can be affected by varying the threshold distances used in classification. A target value for

---

Algorithm 8.2 AOAR algorithm

1: Given c_1, c_2, N, and L as input
2: Given initial dataset
3: Train *event classification system (ECS)*
4: **repeat**
5:     Update estimates of N and L
6:     Solve the optimization problem (8.3) to obtain x_1 and x_2.
7:     **for** each incident i **do**
8:         Forward i to sysadmin with probability x_2
9:         Send i to ECS with probability x_1
10:         **for** each incident j sent to ECS **do**
11:             **if** incident j is outlier **then**
12:                 Forward j to sysadmin
13:             **else**
14:                 Handle j by automated response
15:             **end if**
16:         **end for**
17:     **end for**
18:     Given estimate of λ, update ECS parameters
19:     **if** ECS performance $! =$ target **then**
20:         Adjust λ
21:     **end if**
22: **until**

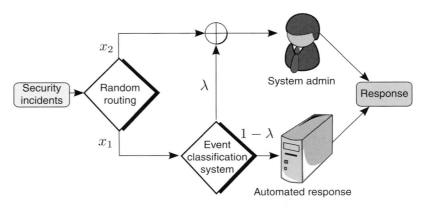

Figure 8.5 Graphical depiction of AOAR algorithm.

this parameter can be selected based on how well the classification algorithm functions in practice. Then, threshold values (parameters) of the classification algorithm can be adjusted in an adaptive manner to keep the value of λ stable.

Forwarding a portion (x_2) of incidents randomly to the system administrator is analogous to random searches of passengers at airport security checks, and it is essentially another layer of protection that adds robustness to the system. If λ is set to zero, then the event classification system is offline and the algorithm relies purely on "random search." In contrast, if λ is set to one, then there is no random search, which means that the classification system is fully trusted.

8.2   Digital trust in online communities

Digital identity constitutes one of the building blocks of the World Wide Web for all types of activity ranging from social networking to e-commerce. During the explosive growth phase of the Web, a variety of digital identity and trust management mechanisms were developed organically to satisfy the emerging needs. However, most of these existing solutions have been either ad-hoc or heuristic in nature [82]. An analytical foundation for digital identity and trust is important not only for the growth of Web services but also in terms of security.

Individuals interact and make security decisions based on digital identities in a virtual environment such as the Web. Opening an attachment from an email address, buying a service or good from an online merchant website, or sharing information with someone on a social network are all trust-related decisions with security consequences. In many situations, community effects play a significant role when these decisions are made. The rating and community reviews of a merchant may influence individual trust or referrals from trusted sources are taken into account during interactions on a social network.

Although some of these mechanisms also exist in the physical world, the online environments and digital identities have special distinguishing characteristics that also influence trust mechanisms. On the one hand, new digital identities are often very easy to create. On the other hand, acquiring trust and sustaining it over long periods of time is rather difficult due to fast information dissemination (e.g. news, gossip) and long-term memory (storage) in the digital world.

Digital trust and reputation are two concepts that are closely related to each other. An individual often decides whether or not to trust a digital identity based on the reputation of that identity. Therefore, reputation of a digital identity can be seen as an aggregate metric which is a function of community members' trust in that identity. Online environments allow for quick dissemination and sharing of such trust decisions (user opinions) through rating systems.

This section presents a game-theoretic model of community effects on trust in digital identities. Factors such as peer pressure, personality traits such as timidness or reluctance to pass judgment, and influence of community leaders are investigated in a noncooperative game setting. As an example, the players (users) take part in a digital trust management system where they explicitly share their opinions on an external digital identity (e.g. seller in e-commerce). After a dynamic evaluation process, the resulting opinion is a mixture of their own individual assessment and community influences. The effect of various parameters on the final outcome as well as the equilibrium and convergence properties of the iterative processes are rigorously studied.

It is worth noting that the term "trust" is used here in a social context, in the sense of trusting a digital identity, for example a seller in e-commerce. This should be distinguished from the use of trust in "trusted computing" or "trusted systems," where the term denotes consistent behavior enforced by hardware in the former and reliance upon a system to enforce a specified security policy in the latter.

8.2.1 Community trust game

Consider a set of **agents**, $\mathcal{A} := \{a_1, \ldots, a_i, \ldots, a_N\}$, which can represent users of a social network (e.g. *Facebook* or *Slashdot*) or participants in an e-commerce environment such as the one provided by *Amazon* or *eBay*. For simplicity, each agent is associated with a single digital identity. The agents may be symmetric in their properties, for example, as members of a social network. In other cases, e.g. in e-commerce environments, they can be divided into two main groups of buyers and sellers.

The agents are issued their digital identities by a **digital identity provider** of the given digital environment. This role is customarily played by the respective owner of the social networking or e-commerce site itself as in the case of *Amazon, Facebook,* or *eBay*. In addition, the identity provider usually provides a reputation and trust management service for the users, allowing them to establish trust relationships for their digital identities. The trust in a digital identity may sometimes be almost independent of its real-life owner. Furthermore, a (positive or negative) reputation may not be as transferable between real-life and digital identity as one might expect. This phenomenon can be partly explained with the geographically dispersed and virtual nature of online communities.

Based on the properties of the agents and digital identity providers, digital reputation and trust systems can be classified according to two factors:

1. Symmetric versus asymmetric agents. An example of the former is social networking such as *Facebook*, *Eopinions*, and *Slashdot*. The examples of the latter include *eBay*, *Amazon*, and *Newegg*, where agents are separated into buyers and sellers.
2. Centralized versus decentralized identity providers. Most of the existing e-commerce and social networking sites belong to the former category, whereas decentralized P2P communities such as *bittorrent* are examples of the latter.

Figure 8.6 visualizes a classification of some specific digital reputation and trust schemes based on the factors discussed above.

Game model

The **community trust game** focuses on the asymmetric and centralized digital identity systems (upper left-hand corner of Figure 8.6). However, the approach to be presented is also applicable to the remaining cases. For simplicity, the game is played among N agents in the set $\mathcal{A}$, who evaluate a single given identity or seller s over a certain finite time interval. In what follows, the terms agent, user, and buyer as well as the terms evaluated identity and seller will be used interchangeably without any loss of generality.

It is assumed here that the seller has a stationary *initial reputation* over this time window. The perceived *initial image* of the seller by individual agents may, however, vary according to personal experiences and observations. The digital trust game allows agents to form new *opinions* on the seller by sharing their evaluations and may result in a *community reputation* (aggregate trust) that differs from the *initial reputation* or trust.

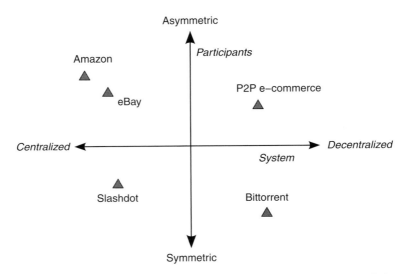

Figure 8.6 Simple overview of example digital identity and trust systems based on their system implementation and the nature of participants.

Given the initial reputation of the seller, $r_s \in \mathbb{R}$, the initial image (or trust level), $e_i \in \mathbb{R}$, perceived by an agent a_i can be considered as a noisy measurement of r_s and is defined by

$$e_i = r_s + n_i. \tag{8.4}$$

The *bias term*, n_i, captures the individual variation in initial opinion of agent i on the seller. This may be a result of varying personal experiences or observational limitations and distortions. Depending on the specific system, the vector $n = [n_1, \ldots, n_N]$ can be modeled as additive (zero-mean) Gaussian noise.

Using the initial image e_i as a starting point, an agent a_i forms an **opinion (trust)**, $x_i \in \mathbb{R}$, of the seller after exchanging information with the rest of the community. The individual opinion or trust, x_i, is influenced by various community effects as well as individual properties of the agent. The opinions of all the agents represented by the vector

$$x = [x_1, \ldots, x_N] \in \mathcal{X} \subset \mathbb{R}^N$$

define the decision space of the digital trust game. The opinions are time-dependent as they are formed over time through an iterative update process.

In the game, $x_i = 0$ corresponds to a neutral or default opinion of agent a_i on the seller. Consequently, the positive values, $x_i > 0$, represent a positive opinion and negative ones, $x_i < 0$, a negative opinion. The same convention also applies to the variables r_s and e, which admit similar interpretations.

The agents' opinions are not only a function of the initial reputation and image but also of factors capturing community influences. The decision process of an agent a_i can be modeled by the minimization of a well-defined cost function that quantifies the factors affecting the opinion of the agent. The cost function of agent a_i is defined as

$$J_i(x_i, \mathbf{x}_{-i}) = \frac{\alpha_i}{2} x_i^2 + \frac{\beta_i}{2} \left(x_i - \frac{1}{N-1} \sum_{j \neq i} x_j \right)^2 + \frac{\gamma_i}{2} (x_i - e_i)^2, \tag{8.5}$$

where $0 \leq \alpha_i, \beta_i, \gamma_i \leq 1$, $\alpha_i + \beta_i + \gamma_i = 1$ $\forall i$, and $x_{-i} := [x_1, \ldots, x_{i-1}, x_{i+1}, \ldots, x_N]$. It is naturally possible to define different cost functions that capture the paradigms discussed. The specific one in (8.5) is mainly chosen due to its analytical tractability.

The **first term**, $\alpha_i x_i^2$, in the cost function (8.5) quantifies the timidness of agent a_i. The term quadratically penalizes any positive or negative opinion of the agent, forcing it to the neutral or zero opinion. Agents with different properties can be represented by choosing the weighting parameter α appropriately. A *timid* agent, who is reluctant to pass judgment, is expected to have a high value of α whereas a *self-assertive* or opinionated one is captured by a small α parameter value.

The **second term** in the cost function quantifies the influence of *peer pressure* on the agent. Here, peer pressure is modeled using a quadratic cost on any opinion deviating from the mean value of others. An individualistic or independent agent is represented with a small β value. On the other hand, an agent who follows the crowd is expected to have a high-valued β parameter.

The **third term**, $\gamma_i(x_i - e_i)^2$, captures the effect of the initial image e_i of an agent a_i on the final opinion x_i. A *steadfast* agent who does not change its own opinion as a result of community interactions or sharing is represented by a high γ value. On the other hand, an agent who updates its opinion easily has a small γ parameter in the respective cost function.

Notice that the weighting parameters α, β, γ are normalized in such a way that the factors discussed above are balanced against each other. Hence, the inherent tradeoffs between the factors are captured by the cost function and the game.

The set of players or agents $\mathcal{A}$, the decision space X, and the cost functions J_i $\forall i$ define together the digital trust game, $G_1(\mathcal{A}, X, J)$. In this noncooperative game each individual agent a_i minimizes its own cost J_i by choosing its own opinion (trust decision), $x_i \in \mathbb{R}$, given the opinions (trust decisions) of others, $\mathbf{x}_{-i}$, i.e.

$$x_i = \arg\min_{x_i} J_i(x_i, \mathbf{x}_{-i}). \tag{8.6}$$

An overview of methods for analyzing such games is provided in Appendix A.2 as well as in references [4] and [31].

Equilibrium analysis

Nash equilibrium (NE) [31] provides an appropriate solution concept for the digital trust game. In this context, NE is defined as a set of agent opinions $\mathbf{x}^*$ of a given seller (and the corresponding one costs J^*), with the property that no agent has any incentive for modifying its own opinion while the other agents keep theirs fixed. The questions of whether the game G_1 admits one or multiple NE solutions and how to compute them are discussed next.

The opinion of an agent given the opinions of others is uniquely determined by the best-response function defined in (8.6). The first and second derivatives of the agent i's cost $J_i(x_i, \mathbf{x}_{-i})$ with respect to x_i are

$$\frac{\partial J_i}{\partial x_i} = x_i - \left(\frac{\beta_i}{N-1} \sum_{j \neq i} x_j + \gamma_i e_i \right) \tag{8.7}$$

and

$$\frac{\partial^2 J_i}{\partial x_i^2} = 1 > 0. \tag{8.8}$$

Hence, J_i is a quadratic function strictly convex in x_i and the minimization in (8.6) admits a unique globally optimum solution. Consequently, the decision, x_i, of agent a_i is a unique response to any given x_{-i}.

If the agents (players) are **symmetric** in their properties, i.e. $\alpha_i = \alpha$, $\beta_i = \beta$, and $\gamma_i = \gamma$ $\forall i$, then the NE solution of the digital trust game can be explicitly characterized with an analytical expression. Let $\bar{x} = \sum_i x_i$ and $\bar{e} = \sum_i e_i$. Due to the strict convexity of J, it is sufficient to check the first-order necessary condition for optimality

$$\frac{\partial J_i}{\partial x_i} = 0 \Rightarrow x_i^* = \left(\frac{\beta}{N-1} \sum_{j \neq i} x_j^* + \gamma e_i \right) \quad \forall i.$$

Then, simple algebraic manipulations using the symmetry of agents yield

$$\left(1 + \frac{\beta}{N-1} \right) x_i^* = \frac{\beta}{N-1} \bar{x}^* + \gamma e_i,$$

and

$$\bar{x} = \frac{\gamma}{1 - \beta} \bar{e}.$$

Thus, the unique NE of the game $\mathcal{G}_1$ is given by

$$x_i^* = \frac{\gamma}{N-1+\beta} \left(\frac{\beta}{1-\beta} \bar{e} + (N-1) e_i \right) \quad \forall i.$$

Even when the agents are **not symmetric**, the uniqueness of the NE is preserved. The best-response functions of the agents can then be written at the NE, x^*, in matrix form

$$x^* = Ax^* + c, \tag{8.9}$$

where $c_i = \gamma_i e_i \ \forall i$ and

$$A := \begin{pmatrix} 0 & \frac{\beta_1}{N-1} & \cdots & \frac{\beta_1}{N-1} \\ \vdots & \ddots & & \vdots \\ \vdots & & \ddots & \vdots \\ \frac{\beta_N}{N-1} & \frac{\beta_N}{N-1} & \cdots & 0 \end{pmatrix}_{N \times N}.$$

Hence, the NE is

$$x^* = (I - A)^{-1} c,$$

where I is the identity matrix and $(\cdot)^{-1}$ denotes the matrix inverse. Notice that the matrix $I - A$ is diagonally dominant as $A_{ii} = 1 > \sum_j A_{ij} = b_i \ \forall i$. Therefore, it is of full rank and invertible. Consequently, the digital trust game $\mathcal{G}_1$ has a unique NE solution.

8.2.2 Dynamics and convergence

The agents participating in the digital trust game cannot usually reach a stable opinion in a single round. They may also change their decisions dynamically while interacting with each other, unless the system is at the NE. These agent dynamics can be modeled using iterative update algorithms. Update algorithms and their convergence analyses are of practical importance and provide valuable insights into the dynamical aspects of digital reputation systems.

Parallel and round robin update

The parallel update algorithm (PUA) and its sequential variant, the round robin update (RRU), are studied as two basic and relevant examples of iterative update schemes. In the PUA, all agents (players) update their opinions (trust decisions) synchronously and in parallel. In the RRU, the agents update their opinions sequentially (one-by-one) given the opinions of previous agents. In both cases, the updates are done in discrete time instances and the initial reputation of the seller, r_s, as well as its initial image, e_i, perceived by the agents are assumed to be constant.

In the **PUA**, each agent a_i updates its own opinion $x_i(t)$ together (in parallel) with all other agents at the same discrete time instances $t = 1, 2, \ldots$ Therefore, the PUA is also known as the *synchronous update algorithm*. The updates are done according to the best-response function of the agent. Then, the PUA is formally defined as

$$x_i(t+1) = \frac{\beta_i}{N-1} \sum_{j \neq i} x_j(t) + \gamma_i e_i, \ \forall i. \tag{8.10}$$

The PUA iteration can be alternatively written in matrix form

$$x(t+1) = Ax(t) + c. \tag{8.11}$$

Algorithm 8.3 summarizes the steps of the PUA.

From the Perron–Frobenius theorem [78], the eigenvalues, λ, of the matrix $\mathcal{A}$ satisfy

$$\min_i \beta_i \leq |\lambda| \leq \max_i \beta_i, \ i = 1, 2, \ldots, N.$$

Hence, all of the eigenvalues of the linear system in (8.11) are inside the unit circle, and the PUA globally geometrically converges to the unique NE of the game, x^*.

Algorithm 8.3 Parallel update algorithm (PUA)

Input: Individual trust values e, convergence threshold ε.
Initialize trust values $x_i(0) = e_i \ \forall i$ and time step $t = 0$.
while $\|x(t+1) - x(t)\| > \varepsilon$ **do**
  $t = t + 1$
  Compute $s(t) := \sum_i x_i(t)$
  for $i = 1$ to N **do**
    Compute $x_i(t+1) = \dfrac{\beta_i}{N-1}(s(t) - x_i(t)) + \gamma_i e_i$.
  end for
end while

In the RRU, all agents update their decisions one by one in discrete time instances. The difference between the PUA and the RRU is that, in the RRU each agent knows the decisions of all other agents who played before and responds accordingly. After some rounds, the agents converge exactly to the same NE as in the PUA, yet their convergence speeds may vary. The RRU for the i-th agent is formally defined as

$$x_i(t+1) = \frac{\beta_i}{N-1}\left(\sum_{j<i}x_j(t+1) + \sum_{j>i}x_j(t)\right) + \gamma_i e_i, \tag{8.12}$$

where the agents are assumed to play with the order $i = 1,\dots,N$. Algorithm 8.4 summarizes the steps of the RRU.

Algorithm 8.4 Round robin update (RRU)

Input: Individual trust values e, convergence threshold ε.
Initialize trust values $x_i(0) = e_i$ $\forall i$ and time step $t = 0$.
while $\|x(t+1) - x(t)\| > \varepsilon$ **do**
$\quad t = t+1$
$\quad$ **for** $i = 1$ to N **do**
$\quad\quad$ Compute

$$x_i(t+1) = \frac{\beta_i}{N-1}\left(\sum_{j<i}x_j(t+1) + \sum_{j>i}x_j(t)\right) + \gamma_i e_i.$$

$\quad$ **end for**
end while

For a scenario with 20 symmetric agents and parameters, $[\alpha,\beta,\gamma] = [0.2, 0.3, 0.5]$, the iterative evolution of trust under the PUA is shown in Figure 8.7.

Robustness and asynchronous update

In many practical cases, such as in P2P networks or e-commerce, it is not always possible to ensure that all agents update their trust decisions sequentially or synchronously in parallel. For example, some of the agents may be offline or their decision update

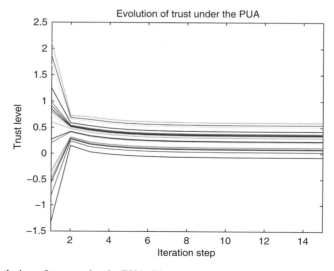

Figure 8.7 Evolution of trust under the PUA. The convergence to the NE point is geometric.

messages may be received with delay. The asynchronous update algorithm (AUA), where only a random subset of agents update their opinions at a given time instance, provides a realistic alternative scheme for such settings.

The **AUA** can be seen as a natural generalization of the PUA due to its parallel nature. The AUA is a more suitable scheme for practical scenarios and P2P networks when it is difficult for the agents to synchronize their exact update instances or sequence. The AUA is defined as

$$x_i(t+1) = \begin{cases} \dfrac{\beta_i}{N-1}\sum_{j\neq i}x_j(t)+\gamma_i e_i, & \text{if } a_i \in \mathcal{U}(t) \\ x_i(t), & \text{if } a_i \in \bar{\mathcal{U}}(t) \end{cases} \tag{8.13}$$

where the random set $U(t)$ represents the updating agents at time t and $\bar{\mathcal{U}}(t)$ the non-updating agents. Naturally, $U(t) \cup \bar{\mathcal{U}}(t) = \mathcal{A}$. Algorithm 8.5 summarizes the steps of the ASU.

Algorithm 8.5 Asynchronous update algorithm (AUA)

Input: Individual trust values e, convergence threshold ε.
Initialize trust values $x_i(0) = e_i \; \forall i$ and time step $t = 0$.
while $\|x(t+1) - x(t)\| > \varepsilon$ **do**
  $t = t + 1$
  Compute $s(t) := \sum_i x_i(t)$
  for $i = 1$ to N **do**
    if agent i updates **then**
      Compute $x_i(t+1) = \dfrac{\beta_i}{N-1}\left(s(t) - x_i(t)\right) + \gamma_i e_i$.
    else
      No change in decision, $x_i(t+1) = x_i(t)$.
    end if
  end for
end while

The ASU converges to the unique NE of the trust game as it satisfies the synchronous convergence condition, which follows from the spectral radius of the matrix $|A|$ being less than one, $\rho(|A|) < 1$, and the box condition. Hence, global geometric convergence of the ASU is established by Proposition 3.1 [38, p. 435]. Figure 8.8 shows the evolution of trust under the AUA again for a scenario with 20 symmetric agents and parameters, $[\alpha, \beta, \gamma] = [0.2, 0.3, 0.5]$. The convergence speed is, although geometric, slower than those of both the PUA and the RRU.

8.2.3   Numerical analysis

We now present a numerical analysis of the digital trust game based on three example scenarios. These scenarios illustrate the underlying concepts discussed, such as community effects and agent properties. They also facilitate a basic exploration of the

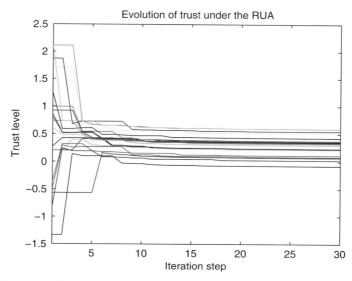

Figure 8.8 Evolution of trust under the AUA where the agents update their decisions with probability 0.5 in an iteration. The speed of convergence to NE, although geometric, is slower than those of both the PUA and the RRU.

game parameter space, which is large due to the general nature of the cost function (8.5). In each of the following scenarios, the digital trust game is played among 20 agents, who have a random initial trust level (image) of the seller, e_i, $i = 1, \ldots, 20$. The same initial values are used for all tests. Since the convergence properties of various update schemes are already established, the focus here is on the initial and final (NE) trust values of the agents, which are depicted with dark and light bars, respectively.

Effects of peer pressure

This scenario studies the effects of peer pressure on agents, for example, in an online community. If the term β, which quantifies the influence of peer pressure on the agent is dominant in the cost function (8.5), then the agents have a strong incentive not to deviate from the mean trust value of others. The cost parameters are

$$[\alpha, \beta, \gamma] = [0.2, 0.6, 0.2]$$

for all agents. The results in Figure 8.9 show that the trust levels of all agents converge close to a common value under strong peer influence, which can be interpreted as a community opinion.

Timid versus self-assertive agents

The case when the agents are timid, i.e. undecided or reluctant to trust or mistrust, is captured by a dominant α value in the cost function. Such agents are hesitant to trust or mistrust a digital identity which causes the trust decisions to converge to values

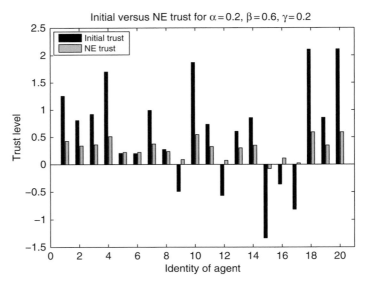

Figure 8.9 Initial and NE trust values for agents with cost parameters $[\alpha, \beta, \gamma] = [0.2, 0.6, 0.2]$ representing strong peer pressure.

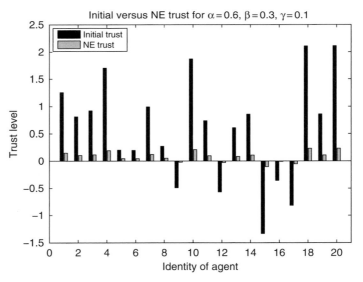

Figure 8.10 Initial and NE trust values for timid agents represented by cost parameters $[\alpha, \beta, \gamma] = [0.6, 0.3, 0.1]$.

close to zero (neutral opinion). The initial and final NE values for timid agents with the parameter set

$$[\alpha, \beta, \gamma] = [0.6, 0.3, 0.1]$$

are depicted in Figure 8.10.

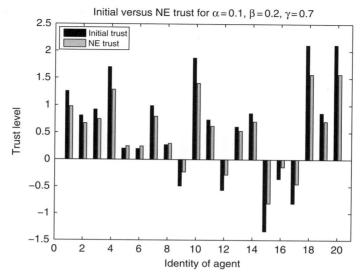

Figure 8.11 Initial and NE trust values for self assertive agents represented by cost parameters $[\alpha, \beta, \gamma] = [0.1, 0.2, 0.7]$.

On the other hand, if the agents are self-assertive (opinionated), which is captured by having a dominant γ value, they do not deviate much from their initial opinion. The results of a numerical analysis with self-assertive agents and the parameter set $[\alpha, \beta, \gamma] = [0.1, 0.2, 0.7]$ are illustrated in Figure 8.11.

Influence of a community leader

In many online communities there are individuals who can influence others. The opinions of such "community leaders" can widely affect the decisions of people who trust them. The effect of community leaders can be taken into account by adding a new term to the cost function (8.5), which now reads:

$$\tilde{J}_i(x_i, \mathbf{x}_{-i}) = \frac{\alpha_i}{2}x_i^2 + \frac{\beta_i}{2}\left(x_i - \frac{1}{N-1}\sum_{j \neq i} x_j\right)^2 + \frac{\gamma_i}{2}(x_i - e_i)^2 + \frac{\delta_i}{2}(x_i - x_k)^2,$$

where agent a_k is a community leader, $0 \leq \alpha_i, \beta_i, \gamma_i, \delta_i \leq 1$, and $\alpha_i + \beta_i + \gamma_i + \delta_i = 1$ $\forall i$. Here, only the effect of a single community leader is modeled. However, the function can be further modified to capture the effects of several leaders who may be of different opinions. It is assumed here that the community leader does not follow the same cost-based behavior model and maintains its own fixed opinion. The fourth term in the equation, $\delta_i(x_i - x_k)^2$, captures the effect of the community leader on the trust decision of agent i, x_i. An agent who faithfully follows the leader is represented by a large δ value.

The influence of the community leader is numerically analyzed under the parameter set $[\alpha, \beta, \gamma, \delta] = [0.1, 0.1, 0.1, 0.7]$. When the agents faithfully follow a community

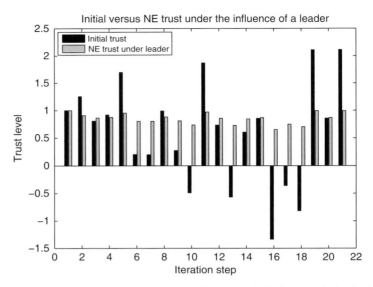

Figure 8.12 Initial and NE trust values in a community under the influence of a leader (agent 1).

leader, the final value is close to that of the leader. The results are shown in Figure 8.12, with agent 1 in the figure being the leader.

8.3   Location privacy in mobile networks

Privacy, just like usability discussed in Section 8.1, is a complex problem with social, legal, and psychological dimensions. As a simple definition, it deals with issues related to the dissemination and control of information that is specific to an individual or group. Privacy is an old problem that is exacerbated by recent technological developments, which make it very easy to communicate, store, and analyze (personal) information. New technologies have sometimes unintended consequences, as here. For example, a piece of personal information such as a joke posted to a social networking site can be read decades later by a potential employer and can lead to significant problems for the individual. At the same time, social networks can facilitate new ways of establishing personal and business contacts with many positive effects.

The problem of privacy in a digital and networked world shares many common aspects with other security issues discussed in previous chapters. One aspect is *observability*, which in this context refers to knowing who has access to personal information and how they use or abuse it. Another issue is how much *control* individuals have of disclosing personal information, i.e. how much information is revealed and in which manner.

The same fundamental reasons that complicate other security problems are also in play when it comes to privacy: the computational (hardware and software) complexity along with global networking makes it difficult for individuals to control the dissemination of their own private information as well as observing how it is used by others. Due to the complexity and scale of networked systems, the individuals need all the technological assistance they can get (along with social and legal ones) to obtain a reasonable level of privacy and to prevent the abuse of personal information in a digital world. Analytical models and quantitative analyses constitute the first step towards providing such assistance.

This section presents an example analytical model, which is based on game theory, and provides a setting in which to analyze and address the problem of **location privacy** in wireless networks with mobile nodes. Location is an important piece of information that compromises the privacy of an individual in a significant way, especially when revealed to an adversary. With proper equipment, a malicious attacker can track individual mobile nodes by monitoring their identity and correlating it with their location information. A frequently proposed solution to protect location privacy involves mobile nodes using pseudonyms and collectively changing them in special regions called *mix zones*. A mix zone is a special place where the attacker cannot distinguish the individual locations of mobile nodes, and hence confuses the location–pseudonym pairs of mobile nodes. The model, which will be discussed next, studies the noncooperative behavior of mobile nodes in a setting where each player aims at maximizing its location privacy at a minimum cost.

8.3.1 A location privacy model

A specific location privacy model is presented based on *multiple pseudonyms* as one of several techniques used to mitigate the tracking of mobile nodes [66]. In this technique, mobile nodes change their pseudonym, with which they sign messages, over time. Thus, they reduce the long-term linkability of name and location. As is commonly done in such networks, the existence of an offline CA is assumed. The CA is run by an independent trusted third party and pre-establishes the credentials, e.g. public–private key pairs for the devices. Here, a public key serves as the pseudonym (identifier) of the owner node. In order to thwart Sybil attacks, as soon as a node changes pseudonym, the old pseudonym expires.

In order to avoid the spatial correlation of their location, mobile nodes can coordinate pseudonym changes with their neighbors, especially in mix-zone regions, which may be fixed and part of the infrastructure or user-generated and ad hoc. Mix zones can also conceal the trajectory of mobile nodes to protect against the spatial correlation of location traces, e.g. by using silent mix zones [79, 96], a mobile proxy [156], or regions where the adversary has no coverage [49]. Without loss of generality, the proposed model assumes silent mix zones where the mobile nodes turn off their transceivers and stop sending messages for a certain period of time. If at least two nodes change pseudonyms in a silent mix zone, a mixing of their whereabouts occurs and the mix zone becomes a *confusion point* for the attacker. This flexible approach has obvious advantages over creating fixed "clean regions."

The (location) privacy needs of individual users (mobile nodes) may vary depending on time and location. Hence, it is desirable to adopt a user-centric approach that allows each user to decide when and where to protect its location privacy. In **user-centric location privacy**, each mobile node locally monitors its location privacy level over time and takes individual actions. This opens the door to scalable, fine-grained, and distributed schemes.

Consider a mobile network composed of a set of mobile nodes, $\mathcal{P}^D$, with cardinality N. When, at time t, a group of $n(t)$ mobile nodes are in close proximity, one of the nodes can initiate the pseudonym change using the one-round protocol suggested in reference [96] (i.e. the Swing protocol): a mobile node broadcasts an initiation message to start the pseudonym change. The $n(t) - 1$ mobile nodes in proximity receive the message and enter a silent period during which they decide whether or not to change their pseudonyms. During the silent period, nodes cannot observe each other's messages. At the end of the silent period, it appears to the attacker that all pseudonym changes occur simultaneously.

The attacker $\mathcal{P}^A$ observes the set of $n\left(T_i^\ell\right)$ nodes changing pseudonyms, where T_i^ℓ is the (last) time at which a successful pseudonym change occurs for node i. Subsequently, the attacker compares the set $\mathcal{B}$ of pseudonyms before the change with the set $\mathcal{D}$ of pseudonyms after the change, in order to predict the most probable matching [33, 96]. Let $p_{d|b}$ denote the probability that a new pseudonym $d \in \mathcal{D}$ corresponds to an old pseudonym $b \in \mathcal{B}$. The uncertainty of the adversary is defined using the entropy term

$$A_i(T^\ell) = - \sum_{d=1}^{n(T^\ell)} p_{d|b} \log_2(p_{d|b}), \tag{8.14}$$

which concurrently quantifies the location privacy level of a node i involved in a successful pseudonym change in a mix zone.

The achievable location privacy depends on both the number of nodes, $n\left(T_i^\ell\right)$, and the unpredictability of their whereabouts in the mix zone, $p_{d|b}$. If a node i is the only one to change its pseudonym, then its identity is known to the adversary and its location privacy level is defined to be $A_i\left(T_i^\ell\right) = 0$. The upper-bound on the entropy is obtained for a uniform probability distribution $p_{d|b}$, which would provide node i with a location privacy level of $log_2\left(n\left(T_i^\ell\right)\right)$. This can only happen after a coordinated pseudonym change by all players.

Assume a linear loss of privacy with time. This can be interpreted in multiple ways such as the amount of information collected by an adversary or the perceived risk increasing with time. A linear function with a user-specific parameter is chosen as a first approximation and the analysis can easily be extended to nonlinear loss functions. Then, the location privacy function of a mobile node i is

$$A_i(t) = \max \left(A_i\left(T_i^\ell\right) - \lambda \left[t - T_i^\ell \right], 0 \right), \quad T_i^\ell \leq t < T_i^f, \tag{8.15}$$

in the time period between the last successful pseudonym change at $t = T_i^\ell$ which has provided a location privacy level of $A_i\left(T_i^\ell\right)$ and the next one at $t = T_i^f$. Here, the parameter λ denotes the privacy loss rate and is chosen to be symmetric for all nodes for simplicity. Notice that the location privacy level provided by pseudonym changes depends on various factors and may be different each time, i.e. $A_i\left(T_i^\ell\right) \neq A_i\left(T_i^f\right)$. The privacy level dropping to zero means the successful identification and monitoring of the mobile node identity and position by an attacker. Thus, the resulting location privacy function of a mobile node resembles a sawtooth pattern over the x-axis.

8.3.2   Location privacy games

Based on the model in the previous section, a noncooperative pseudonym change or location privacy game is defined among $n(t)$ mobile nodes, who are in transmission range of each other at time t. In this section, full information games are studied, where each node or player $\mathcal{P}_i^D$ knows the number and properties of other players in the mix zone. This information can be obtained, for example, by running a special protocol such as neighbor discovery [182]. A Bayesian extension of location privacy games is discussed in reference [66].

The location privacy game considered is as a succession of one-shot or static games since all nodes make a single decision in parallel during the silent period of the mix zone on whether or not to change their pseudonyms. Thus, the action (strategy), s_i, of a player i can be one of the two moves $s_i \in \{C,D\}$, cooperate (C) or defect (D). The actions of all other players are denoted by s_{-i} as is customary.

The cost function of a player is the difference between the cost of a pseudonym change, γ, [65] and the privacy benefit obtained from the change. For simplicity, the cost γ is assumed to be symmetric for all users. A possible cost function is then defined as

$$J_i(s_i, s_{-i}) = \begin{cases} \gamma - \max\left(A_i, \hat{A}_i(s_{-i})\right), & \text{if } s_i = C \\ -A_i, & \text{if } s_i = D, \end{cases} \qquad (8.16)$$

where A_i is the privacy level of the player at the time of play, $\hat{A}_i(s_{-i})$ is the new privacy level as a function of the number of other players cooperating, i.e. joining the pseudonym change. If no other player cooperates, then the pseudonym change fails and $\hat{A}_i = 0$. Let $\bar{n}$ be the total number of nodes in the system. Then, by definition, $A_i \leq \log_2(\bar{n})$ and $\hat{A}_i \leq \log_2(\bar{n})$ hold.

In order to gain further insight into the location privacy game defined, first a two-player version is investigated. Subsequently, an n-player extension is discussed.

Two-player game
Consider a full information location privacy game between two mobile nodes or players, $\mathcal{P}_1^D$ and $\mathcal{P}_2^D$. If upon a successful pseudonym change, each node achieves the same level of privacy, then all cooperative players receive a new privacy level of $\hat{A} = \log_2(2) = 1$. As a special case of (8.16), the game matrices of players $\mathcal{P}_1^D$ and $\mathcal{P}_2^D$ are

$$\begin{matrix} & \begin{matrix} (C) & (D) \end{matrix} \\ G^1 = & \begin{bmatrix} \gamma - 1 & \gamma - A_1 \\ -A_1 & -A_1 \end{bmatrix} \end{matrix} \begin{matrix} (C) \\ (D) \end{matrix}, \quad \begin{matrix} & \begin{matrix} (C) & (D) \end{matrix} \\ G^2 = & \begin{bmatrix} \gamma - 1 & -A_2 \\ \gamma - A_2 & -A_2 \end{bmatrix} \end{matrix} \begin{matrix} (C) \\ (D) \end{matrix}, \qquad (8.17)$$

where A_1 and A_2 are player privacy levels, respectively. Note that $A_1, A_2 < 1$ from (8.15) and assuming a minimum amount of time between pseudonym change attempts or games, the next theorem characterizes NE strategies under various conditions on game parameters.

Theorem 8.6 *The two-player location privacy game admits the following NE solutions:*

(a) the pure strategy (D,D), regardless of game parameters,
(b) the pure strategy (C,C), if $\gamma < \max(1-A_1, 1-A_2)$,
(c) the mixed strategy (p^*, q^*), given by

$$p^* = \frac{\gamma}{1-A_2}, \quad q^* = \frac{\gamma}{1-A_1}, \quad \gamma < \max(1-A_1, 1-A_2),$$

where p and q are the probabilities of the collaboration of players $\mathcal{P}_1^D$ and $\mathcal{P}_2^D$, respectively.

Proof The pure strategy equilibria follow directly from the definition of NE. The mixed equilibrium is obtained directly from the indifference condition

$$J_1(p, q^*) = const., \quad J_2(p^*, q) = const. \; \forall p, q,$$

similarly to the cases studied in Chapter 3. □

If the condition $\gamma < \max(1-A_1, 1-A_2)$ is satisfied, i.e. the cost γ or both A_1, A_2 are sufficiently low that collaboration is feasible, then the pure NE strategy (C,C) is the dominant one and preferred by both players over other equilibria. The players then have an incentive to collaborate and change their pseudonyms when they are in proximity of each other. This is naturally not the case if the condition is violated, for example, when one of the players has just changed its own pseudonym (i.e. A_i high) or when the name change cost γ is prohibitively high.

N-player game

Consider now an *n*-player location privacy game as an extension of the two-player version. It is useful to reorder (relabel) players, $\mathcal{P}_i^D \in \mathcal{A}^D$, $i = 1, \ldots, n$, based on their current privacy levels such that $A_1 \leq A_2 \leq \cdots \leq A_n$. Then, define the set $\mathcal{A}^D(k) \subset \mathcal{A}^D$ as one of the first $k < n$ players.

If k players (collaborate) participate in a successful pseudonym change, then each of them obtains a new privacy level, e.g. $\hat{A} = \log_2(k)$. Each node minimizes its own cost in (8.16) as before. The equilibrium solutions of this game are characterized in the next theorem.

Theorem 8.7 *Consider an n-player location privacy game between the set of players $\mathcal{A}^D$, where the players are labeled in increasing order based on their current privacy levels such that $A_1 \leq A_2 \leq \cdots \leq A_n$. The game admits the following pure-strategy NE solutions under the respective conditions.*

(i) The strategy where all players defect, $(D, \ldots, D)$, is an NE.
(ii) The pure strategy where first $2 \leq k < n$ players, $\mathcal{A}^D(k)$, collaborate (C) and the remaining, $\mathcal{A}^D \backslash \mathcal{A}^D(k)$, defect (D), i.e.

$$\underbrace{(C, \ldots, C_k,}_{k} \underbrace{D, \ldots, D)}_{n-k}, \ 2 \le k < n,$$

is an NE, if

$$A_{k+1} + \gamma > \hat{A}(k) > A_k + \gamma.$$

(iii) The pure strategy where all players collaborate, $(C, \ldots, C)$, is an NE, if

$$\hat{A}(n) > A_n + \gamma.$$

Proof The cases of (i) and (iii) follow directly from the definition of NE. In order to prove case (ii), it is sufficient to show that no player from the set of $2 \le k < n$ collaborating players, $\mathcal{A}^D(k)$, has an incentive to defect and from the set $\mathcal{A}^D \backslash \mathcal{A}^D(k)$ to collaborate. This follows immediately from (8.16), since for any player $i \in \mathcal{A}^D(k)$ the condition in the theorem states that $\hat{A}(k) > A_i + \gamma$, and hence $J_i(D, s_{-i}) > J_i(C, s_{-i})$. Likewise for any player $i \in (\mathcal{A}^D \backslash \mathcal{A}^D(k))$, the condition leads to $A_i + \gamma > \hat{A}(k)$, and hence $J_i(D, s_{-i}) < J_i(C, s_{-i})$. □

The results in the theorem can be intuitively interpreted as follows, assuming that pseudonym change cost γ is reasonably low. If all the players have high privacy levels, then nobody has an incentive for collaboration and all nodes defect. If two or more players have sufficiently low privacy levels, then they collaborate to create a group for successful pseudonym change. Finally, if all players have low privacy levels, then all in the mix zone collaborate.

8.4 Discussion and further reading

Reference [58] contains a comprehensive discussion on security and usability from multiple perspectives and dimensions. The relation between security and usability is further studied in reference [23, Chap. 2] and reference [149]. The security alert dissemination system [16] in Section 8.1.1 is one example application of decision and the game-theoretic approach for improving usability in network security. The effective administrator response scheme in Section 8.1.2 is another example that utilizes an optimization framework which was first introduced in reference [41].

The community trust game presented in Section 8.2 extends reference [17] and studies community effects on trust decisions for digital identities and in virtual environments. For a survey on trust and reputation systems, see reference [82]. Earlier game-theoretic approaches to the topic include references [63, 118, 133, 194]. Trust and reputation on the Internet is investigated in references [143, 144]. Other studies, such as reference [47, 48], have applied reputation systems to P2P and mobile ad-hoc networks.

Section 8.3 discusses location privacy in mobile networks and presents a summary of the model in reference [66]. The location privacy approach here is based on mix zones and pseudonym changes [79, 96] building upon the intuitive idea of "disappearing in the crowd." An incomplete information extension and a protocol implementation of the same privacy scheme are also investigated in references [65] and [66].

Part IV

Security attack and intrusion detection

9 Machine learning for intrusion and anomaly detection

Chapter overview

1. Intrusion and anomaly detection
   - intrusion detection and prevention systems (IDPSs)
   - open problems in intrusion detection
2. Machine learning (ML) for security
   - overview of ML methods
   - open problems in ML for security
3. Distributed machine learning
   - support-vector-machine (SVM) classification
   - parallel update algorithms
   - asynchronous and stochastic algorithms
   - active set-based algorithm and a numerical example
   - behavioral malware detection in mobile devices

Chapter summary

An overview of anomaly and intrusion detection (prevention) as well as machine learning is presented along with a discussion of open research problems. Machine learning provides a scalable and decentralized framework for detection, analysis, and decision making in network security. As an example of a distributed machine-learning (ML) scheme, a distributed binary SVM classification algorithm is analyzed. It is then applied to the problem of malware detection in mobile devices using behavioral personalized signatures.

9.1   Intrusion and anomaly detection

Networked systems are difficult to **observe and control** even by their owners due to their **complexity**. The complexity is a result of three factors: powerful and ubiquitous computing hardware, complex software running on the hardware, and interconnectivity between separate computing systems. The same factors also make it difficult to observe the multitude of processes on modern networked systems. As a simple example, it is practically impossible for users to continuously be conscious of which software and communication processes are running on their mobile phone and to control their functionality.

An intuitive result, supported also by the classical control theory, states that it is very difficult to control any system that is not observable. Therefore, **observation** capabilities should be seen as the first step towards controlling networked systems by their owners. Increasing observability of systems (by their owners) enables building better defenses against unwanted behavior and malicious security compromises. Nevertheless, observation capabilities are often not a built-in feature of networked systems. Furthermore, collecting, interpreting, and storing such data brings additional overhead and requires a nontrivial investment.

It is useful to briefly review the history of network security in order to better understand the current problems. When networked systems started playing an increasingly important role in almost all aspects of modern life, security threats against them both increased and were taken more seriously by the owners of the systems. Ever since, the potential and actual costs of security compromises have been naturally motivating defensive measures. The first defenses were mostly static such as firewalls establishing a perimeter and protecting it or reactive such as the signature-based antivirus software. Next, dynamic intrusion detection systems emerged, which increased system observation capabilities. More recently, the trend is towards dynamic and preventive measures with the emergence of intrusion prevention systems augmenting detection capabilities. This is partly a result of technological developments making such preventive measures feasible on the network and hosts.

On the other side, the attacks have also become more sophisticated. The security threats have evolved from simple ones driven by scientific curiosity of renegade enthusiasts to complex and distributed attacks for financial gain, sometimes in connection with organized crime. The botnets which launch hybrid worm-Trojan malware from thousands of computers (without the knowledge of their owners) exemplify the level of sophistication of recent attacks.

The **castle analogy** provides a useful and fun model for visualizing the fundamental security concepts discussed above [69]. If the networked system to be secured is compared to a medieval castle, then static security measures such as firewalls, authentication, and access control correspond to the walls and gates of the castle, respectively. Reactive measures such as antivirus software can be similarly compared to antisiege weapons. All these semistatic security measures naturally require maintenance and upgrades in the face of new technologies. On the other hand, no matter how strong the walls are, the castle is not secure without guards defending it. Without dynamic measures the castle, however well protected, would be vulnerable and "blind" to security violations and attacks. The direct counterpart of these guards in the IT context is intrusion detection systems. Similarly, scouts and guards patrolling the vicinity of the castle can be compared to intrusion prevention systems which detect and try to address threats before they arrive at the gates.

The characteristics of attackers and defenders can also be modeled within the same castle analogy. Over time, the defensive guards have evolved from simple people or owners defending the castle to professional soldiers protecting it. The attackers, likewise, turned into well-organized and equipped opponent armies from their humble beginnings of disorganized roving bands of outlaws.

This chapter focuses on computer-assisted security attack detection and analysis using (statistical) ML methods from the perspective of the defense side. As already mentioned, observation capabilities provide the necessary basis for attack prevention and security measures. It is obvious yet worth mentioning that observation and detection in the context of network security cannot be done manually by human beings and require computer assistance at the minimum. Therefore, *state-of-the-art ML techniques are potentially an indispensable component of computer assistance systems for the detection, analysis, and prevention of security threats.*

9.1.1 Intrusion detection and prevention systems

Intrusion detection can be defined as the process of monitoring the events occurring in a computer system or network and analyzing them for signs of intrusions [25, 26]. This classical definition is somewhat limited and focuses mainly on detecting attacks after they occur and reacting to them. Recent developments in the field have resulted in a more extended approach of intrusion prevention where monitoring capabilities are utilized to take preventive actions to defend against malicious attacks before or as they occur.

The evolution of intrusion detection systems to hybrid intrusion detection and prevention systems is not very surprising in light of the close relationship between observation and control as discussed above. While intrusion detection focuses more on detection and reporting aspects of the problem, prevention systems tend to be more action- and response-oriented. Responses such as filtering of malicious packets at the perimeter of the network aim to prevent attacks proactively. This shift can be seen as a result of continuous technological improvements, which enable better and economically feasible defensive solutions. Consequently, once monitoring capabilities are developed and available, it makes sense to deploy them both inside a networked system and at the periphery.

Although IDPSs are constantly evolving, it is useful to study the basic concepts and building blocks, which mostly remain invariant. An IDPS consists of three main components: *information sources*, *analysis*, and *response*.

The **information sources** in an IDPS observe the networked system and collect data that will help to detect and prevent attacks. They can be implemented, for example, as software agents running as virtual sensors or on dedicated hardware for packet inspection. The data collected is often sequential and heterogeneous in nature. Depending on the type and number of sources, the amount of data collected can be quite large, which may require significant storage and processing capabilities.

A set of (virtual) sensors network can be deployed as part of an IDPS in order to collect information and detect malicious attacks on the network (see Figure 9.1). A virtual sensor network is defined as a collection of autonomous software agents that monitors the system and collects data for detection purposes. The sensors report possible intrusions or anomalies occurring in a subsystem using common pattern recognition and statistical analysis techniques. Some of the desired properties of sensors can be summarized [191] as

- **completeness:** a sensor should cover the assigned subsystem in the sense that it collects all the information necessary for detection. Structural differences between the information sources such as log files, packet headers, etc. and resource constraints increase the difficulty of this task;
- **correctness:** the integrity of the collected information by the sensor should be ascertained against any tampering by the attacker. Misinformation is much more harmful to the IDS than partial information;
- **resource usage:** especially host-based sensors use a portion of the system resources, which include memory, storage space, processing power, and bandwidth. Hence, they are naturally bounded by them;
- **reconfigurability:** sensors should be reconfigurable both in terms of operating parameters and deployment.

It is informative to compare and contrast the virtual sensors, including those in dedicated appliance devices, with physical hardware-based sensors (motes). Motion, temperature, and biometric sensor motes are, for example, utilized for the physical security of buildings and rooms. While both classes of sensor are functionally similar with the common goal of collecting information on a system, there are some fundamental differences in terms of resource constraints. For example, there is no limit to the number of virtual sensors to be deployed in a system except from the communication and computation overhead. Clearly, this is not the case with physical sensors. Moreover, virtual sensors do not have power constraints. On the other hand, communication overhead is a problem for both networks. The unique characteristics of the virtual sensor network have to be taken into account in the IDS deployment process as well as in addressing the related resource allocation problems.

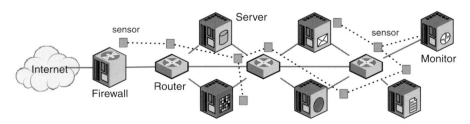

Figure 9.1 Graphical representation of a virtual sensor network for intrusion detection.

According to their deployment and source of information, IDPSs are categorized as host-based or network-based. **Network-based** IDPSs detect attacks by capturing and analyzing individual packets on the network. They can be deployed at the periphery or gateway, monitoring the entire traffic or at individual switches. They can provide blanket protection to the entire network and potentially stop attacks before they reach the protected system. On the other hand, large data volumes or encrypted traffic pose significant challenges for such systems.

A concept related to information collection is a **honeypot**. A honeypot is an isolated, unprotected, and monitored component of the protected system, e.g. a computer. Its sole purpose is to collect attack data but it has no other production value. The information collected by honeypots is valuable as a surveillance and early warning tool. Honeypots can be deployed in production environments to assess the actual threats faced as well as for research purposes.

Host-based IDPSs are deployed in individual computer systems and utilize, for example, operating system calls and system logs to detect attacks. They can also check file system integrity. Although they are immune to the issues faced by network-based variants, host-based IDPSs have their own problems. They are often difficult to manage, have nontrivial computational overhead, and can themselves be targeted by attacks.

Once the observational data is collected, it needs to be analyzed for signs of security threats. There are two main approaches for the **analysis** of the collected data: misuse and anomaly detection. **Misuse detectors** analyze the data, looking for events that match a predefined pattern of a known attack. Since the patterns corresponding to known attacks are often called (attack) signatures, the terms *"signature-based detection"* and "misuse detection" are used interchangeably. The most well-known and widely used example of signature-based detection is that used in antivirus software. Another example is the rules in the Snort IDPS software.[1] The signatures have to be written manually by experts one by one for each threat. The updated signatures are then sent to individual IDPS software periodically. As a countermeasure the attackers create multiple variants of the same malware as well as malware that self-mutates. Zero-day attacks for which no signature is generated yet and the huge number of threat types make signature-based detection a limited solution.

The second analysis approach is **anomaly detection** where "abnormal" or unusual behavior (anomalies) on the networked system are identified using various pattern recognition and (statistical) machine-learning techniques. The underlying assumption here is that attacks will be observable in the data collected by (legitimate) activity differing from "normal." This approach has the potentially desirable properties of scalability and detection of previously unknown attacks for which there is no manually generated signature. In practice, however, anomaly detectors have often very high false positive rates, i.e. mistake legitimate events for anomalies. This is maybe not surprising considering that detecting an anomaly in a large dataset is roughly analogous to

---

[1] http://www.snort.org/

Table 9.1 Comparison of signature-based and anomaly detection

Signature-based	Anomaly
Effective in detecting known attacks	Detects previously unknown attacks
Quick and reliable	Many false-alarms
Manual signature generation	Often requires training sets
Not scalable with threat types	Potentially scalable

finding a needle in a haystack. As a further complication, most ML techniques require a "clean" training set to learn normal behavior, which may not exist at all. A summary of signature-based and anomaly detection techniques' properties is shown in Table 9.1.

Once the data is collected and analyzed, the **response** to detected incidents can be active or passive. **Passive response** mainly involves alerting the system administrators and logging the suspected events for later analysis or network forensics. The problems of prioritizing alerts and how to present them to the administrators can be addressed within an optimization framework as discussed in Section 8.1.2. Given the limited time and attention of security officers, these tasks have to take usability into account as an important factor.

Active response schemes are one of the main distinguishing characteristics of intrusion prevention systems. One possible active response approach is to drop suspected packets and flows to prevent attacks before they reach the protected system. If not carefully configured, however, such measures can have significant detrimental side effects. A less drastic approach is to initiate additional investigation on the incident, for example, by collecting more data than usual via additional sensors. Another alternative is to reconfigure the environment, e.g. by increasing access restrictions and isolating parts of the protected system. Active response is crucial especially in cases where *response time* is an important factor such as fast-spreading worms or viruses which can paralyze the whole networked system within minutes.

9.1.2   Open problems and challenges

Although IDPSs are important components of network security, they are far from perfect and face open problems. One problem is the attacks targeting IDPSs and their capabilities directly. A second one is usability, deployment, and configuration of IDPSs. A third problem is the inherent difficulty of the detection task itself, which, using the earlier analogy, is similar to "searching for a needle in a haystack."

Just like the system it protects, the **IDPS can itself be targeted** by attackers. Carefully constructed packets or behavior patterns may function as a DoS attack against the IDPS and render it ineffective. Alternatively, the attackers can use clever stealth techniques to avoid detection by IDPS. One example is polymorphic worms that change their representation as they spread throughout networks and can evade especially signature-based worm detection systems which rely on fixed descriptions. Another example is encrypted network traffic that is becoming more and more common. If an attacker encrypts the command and control mechanisms of a botnet, the challenge for an IDPS to detect and disrupt it becomes an order of magnitude more difficult. Another class of attacks targets the learning phase of anomaly detection schemes. Here, a malicious adversary constructs labeled samples artificially on purpose and lets them be observed by the IDPS. If these samples are used by the detection algorithm as data points during training, the accuracy of the classifier is compromised for the benefit of the attacker.

Usability of IDPSs is an important problem that is usually overlooked by security researchers, partly owing to its nontechnical nature. Since each networked system is different, the IDPS has to be configured according to the deployment environment. Consequently, the initial deployment and configuration of an IDPS in a medium- to large-scale networked system can be so complex that it becomes a barrier itself. Recently, a variety of security appliances have been developed as a solution to address the usability issue. They are, however, usually only network-based and focus mainly on packet filtering.

A third problem, which may well be the most significant of all, is the **base-rate fallacy** and its implications for intrusion detection [24]. It is a direct consequence of the basic Bayesian theorem and is best explained by means of a numerical example. If an IDPS generates one million records a day and only 20 of them indicate real intrusions, then to reach a true positive (attack detection) rate of 0.7 and a Bayesian detection rate (the rate an alarm really indicates an intrusion) of 0.58, the false-alarm probability has to be as low as 0.00001.

This rather unexpected result puts significant constraints on the desired effectiveness of an IDPS. One of the underlying issues here is the rarity of intrusion-related events within a significant amount of data. Another is the adverse effects of false-alarms, which make the job of security officers more difficult and potentially waste resources. In other words, the situation is worse than looking for a "needle in a haystack" since one has not only to go through a big haystack to find the needle but also have a very sharp eye distinguishing each straw from the needle very accurately.

9.2 Machine learning for security: an overview

Complex networked systems cannot be continuously observed, controlled, and defended against security compromises *manually* by their users or administrators. **Computer assistance** is absolutely necessary to facilitate all these processes. This requires certain decisions to be taken automatically without human intervention. Machine (statistical) learning studies such decision-making problems in a variety of fields ranging from computer vision to speech recognition. Hence, machine learning has significant potential and has been increasingly used in the security domain.

The application of ML to network security has interesting implications for both fields. While ML creates a foundation for computer-assisted decision making in network security, it in turn benefits from unique aspects of this relatively novel application domain. This opens up interesting research directions such as data mining for security, distributed ML, and adversarial ML. Among these, distributed machine learning in networked environments will be discussed in Section 9.3 in more detail.

As we have already seen, intrusion and anomaly detection involves collecting a significant amount of observation data and its analysis. Therefore, it is natural to adopt existing statistical and pattern recognition schemes in the analysis of the collected data to classify anomalous and potentially malicious behavior. One way to approach this problem is to characterize the data within a parametric statistical model and estimate its parameters from the dataset. The anomalies potentially indicating malicious behavior can then be distinguished using hypothesis testing. There exists a rich literature on detection and estimation theory which approaches this problem from a signal processing and information theoretic perspective. This point of view will be presented and discussed in Chapter 10.

On the other hand, it is often difficult to characterize the IDPS-collected data by a probability distribution or an independent identically distributed random process. In such cases nonparametric ML and data-mining algorithms such as clustering, support vector machines (SVMs), or kernel-based methods become more relevant. It is worth noting that the distinction between parametric (statistical) and nonparametric (ML) methods is not as crisp as it may first seem. Both approaches aim to solve similar problems, share fundamental properties, and usually borrow many techniques from each other.

Although it is not feasible to provide here a comprehensive treatment of ML methods, a brief overview is beneficial to better understand their advantages and limitations within network security context. The basic properties of various ML algorithms [40] and their relevancy to decision making in the security domain are discussed next.

9.2.1   Overview of machine-learning methods

We provide in this section an overview of preprocessing, unsupervised and supervised learning, and reinforcement learning, which are the salient ML methods.

Preprocessing

The first step of data analysis is *preprocessing* where original input variables are transformed into a new variable space and – if possible – undergo a dimension reduction to facilitate subsequent computations. This may involve *feature extraction* where a subset of features that preserves useful discriminatory information are selected. The data transformation and feature selection may be done in such a way that variance of the projected feature set is maximized as in the well-known principle component analysis (PCA). Preprocessing is one of the most important yet sometimes overlooked aspects of ML and plays a crucial role in the success of any subsequent data analysis method.

In the case of network security problems, the amount of data collected by IDPS is often large. Furthermore, the data is usually noisy and redundant due to its observational nature. Proper application of preprocessing techniques in combination with problem-dependent domain knowledge helps to reduce the problem dimension and decrease the computational costs in subsequent steps significantly. However, proper care must be shown in preprocessing in order *not* to discard potentially useful information from original input data.

Unsupervised learning

One of the two main approaches in the analysis of IDPS observations is **unsupervised learning**. Clustering, density estimation, and visualization are among the well-known unsupervised learning techniques. *Clustering* discovers patterns in the dataset by grouping data points with similar features, *density estimation* studies the distribution of the data within the input space, and *visualization* projects the data into two or three dimensions to facilitate its human readability.

Security visualization[2] aims to process and transform complex security data collected into human-readable visual representations. Hence, the actual tasks of intrusion and anomaly detection are left to system administrators to be handled manually. The main idea is to utilize the superb pattern recognition capabilities of the human brain instead of relying on various ML algorithms. Thus, security visualization can be seen both as a computer-assistance scheme for security monitoring and as an alternative approach that addresses the disadvantages of fully automated anomaly detection methods. On the other hand, this approach has its own potential disadvantages such as imperfect data aggregation and scalability issues owing to its manual nature.

Density estimation assumes a probabilistic model of the input data and estimates the parameters of the model, e.g. probability distribution, from the given dataset which may be incomplete. Therefore, it can be categorized as a parametric or a statistical technique. As an example, the expectation–maximization algorithm is a well-known formal statistical technique for finding the maximum likelihood estimates of the model parameters. It can be used in the security context, for example, to characterize network traffic, and hence detect flow anomalies which may indicate DoS attacks.

Clustering is probably the most well-known and widely used unsupervised learning method in ML and data mining. Clustering techniques categorize data points in the input set to subsets or categories (clusters) based on a similarity measure defined over their feature space. A well-known variant called *K-means clustering* partitions n given points into a predefined k cluster in which each point is assigned to the cluster with the nearest mean value. Expectation–maximization is another widely used clustering method. While neither algorithm guarantees a globally optimal solution, they are both fast and easy to implement.

Clustering techniques have been utilized in network security especially for anomaly detection. The common approach is as follows: the data collected by IDPS is clustered based on a chosen feature vector and distance metric on the feature space. The computed clusters are then used to characterize the data. For example, the "neurons" of a SOM are a discretized representation of the input data. After this analysis step, subsequent

---

[2] http://www.secviz.org/

IDPS observations can be compared with the computed representation using a distance metric. If it can be ensured that the initial dataset represents "normal" system behavior, then any future data that significantly deviates from the computed representation can be interpreted as an "anomaly."

Supervised learning

ML methods that aim to "learn" a function from a given *training data* set constitute the second main approach, called **supervised learning**, for the analysis of IDPS observations. The objective of *classification* problems is to learn a function that assigns a label from a discrete set such as {*clear, malicious*} to each data point. If the label set has only two elements, then the problem is called binary classification. On the other hand, if the function to be learned is real-valued, then the problem is called *regression*. For example, each packet analyzed by the IDPS can be assigned a real-valued suspicion level between zero and one.

Two main criteria for the success of supervised learning algorithms are *prediction* and *generalization*. The learned function should be able to predict successfully the output for data other than in the training set used. At the same time, it should act as a model that captures the underlying characteristics of the training data and generalize to new data points. The opposite behavior is called *overfitting* where the function describes the training data perfectly but has poor predictive power. The performance of a supervised learning algorithm is quantitatively assessed using a *test data* set and *cross-validation* techniques.

Supervised ML techniques offer an invaluable set of formalized computing methods to develop computer-assisted detection, analysis, and decision systems for network security. These methods have been successfully applied to a wide variety of fields ranging from image recognition, speech processing, and data mining. In the network security domain, whether the objective is to distinguish whether a packet contains malicious payload or to assess how suspicious is certain behavior (e.g. flow pattern), the problem can be naturally formulated within the ML framework. IDPSs benefit from the scalability and computational power of ML algorithms as opposed to manual processing by system administrators. Furthermore, an IDPS usually generates significant amounts of data which improves the performance of many ML algorithms.

9.2.2   Open problems and challenges

In addition to its potential, the application of machine learning to network security gives rise to multiple problems and research directions. The first issue is the lack of good quality datasets to assess the performance of ML algorithms in the security domain. This lack of data hinders research efforts and is in stark contrast to other application fields of ML such as computer vision. The second problem is the fact that attackers can (secretly) influence the learning and decision processes in ML, especially in the training phase. This opens up an interesting new research direction for ML called *adversarial machine learning*. The adversarial ML algorithms, which adopt the role of the defender, try to accomplish their objectives while facing a malicious attacker who aims to undermine their efforts. The third problem is *distributed machine learning*, which extends ML to networked environments where the training set itself may be distributed and the communication between distributed algorithm subroutines may be imperfect.

The current **lack of good quality datasets** in network security is an important yet hopefully temporary problem. The problem was acknowledged a decade ago and the famous IDS evaluation dataset was published by the Defense Advanced Research Projects Agency (DARPA)[3] in 1998, which was then used in 1999 at the knowledge discovery and data mining (KDD) cup. Since then, the KDD dataset has been used by numerous papers, partly owing to lack of alternatives and partly for convenience. Recently, various flaws with this dataset have been discovered. Its usage as an evaluation tool is discouraged by many experts. Disappointingly, there is almost no other dataset proposed as a replacement until now.

Some security researchers claim that the only way of checking a security mechanism is to test it experimentally in a realistic setting. Although it has its obvious merits, this approach can also be potentially harmful to research efforts in the field. Building an experimental IDPS environment is a costly endeavor in terms of both manpower and monetary expenses. Therefore, such an approach creates a significant barrier of entry for cross-disciplinary efforts and prevents knowledge transfer from the ML community to security researchers and vice versa. It is unrealistic to expect every ML researcher, who is interested in security as an application field, to have a full-fledged and expensive security laboratory. Considering that quantitative evaluation using training and testing datasets is an indispensable aspect of ML algorithms, lack of high-quality datasets shared by the whole community is potentially harmful to the healthy development of ML in the security domain.

Adversarial machine learning constitutes a second open problem and potentially promising research direction. In classical ML, the training data can be noisy; yet there is no conscious attacker trying to manipulate it. The presence of an attacker, who either tries to hide attack symptoms or poison the training data to mislead the algorithm, brings a novel and interesting dimension to existing ML approaches discussed above. Given the widespread use of optimization formulations in classical ML, *game theory*, which can be seen as the multi-person decision-making counterpart of optimization, can provide the needed framework to investigate adversarial ML problems. Security game formulations discussed in Chapter 3 may be especially relevant here.

Recent advances in cloud computing, multiprocessor systems, and multicore processors make **distributed machine learning** an important research field relevant not only to the security domain but also to many others. The widespread and cheap availability of powerful parallel and distributed computing resources already provides a motivation for distributed and parallel ML schemes. They find a natural application domain in network security due to the inherently distributed nature of networked systems. The data collection, analysis, and storage processes of a networked IDPS are often decentralized. Distributed ML can easily be deployed within such an IDPS and remove the need

---

[3] DARPA is an agency of the United States Department of Defense.

for forced centralization at a security center within the network. A distributed binary classification algorithm will be discussed in the next section as an example.

9.3   Distributed machine learning

Powerful parallel and distributed computing systems have recently become widely available in the form of multicore processors and cloud computing platforms. Motivated by these developments, **distributed machine learning** algorithms can be naturally applied to address various decision-making problems in network security, where data collection and storage processes are inherently distributed. Both the efficiency and robustness of IDPSs are improved by taking advantage of computing capabilities distributed over the whole networked system. Furthermore, overall communication overhead is decreased by not sending all of the observation data to a single information fusion and decision-making center. A similar approach is also applicable in multicore processors or GPU programming, albeit in a smaller scale.

This section discusses a **distributed SVM-based binary classification framework** as an exemplary distributed ML scheme. First, the quadratic SVM binary classification problem is divided into multiple separate subproblems by relaxing it using a penalty function. Then, distributed continuous- and discrete-time gradient algorithms are analyzed that solve the relaxed problem iteratively and in parallel. A sufficient condition is derived under which the synchronous parallel update converges to the approximate solution geometrically. Next, an asynchronous algorithm is investigated where only a random subset of processing units update at a given time instance and show its convergence under the same condition. Subsequently, stochastic update algorithms are studied which may arise due to imperfect information flow between units or distortions in parameters. Sufficient conditions are derived under which a broad class of stochastic algorithms converge to the solution.

Unlike sequential or centralized approaches in the classical ML literature, the focus here is exclusively on **parallel update schemes** to address classification problems such as detecting malware and suspicious activity given data obtained by an IDPS. The approach introduced allows individual processing units to do simultaneous computations. This is in contrast to training SVMs either sequentially or in parallel first and then fusing them into a centralized classifier for intrusion detection. The distributed SVM classifier is especially useful when each unit has access to a different subset of the overall dataset, which may be time-varying, and has its own computing resources.

In practice, the communication overhead and the number of support vectors resulting from the relaxation of the original binary SVM classification problem can be unacceptably high, regardless of the update algorithm used. To address this issue, **active set methods** are utilized, which have been widely used in solving general quadratic problems as well as in centralized SVM formulations. The resulting algorithm is greedy in nature and has been observed to converge to a solution in a low number of rounds. In addition, the framework is suitable for online learning and gives a choice on the upper-bound of the resulting support vectors.

9.3.1 SVM classification and decomposition

First, an overview of the classical SVM-based binary classification problem is provided. Subsequently, a relaxation of the centralized formulation and its decomposition into parallel subproblems are discussed.

Centralized classification problem

Assume that a set of labeled data,

$$S := \{(x_1, y_1), \ldots, (x_N, y_N)\},$$

is given where $x_d \in \mathbb{R}^L$ and $y_d \in \{\pm 1\}$, $d = 1, \ldots, N$. A *classification problem* is considered with the objective of deriving a generalized rule from this *training data* that associates an input x with a label y as accurately as possible. It is important to note that no assumption is made on the nature of the training or test data. A well-known method for addressing this binary classification problem involves the representation of input vectors in a high(er)-dimensional feature space through a (nonlinear) transformation. Define the dot product of two feature vectors, say x_d and x_e, as a *kernel*, $k(x_d, x_e)$. In many cases this dot product can be computed without actually calculating the individual transformations, which is known as the *kernel trick*.

The optimal margin nonlinear binary **SVM classification** problem [162] described above is formalized using the quadratic programming problem

$$\max_{\alpha_d} \sum_{d=1}^{N} \alpha_d - \frac{1}{2} \sum_{d=1}^{N} \sum_{e=1}^{N} \alpha_d \alpha_e q_{d,e}$$

$$\text{such that } \alpha_d \geq 0, \quad d = 1, \ldots, N \tag{9.1}$$

$$\text{and } \sum_{d=1}^{N} \alpha_d y_d = 0, \tag{9.2}$$

where the α_d are the Lagrange multipliers of the corresponding *support vectors* (SVs) and $q_{d,e}$ are the entries of the positive definite matrix

$$Q := \left[(y_d y_e \, k(x_d, x_e))_{d,e} \right]_{N \times N}. \tag{9.3}$$

The positive definiteness of Q simply follows from the assumption that the kernel matrix $K_{d,e} := [k(x_d, x_e)]_{N \times N}$ is positive definite [162]. The decision function classifying an input x is then

$$f(x) = \text{sgn}\left(\sum_{l=1}^{N_D} \alpha_l y_l k(x, x_l) + b \right),$$

where b is the bias term and $N_D < N$ is the number of support vectors. Here, the well-known *representer theorem* [162] enables a finite solution to the infinite-dimensional optimization theorem in the span of N particular kernels, $k(x, x_d)$, centered on the training points x_d, $\forall d$.

Decomposition into subproblems

In order to decompose the centralized classification problem into **subproblems**, define a set $\mathcal{M} := \{1, 2, \ldots, M\}$ of separate *processing units* with access to different (possibly overlapping) subsets, S_i, $i \in \mathcal{M}$, of the labeled training data such that $S = \bigcup_{i=1}^{M} S_i$. The initial partition of the data S can be due to the nature of the specific problem at hand or as a result of a partitioning scheme at the preprocessing stage. Given the partition, define the vectors $\{\alpha^{(1)}, \alpha^{(2)}, \ldots, \alpha^{(M)}\}$ where the i-th one is $\alpha^{(i)} := \left[\alpha_1^{(i)}, \ldots, \alpha_{N_i}^{(i)}\right]$. In order to devise a distributed algorithm that solves the optimization problem (9.1), it is relaxed by substituting the constraint $\sum_{d=1}^{N} \alpha_d y_d = 0$ by a quadratic penalty function, $0.5M\beta \left(\sum_{d=1}^{N} \alpha_d y_d\right)^2$, where $\beta > 0$. Next, an upper-bound α_{max} on α_d is imposed such that $\alpha_d \leq \alpha_{max} \; \forall d$. This upper-bound can be chosen to derive a soft margin hyperplane (i.e. maximizing the margin). Alternatively, it can be chosen as very large to minimize the training error ignoring the margin. Thus, the following constrained optimization problem

$$\max_{\alpha_d \in [0, \alpha_{max}]} F(\alpha) = \sum_{d=1}^{N} \alpha_d - \frac{1}{2} \sum_{d=1}^{N} \sum_{e=1}^{N} \alpha_d \alpha_e q_{d,e} - \frac{M\beta}{2} \left(\sum_{l=1}^{N} \alpha_l y_l\right)^2 \tag{9.4}$$

approximates (9.1). Note that the objective function $F(\alpha)$ is strictly concave in all its arguments due to Q in (9.3) being positive definite, and the constraint set $X := [0, \alpha_{max}]^N$ is convex, compact, and nonempty.

The convex optimization problem (9.4) is next **partitioned** into M subproblems. Hence, the i-th unit's optimization problem is

$$\underset{\alpha_d \in [0, \alpha_{max}], \, d \in S_i}{\text{maximize}} \; F_i(\alpha) = \sum_{d \in S_i} \alpha_d - \frac{1}{2} \sum_{d \in S_i} \sum_{e=1}^{N} \alpha_d \alpha_e q_{d,e} - \frac{\beta}{2} \left(\sum_{l=1}^{N} \alpha_l y_l\right)^2. \tag{9.5}$$

Again, the individual objective functions $F_i(\alpha)$ are strictly concave in all $\alpha_d \in S_i$, and the respective constraint sets are convex, compact, and nonempty for all i.

Remark 9.1 *The individual optimization problems of the units are interdependent. Hence they cannot be solved individually without deployment of an information exchange scheme between the processing units.*

9.3.2   Parallel update algorithms

Continuous-time gradient algorithm

A distributed continuous-time algorithm, similar to that in reference [15], solves the problem (9.4). Clearly, solving all unit problems at the same time is equivalent to finding the solution of the relaxed problem (9.4). One possible way of achieving this objective is to utilize a gradient algorithm that converges to the unique maximum, α^*, of (9.4) which closely approximates that of the original optimization problem. Then, every unit implements the following **gradient algorithm** for each of its training samples

$$\frac{d\alpha_d}{dt} = \kappa_d \left[1 - \frac{\alpha_d q_{dd} - \sum_e \alpha_e q_{d,e}}{2} - \beta y_d \sum_{l=1}^{N} \alpha_l y_l \right]^P \tag{9.6}$$

for all $d \in S_i$ and all i where $\kappa_d > 0$ is a unit-specific step-size constant. Here, the shorthand $[\cdot]^P$ refers to the following projection of the $\dot{a}_d := d\alpha_d/dt$:

$$\dot{\alpha}_d = \begin{cases} \dot{\alpha}_d & \text{if } 0 < \alpha_d < \alpha_{max} \\ \dot{\alpha}_d & \text{if } \alpha_d = 0, \dot{\alpha}_d > 0 \text{ or } \alpha_d = \alpha_{max}, \dot{\alpha}_d < 0 \\ 0 & \text{otherwise} \end{cases}$$

Each unit i has access to only its own (possibly overlapping) **training data** S_i whereas the algorithm (9.6) requires more information to be shared between units. One possible solution to this communication problem is to define a *system node* facilitating the information flow between individual units by collecting and sending back all the necessary information from and to each unit, respectively. Every unit sends to the central node its own set of SVs $D^{(i)} = \left\{ (x_d, y_d, \alpha_d) | x_d, y_d \in S_i, \forall d \text{ s.t. } \alpha_d^{(i)} > 0 \right\}$. The system node aggregates this information $D := \bigcup_{i=1}^{M} D^{(i)}$ to obtain

$$x_D = \{x_d \in S | \alpha_d > 0\}$$
$$y_D = \{y_d \in S | \alpha_d > 0\}$$
$$\alpha_D = \{\alpha_d > 0\}.$$

Subsequently, the system node broadcasts the triple (x_D, y_D, α_D) back to all units $\{1, \dots, M\}$ after which each unit i locally computes the terms

$$t^{(i)} = Q_{N_i \times N_D}^{(i)} \alpha_D$$
$$u = \alpha_D^T y_D, \tag{9.7}$$

where $[\cdot]^T$ denotes transpose operation, N_D is the number of elements of x_D, and $Q_{N_i \times N}^{(i)}$ is the portion of the matrix Q relevant to the unit. Thus, the **unit update algorithm** (9.6) is redefined as

$$\dot{\alpha}_d = \kappa \left[1 - \frac{1}{2} \alpha_d q_{dd} - \frac{1}{2} t_d^{(i)} - \beta y_d u \right]^P, \forall d \in S_i, \forall i. \tag{9.8}$$

Theorem 9.2 *The distributed scheme (9.6) globally asymptotically converges to the unique maximum of the centralized problem (9.4).*

Proof In order to investigate the convergence properties of the unit algorithms (9.6), first define a Lyapunov function V on the compact and convex constraint set $X = [0, \alpha_{max}]^N$:

$$V(\alpha) = \sum_{d=1}^{N} \alpha_d - \frac{1}{2} \sum_{d=1}^{N} \sum_{e=1}^{N} \alpha_d \alpha_e q_{d,e} - \frac{\beta}{2} \left(\sum_{l=1}^{N} \alpha_l y_l \right)^2. \tag{9.9}$$

This function is strictly concave and admits its global maximum at the solution of (9.4). Furthermore, its unique maximum, $\alpha^* \in X$, coincides with that of the centralized problem (9.4).

Taking the derivative of the **Lyapunov function** with respect to α_d and time t, respectively, yields

$$\frac{\partial V(\alpha)}{\partial \alpha_d} = 1 - \alpha_d q_{dd} - \frac{1}{2} \sum_{e \neq d} \alpha_e q_{d,e} - \beta y_d \sum_{l=1}^{N} \alpha_l y_l$$

and

$$\frac{dV(\alpha)}{dt} = \sum_{d=1}^{N} \frac{\partial V(\alpha)}{\partial \alpha_d} (\dot{\alpha}_d).$$

If $\alpha_{max} > \alpha_d > 0$ for some d, then

$$\frac{\partial V(\alpha)}{\partial \alpha_d} \dot{\alpha}_d = \kappa_d \left[1 - \alpha_d q_{dd} - \frac{1}{2} \sum_{e \neq d} \alpha_e q_{d,e} - \beta y_d \sum_{l=1}^{N} \alpha_l y_l \right]^2 > 0.$$

In the case of $\alpha_d = 0$ for a data point d, either $\dot{\alpha}_d > 0$ and $(\partial V / \partial \alpha_d) \dot{\alpha}_d > 0$ as above or $\dot{\alpha}_d = 0$ leading to $(\partial V / \partial \alpha_d) \dot{\alpha}_d = 0$. The cases where $\alpha_d = \alpha_{max}$ (i.e. upper-bound) are handled similarly.

If the **trajectory** hits the boundary (i.e. $a_d = 0$ or $\alpha_d = \alpha_{max}$ for some d) at a point other than the unique maximum $\alpha^* \in X$, then $dV(\alpha)/dt > 0$ until the trajectory reaches α^*, where $[dV(\alpha)/dt]^P = 0$. Assume otherwise, i.e. that there exists a point $\tilde{\alpha}$ on the boundary of the set X and $\tilde{\alpha} \neq \alpha^*$. Then, the projection of the gradient $[dV/dt]^P$ is zero at the point $\tilde{\alpha} \in X$. However, due to the strict concavity of V, there exists at least one term d such that $(\partial V / \partial \tilde{\alpha}_d) \dot{\tilde{\alpha}}_d > 0$ resulting in $[dV/dt]^P > 0$ which leads to a contradiction. Thus, the system convergences globally asymptotically to the unique maximum α^*.   □

Discrete-time gradient projection algorithm

We discuss a discrete-time counterpart of the parallel update scheme (9.6). In this case, every processing unit implements the following **discrete-time gradient projection** algorithm for each of its training samples

$$\alpha_d(n+1) = [\alpha_d(n) + \kappa_d G_d(\alpha)]^+ \ \forall d, \tag{9.10}$$

where

$$G_d(\alpha(n)) := 1 - \frac{1}{2} \left(\alpha_d(n) q_{dd} - \sum_{e=1}^{N} \alpha_e(n) q_{d,e} \right) - \beta y_d \sum_{l=1}^{N} \alpha_l(n) y_l.$$

Here, n denotes the update instances and the notation $[\cdot]^+$ represents the orthogonal projection of a vector onto the convex set X defined by $[\alpha]^+ := \arg\min_{z \in X} \|z - \alpha\|_2$, where $\|\cdot\|_2$ is the Euclidean norm. In this special case, the projection of α_d onto X can be computed by mapping α_d onto $[0, \alpha_{max}]$ for each d. This facilitates an easy implementation of the parallel algorithm.

The function $F(\alpha)$ in (9.4) is a polynomial and hence continuously differentiable in its arguments. Furthermore, there exists a scalar constant C such that

$$\|\nabla F(\gamma) - \nabla F(\delta)\|_2 \le C \|\gamma - \delta\|_2, \ \forall \gamma, \delta \in X,$$

where ∇F is the gradient operator. Define $z := \gamma - \delta$. Then,

$$\|\nabla F(\gamma) - \nabla F(\delta)\|_2^2 = z^T A^T A z,$$

where

$$A := \frac{1}{2}\text{diag}(Q) + \frac{1}{2}Q + \beta y y^T. \tag{9.11}$$

The matrix $\text{diag}(Q)$ contains the diagonal elements of Q with all its off-diagonal elements set to zero. The scalar constant C is then given by

$$C = \sqrt{\max \lambda(A^T A)},$$

where $\max \lambda(\cdot)$ is the maximum eigenvalue. Therefore, the gradient of the objective function $F(\alpha)$, ∇F, is Lipschitz continuous. Moreover, it is bounded from above on X.

Theorem 9.3 *The gradient projection algorithm (9.10) converges to the unique maximum, α^*, of the objective function F in (9.4), if the step-size constant κ_d satisfies*

$$0 < \kappa_d < \frac{2}{\sqrt{\max \lambda(A^T A)}}, \ \forall d.$$

Proof The proof follows directly from the upper-boundedness and strict concavity of the function F and Lipschitz continuity of its gradient ∇F. Propositions 3.3 and 3.4 in reference [38, p. 213] contain the details. □

Theorem 9.4 *Assume that the conditions in Theorem 9.3 hold. Then, the gradient projection algorithm (9.10) converges to the unique maximum α^* of (9.4) geometrically.*

Proof This result follows directly from Proposition 3.5 of [38, p. 215], if there exists a constant $c > 0$ such that

$$(\nabla F(\gamma) - \nabla F(\delta))^T (\gamma - \delta) \ge c \|\gamma - \delta\|_2^2, \ \forall \gamma, \delta \in X.$$

Let us again define $z := \gamma - \delta$. Then,

$$(\nabla F(\gamma) - \nabla F(\delta))^T (\gamma - \delta) = z^T A^T z$$
$$\ge z^T \max \lambda(A^T) z,$$

$\forall \gamma, \delta \in X$, where the matrix A is defined in (9.11). We note that the matrix A is the sum of two positive definite matrices, $\text{diag}(Q)$ and Q, and a positive semidefinite one, $y y^T$. It is hence positive definite and all of its eigenvalues are positive. Hence, there exists a positive constant c which satisfies the sufficient condition for the theorem to hold. □

Asynchronous update algorithm

A natural generalization of the parallel update algorithms presented is the asynchronous update scheme where only a random subset of processing units update their α values at a given time instance. Notice that the synchronous update scheme can be thought of as a limiting case of this more general version where all units participate in a random subset update. Define the set of units that update at a given instance n as $\mathcal{M}_u^{(n)}$ and the rest as $\mathcal{M}_{no}^{(n)}$, such that $\mathcal{M}_u^{(n)} \cup \mathcal{M}_{no}^{(n)} = \mathcal{M} \; \forall n$. Then, the update algorithm for the i-th unit is:

$$\alpha_d(n+1) = \begin{cases} [\alpha_d(n) + \kappa_d G_d(\alpha)]^+, \forall d \in S_i, \text{ if } i \in \mathcal{M}_u^{(n)} \\ \alpha_d(n), \forall d \in S_i, \text{ if } i \in \mathcal{M}_{no}^{(n)}, \end{cases} \tag{9.12}$$

where $G_d(\alpha)$ is defined in (9.10).

Asynchronous update schemes are in fact more relevant in practical implementations, since it is usually difficult for the units to synchronize their exact update instances. Two well-known conditions, which are given below, are together sufficient for the asynchronous convergence of a nonlinear iterative mapping $\mathbf{x}(n+1) = T(\mathbf{x})$ [38, p. 431].

Definition 9.5 (*Synchronous convergence condition*) *For a sequence of nonempty sets* $\{X(k)\}$ *with* $\ldots \subset X(k+1) \subset X(k) \subset \ldots X$, *we have* $T(\mathbf{x}) \in X(k+1)$, $\forall k$, *and* $\mathbf{x} \in X(k)$. *Furthermore, if* $\{y^k\}$ *is a sequence such that* $y^k \in X(k)$ *for every* k, *then every limit point of* $\{y^k\}$ *is a fixed point of* T.

Definition 9.6 (*Box condition*) *Given a closed and bounded set* X *in* $\mathbb{R}$, *for every* k, *there exist sets* $X_i(k) \subset X$ *such that*

$$X(k) := X_1(k) \times X_2(k) \times \cdots \times X_M(k).$$

For the gradient projection algorithm (9.10), it is straightforward to apply the results of the convergence analysis above to the asynchronous update case. Towards this end, define the sequence of nonempty, convex, and compact sets

$$X(k) := X_1 \times X_2 \times \cdots X_M,$$

where $X_i := [x_i^* - \delta(k), x_i^* + \delta(k)] \; \forall i$ and $\delta(k) := \|\alpha(k) - \alpha^*\|$. Since $\delta(k+1) < \delta(k)$ by Theorem 9.4, we obtain

$$\cdots \subset X(k+1) \subset X(k) \subset \cdots X.$$

Here, X is defined as the interval $[0, \alpha_{max}]$ where a sufficiently large fixed α_{max} is chosen without any loss of generality due to the existence of a finite equilibrium point and geometric convergence. Hence, the box condition is satisfied by the definition of $X(k)$. Since $\delta(k)$ is monotonically decreasing in k by Theorem 9.4, the synchronous convergence condition also holds. Therefore, the next convergence result for the asynchronous counterpart of the parallel update algorithm (9.10) immediately follows from the asynchronous convergence theorem [38, p. 431]:

Theorem 9.7 *Assume that the conditions in Theorem 9.3 hold. If a random subset of the units update their α values at each iteration according to (9.10) while others keep theirs fixed, then the resulting (totally) asynchronous update algorithm defined in (9.12) converges to the unique maximum α^* of (9.4).*

Stochastic update algorithm

All of the update schemes described heretofore require an information exchange system (see Remark 9.1) to function properly. Also, the information flow within the system has been assumed to be perfect, i.e. the units have access to all the parameters needed by the update algorithms. However, this may not be the case in practice for a variety of reasons such as **communication errors**, i.e. if the units are not collocated or approximations are imposed in order to reduce the communication load between the units. To accommodate this, consider the update algorithm

$$\alpha(n+1) = [\alpha(n) + \kappa s(n)]^+. \tag{9.13}$$

Define $s(n) := G(\alpha, n) + \beta(n)$, where $G(\alpha, n)$ is defined componentwise in (9.10). Here, $\beta(n)$ is the random *distortion* at time instance n. It is possible to establish convergence of the algorithm (9.13) which can be interpreted as a parallel update scheme under **stochastic distortions**, if these random perturbations satisfy some conditions. Toward this end, let us characterize the relationship between the distortion term $\beta(n)$ and the real gradient $G(\alpha, n)$ through the variables $\rho(n) > 0$ and $-\pi \leq \theta(n) \leq \pi$:

$$|\beta(n)| = \rho(n)|G(\alpha, n)|, \quad \cos(\theta(n)) = \frac{\beta^T(n)G(\alpha, n)}{|\beta(n)||G(\alpha, n)|}. \tag{9.14}$$

Theorem 9.8 *Let the stochastic distortion $\beta(n)$ defined in (9.14) through parameters $\rho(n) > 0$ and $-\pi \leq \theta(n) \leq \pi$ satisfy:*

$$\rho^2(n) + 2\rho(n)\cos(\theta(n)) + 1 > 0, \forall n,$$

and

$$\frac{1 + \rho(n)\cos(\theta(n))}{(1 + \rho(n))^2} \geq \bar{K} > 0, \forall n,$$

where $\bar{K}$ is a positive real number. Then, the update algorithm (9.13) converges to the unique maximum, α^*, of the objective function F in (9.4) geometrically, if the elements of the step-size vector, κ_d, satisfy

$$0 < \kappa_d < \frac{2\bar{K}}{\sqrt{\max \lambda(A^T A)}}, \forall d.$$

Proof The proof makes use of the upper-boundedness and strict concavity of the function F and Lipschitz continuity of its gradient $G(\alpha) = \nabla F$ as in the proof of Theorem 9.3. Modify the function F, without any loss of generality, such that it is bounded above by zero. It is straightforward to show that the conditions in the theorem on the distortion, $\beta(n)$, imposed through parameters $\rho(n)$ and $\theta(n)$ ensure that

$$\|s(n)\|_2 \geq k\|G(\alpha,n)\| \ \forall n,$$

where $k > 0$ is a positive constant and

$$s(n)^T G(\alpha,n) \geq \bar{K} \|s(n)\|_2^2 \ \forall n.$$

These conditions, together with that on the positive step-size constant vector k, ensure that the update algorithm (9.13) converges to the unique maximum of the objective function F, which follows immediately from Propositions 2.1 and 2.3 in reference [38, pp. 204–206]. Furthermore, since the objective function is strictly concave, the rate of convergence is geometrical (see Proposition 2.4 in reference [38] for details).   □

The conditions in Theorem 9.8 are now investigated through a couple of **illustrative examples**.

Example 1

Let $-\pi/2 < \theta(n) < \pi/2 \ \forall n$. The theorem holds for any value of ρ as long as the condition on the step-size κ is satisfied, where $\bar{K} \leq (1 + \rho(n)\cos(\theta(n)))/(1 + \rho(n))^2 \ \forall n$. This means in practice that if the angle between the gradient $G(\alpha)$ and distortion β vectors remains less than $90°$, then it is possible to choose a sufficiently small step-size to compensate for the errors and ensure convergence.

Example 2

Let $|\theta(n)| > \pi/2 \ \forall n$. If $\rho(n) < 1$, then the step-size can be chosen sufficiently small and the conditions in the theorem can be satisfied. In this case, the distortion is in the almost opposite direction of the correct gradient, and hence its norm with respect to the gradient needs to be bounded.

9.3.3   Active set method and a numerical example

Although the update algorithms presented in Section 9.3.2 provably converge to the unique solution of (9.4), and hence approximately solve the original binary classification problem (9.1), they often result in a large number of support vectors. This is undesirable not only for efficiency reasons but also due to the communication overhead it brings to the system. To address this problem, *active set* methods are proposed, which have been widely used in solving general quadratic problems. Furthermore, they have also been applied to SVMs in centralized classification formulations [122, 158, 187].

Active set algorithms solve the problem iteratively by changing the set of active constraints, i.e. the inequality constraints that are fulfilled with equality, starting with an initial active set $\mathcal{A}_0$. Since the initial set is in most cases not the correct one, it is modified iteratively by adding and removing constraints according to some criteria and testing if the solution remains feasible [187]. Active set methods are considered to be more robust and better suited for warm starts, i.e. when there is some prior knowledge on the initial set [94]. For the relaxed problem (9.4), there are only non-negativity constraints on α. In this case, the vectors with $\alpha > 0$ constitute the set of SVs. Hence, the set of SVs is the complement of the active set and both sets are mutually exclusive. Furthermore, the union of both sets gives the (universal) set of all feature vectors.

A combinatorial trial-and-error approach is infeasible in practice except from very small constraint sets. Several schemes have been proposed in the literature for **iteratively updating the active set**. However, popular approaches such as *line search* [158] are not suitable for distributed implementation. The selection of active constraint sets has also been studied in the context of the simplex method [106]. In order to minimize the communication overhead and simplify implementations, a greedy approach has been proposed [11] for updating the active set, similar to those in the literature. For many problems, such heuristic methods are known to perform well [162].

The **greedy algorithm** adopted for updating the active set is summarized as follows. At each iteration, the data point with the highest positive gradient (derivative of the objective function F in (9.4)) is removed from the active set as an SV candidate. Next, the problem described in Section 9.3.2 is solved through parallel updates under the resulting active set constraints. Finally, the SV with the lowest α is added to the active set which is a rough approximation to finding the constraint that is violated by the solution.

The last step in the algorithm both speeds up the convergence and ensures a **user-defined upper-bound** on the number of SVs, thus turning it into a tunable parameter of the scheme. The total number of active set updates (iterations) in the algorithm should be at most as many as the desired number of SVs in general unless specific assumptions are made on the initial active set. Due to its iterative nature, this algorithm is also very suitable for online learning where training data is dynamic. Clearly, there is no need to rerun the whole algorithm in order to incorporate new data into the decision function each step.

Communication overhead

The communication overhead of the active set-based approach is significantly lower than the plain algorithm presented in Section 9.3.2. During parallel updates at each active set stage (inner loop), individual nodes exchange only the scalar α parameter values. They send an SV to the system node only when there is a change in the active set. Naturally, the amount of information sent by each unit would have been excessive if there were no active set restriction and all the corresponding data points were transmitted. Therefore, limiting the number of SVs through the active set decreases the communication load significantly. The information flow within the distributed system results, thus, in an efficient communication scheme.

Numerical example

The framework presented is analyzed on a synthetic but complex two-dimensional[4] dataset and compared with the centralized solution. First, the problem is solved centrally using a standard radial-basis-function (RBF) kernel and quadratic programming tools

---

[4] This two-dimensional dataset [76] is selected to facilitate visualization, and not because of any restriction of the algorithm on data dimension.

provided by Matlab. Next, the results of the distributed algorithm are obtained with an imposed upper-bound on the number of support vectors of 40, respectively. The starting point of the algorithm is random and perfect information flow is assumed in the system. The SVs of the centralized classifier and an example solution of the distributed one are depicted in Figure 9.2. It is observed that the distributed algorithm performs almost as well as the centralized one. The increased number of SVs needed for classification and

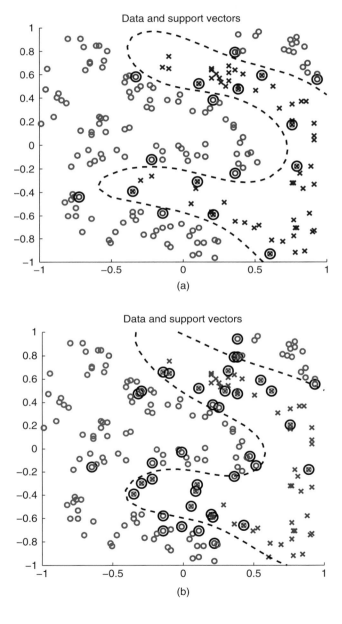

Figure 9.2 Comparison of (a) a centralized classifier with (b) the distributed one.

slight performance degradation in the distributed case are attributed to the approximate nature of the algorithm and the specific penalty function used in the relaxation of the problem. However, the presented framework, especially the active-set algorithm, has a lot of room for improvement in terms of, for example, preprocessing, better search heuristics, and post clustering of obtained SVs.

9.3.4 Behavioral malware detection for mobile devices

Widespread use and general-purpose computing capabilities of next generation smartphones make them the next big targets of malware and security attacks. Given the battery, computing power, and bandwidth limitations inherent in mobile devices, detection of malware on them is a nontrivial research challenge that requires a different approach from those used for desktop or laptop computing. The distributed machine learning and SVM classification framework presented has many suitable characteristics and can be applied to this problem as an alternative or additional defense method.

Mobile device usage patterns such as the number of SMSs sent and call durations can be exploited to collaboratively derive flexible, personalized, and behavioral signatures of malware. For example, a security laboratory can provide the malware behavior data while the participating users join the system with their normal usage data. Once a binary classification function (either specific to malware or for aggregate anomaly detection) is collectively trained, it is used to detect malware and other attacks. The distributed learning approach adopted provides multiple advantages:

1. It does not require the mobiles to send all of their behavior data to a security center. Hence, it is lightweight in terms of bandwidth usage.
2. Due to the abstract nature of feature vectors and not requiring a central repository for all of the user data, it preserves the privacy of the participating users.
3. The classification function learned is essentially an automatically generated behavioral (malware) signature that takes into account the usage patterns of ordinary users.

Given its favorable properties, this scheme provides a promising and low-overhead defensive layer for mobile devices, possibly alongside existing approaches.

Simulations and example implementation

A proof-of-concept illustrative prototype (Figure 9.3) is developed on smartphones with Symbian S60 operating system that communicates to the main server over HTTP. Both the server and clients use Python programming language [180] for ease of implementation. In addition, multiple clients are faithfully emulated on a single computer as part of more extensive simulations thanks to the interpreted nature of the Python language. Hence, the experimental infrastructure requirements are kept at minimum.

A dataset from the MIT reality mining project [60] is used for the simulations. It consists of phone call, SMS, and data communication logs collected via a special application during normal daily usage of volunteers. This usage data is preprocessed to generate histograms with a set of 20 features (see Table 9.2) over 6 h periods. Here,

Table 9.2 Histogram features

Feature (in numbers per 6 h intervals)
Short-duration calls (less than 2 min)
Medium-duration calls (between 2 and 6 min)
Long-duration calls (more than 6 min)
Short intervals between calls (less than 1 h)
Medium-length intervals between calls (between 1 and 3 h)
Long-length intervals between calls (more than 3 h)
Outgoing SMS
Short periods between outgoing SMS
Medium periods between outgoing SMS
Long periods between outgoing SMS
Incoming SMS
Short periods between incoming SMS
Medium periods between incoming SMS
Long periods between incoming SMS
Short-duration packet-sending activities
Medium-duration sending activities
Long-duration sending activities
Short periods between sending activities
Medium periods between sending activities
Long periods between sending activities

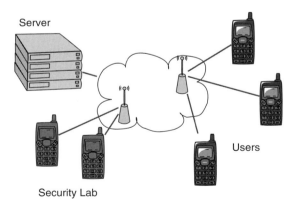

Figure 9.3 Illustration of the distributed malware detection system for smartphones.

the short periods or intervals refer to less than 1 h, medium ones to between 1 and 3 h, and long ones to more than 3 h, respectively. Short-call duration refers to less than 2 min, medium one to between 2 and 6 min, and long one to more than 6 min.

The selected example feature set is clearly statistical in nature, and hence, privacy preserving. The system focuses on aggregate and high-level usage characteristics rather than private data such as lists of people, time, and content. Thus, it is possible to run the algorithm on the server side without intruding into the privacy of the mobile user.

In one experiment, usage data from a subject over 50 days (200 samples) constitutes the training set. A malware is artificially injected into a test set of 25 days. The simulated malware behavior is based on well-known *Viver*[5] and *Beselo*[6] Trojans. These Trojans (malware) exhibit themselves by excessive SMS usage, e.g. sending out up to 20 SMSs in a short duration of time. The results obtained are promising and indicate high accuracy rates typically on the order of 90 percent or more.

9.4 Discussion and further reading

A comprehensive overview of intrusion detection and prevention systems can be found in references [25, 26]. The thesis, reference [191], presents use of internal and virtual sensors for intrusion detection. The base-rate fallacy is defined and discussed in detail in reference [24].

References [40, 162] provide an excellent introduction to machine learning and the kernel methods mentioned in Section 9.2. A recent book focusing on applications of machine learning to security is reference [103].

The distributed machine learning framework in Section 9.3 is based upon references [11, 12]. It has been recently applied to behavioral malware detection for mobile devices (smartphones) as discussed in Section 9.3.4. Another recent application of machine learning (probabilistic diffusion) to malware detection on mobile devices is reported in reference [13], which also contains an overview of the related literature in this area.

---

[5] http://www.f-secure.com/v-descs/trojan_symbos_viver_a.shtml
[6] http://www.f-secure.com/v-descs/worm_symbos_beselo.shtml

10   Hypothesis testing for attack detection

Chapter overview

Chapter summary

Hypothesis testing (signal detection) has been increasingly used in network security to detect attacks, anomalies, and malicious behavior. An overview of various approaches to multiple hypothesis testing is provided, among which are Bayesian, minimax, Neyman–Pearson, sequential, composite, nonparametric, and robust hypothesis testing. Decentralized detection of hypotheses, where observations (measurements) of different sensors are correlated, is discussed under Bayesian and Neyman–Pearson criteria. In addition, a connection with the majority vote is established for some special cases, and an algorithm for the computation of thresholds in likelihood ratio tests is introduced.

10.1   Hypothesis testing and network security

The previous chapter has discussed applications of machine learning in network security, where statistical techniques, among others, are employed to detect and learn the behaviors of friends and foes in a network. This chapter focuses on another statistical method, used more and more widely in network security: that based on hypothesis testing (also called signal detection). In addition to providing an overview of the subject, the chapter will discuss the decentralized detection problem where the observations from different sensors are correlated, a scenario that is prevalent in network intrusion detection problems.

Recently, there has been a surge of research activity in applying hypothesis testing (HT) to network attack and anomaly detection, which is a task often delegated to IDSs as discussed in Section 9.1. Intrusion and attack detection approaches are normally classified into two categories: anomaly detection and misuse detection. In anomaly detection, the IDS characterizes the correct and/or acceptable behavior of the system to detect wrongful behavior. Misuse detection, on the contrary, uses known patterns of attack, called *signatures*, to detect intrusions.

Security approaches using HT can be considered to lie somewhere in between these two categories. As recognized earlier, in today's networks both regular nodes and misbehaving or attacker nodes are much more powerful and smarter than their predecessors. In some situations, canonical approaches based on fixed patterns (of acceptable or intrusive behaviors) may not be effective anymore. Rather, HT-based attack detection differentiates behaviors using the probabilistic models of their observed parameters. In a loose sense, the signatures are now the probability densities of these parameterized behaviors.

The most widely used approaches in hypothesis testing are Bayesian, minimax, and Neyman–Pearson. In Bayesian HT, given the prior densities of the hypotheses and based on the measurements made,[1] the most likely hypothesis is selected using a decision rule obtained through minimization of the average cost or risk. In practical situations where the prior densities are unknown, an alternative design criterion is the minimax approach, where the maximum of all the conditional costs given each hypothesis is minimized. The third criterion, Neyman–Pearson, is used in situations where a cost structure is not available or is not desirable. In this approach, the miss probability is minimized given an upper bound on the false alarm probability. All of these approaches will be introduced in more detail in Section 10.2.

Results from centralized HT do not readily carry over to the decentralized setting, particularly if the measurements at the sensors given each hypothesis are not conditionally independent. However, this scenario is one that is often encountered in network security problems. The existence of optimal solutions for both Bayesian and Neyman–Pearson formulations of the decentralized HT problem, where the sensor observations are conditionally correlated, is established in Section 10.3. Subsequently, Section 10.4 investigates some relationships between the majority vote and the likelihood ratio test

---

[1] We use the terms measurement and observation interchangeably in this chapter.

for a parallel configuration. Building on the theoretical results obtained, Section 10.5 presents an algorithm for computation of the optimal thresholds, provided that the sensors are restricted to use likelihood ratio tests.

10.2   An overview of hypothesis testing

Signal detection or hypothesis testing deals with problems where the goal is to make an accurate decision on the actual hypothesis based on a single measurement or a set of measurements that involve the hypothesis. Its earlier application area has been radar systems, and more recently it has found applications in sensor networks and network security.

Specifically, for a single sensor problem, the sensor is to decide which hypothesis is true among M hypotheses, $H_0, H_1 \ldots, H_{M-1}$. The observation is a random variable Y that take values in a finite or an infinite observation set $\mathcal{Y}$. The conditional density[2] of Y given hypothesis H_i, denoted by $P_i(Y)$, is assumed to be known to the sensor for each hypothesis $i = 0, \ldots, M-1$. A typical example of hypothesis testing in communications is the problem of modulation/demodulation, which entails *binary hypothesis testing*, where $M = 2$. Suppose that the transmitter employs a simple antipodal modulation scheme, where zero and one are modulated as pulses of amplitude -1 volt and $+1$ volt, respectively, where zero stands for hypothesis H_0 and one stands for hypothesis H_1. Suppose further that the signal goes through a channel with zero-mean additive Gaussian noise (AGN) that has noise with variance σ^2. The conditional probability density functions (pdfs) of the observation given each of the two hypotheses, $P_0(Y)$ and $P_1(Y)$, will be Gaussian with variance σ^2 and means -1 and $+1$, respectively. Given an observed value of Y, the receiver has to decide whether the transmitter had sent zero or one. If at the transmitter zero and one are equally likely, it is well known (and totally intuitive) that the decision rule that minimizes the probability of error will be symmetric, and given by:

$$\delta(y) = \begin{cases} 1 & \text{if } y \geq 0 \\ 0 & \text{otherwise}. \end{cases} \tag{10.1}$$

This actually is just a special case of the standard *likelihood ratio test* (LRT), where one picks H_1 if the likelihood ratio $P_1(Y)/P_0(Y)$ exceeds 1, and H_0 otherwise. In this case we also have $P_1(Y)/P_0(Y)$ a monotone increasing function of Y, and hence equivalently $Y = y$ is tested against a threshold, which in this case is 0. It is known that in general the test will be a threshold test based on the likelihood ratio for all three criteria – Bayesian, minimax, and Neyman–Pearson criteria, as discussed below. In the context of network security, hypothesis testing models can be readily applied to detect the state of a network or a node (normal or abnormal), the type of a connection (regular connection or some kind of attack), or a misbehaving node.

---

[2] We will occasionally use "density" to mean *probability density function (pdf)* or *probability mass function (pmf)* depending on whether the random variables are continuous or discrete.

10.2.1 Bayesian hypothesis testing

For Bayesian hypothesis testing, it is assumed that the a priori probabilities of the M hypotheses, $\pi_0, \pi_1, \ldots, \pi_{M-1}$, are known. A cost matrix $\{C_{ij}\}$ is defined, where C_{ij} is the cost of deciding H_i when actually H_j is the true hypothesis (e.g. false alarm or missed attack). The sensor, in the process of arriving at a decision, is to minimize the expected cost (also called the Bayes risk) $\mathbb{E}[C_{ij}]$, which can be written as

$$r(\delta) = \sum_{j=0}^{M-1} \pi_j R_j(\delta), \tag{10.2}$$

where $R_j(\delta)$ is the conditional risk for decision rule δ given H_j, that is

$$R_j(\delta) = \sum_{i=0}^{M-1} C_{ij} P_j(\delta(y) = i), \tag{10.3}$$

where $P_j(\delta(y) = i)$ is the probability of a particular decision rule δ ending up ruling H_i when the actual hypothesis is H_j. For the case of binary hypothesis testing (that is, when $M = 2$), the Bayes decision rule, which minimizes the Bayes risk, is given by

$$\delta(y) = \begin{cases} 1, & \text{if } \dfrac{P_1(y)}{P_0(y)} \geq \dfrac{\pi_0(C_{10} - C_{00})}{\pi_1(C_{01} - C_{11})} \\ 0, & \text{otherwise}, \end{cases} \tag{10.4}$$

where we use the indices of the hypotheses (0, 1) to indicate the hypotheses (H_0, H_1). When the cost parameters are uniform, i.e. $C_{10} = C_{01} = 1$ and $C_{00} = C_{11} = 0$, the Bayes risk becomes the average probability of error, P_e. When there are more than two hypotheses, the Bayesian decision rule is also a threshold rule based on likelihood ratios, but now involving not two but M partitions of the space $\mathcal{Y}$ where y takes values in. For more details, the interested reader is referred to references [138, 181].

10.2.2 Minimax hypothesis testing

In practice, the prior probabilities on the hypotheses may not be available. One approach to overcome this shortcoming is to use Neyman–Pearson detection, which will be introduced in Section 10.2.3. Another approach is *minimax decision making*. It entails finding a decision rule that minimizes the maximum of the conditional risk $R_i(\delta)$, with maximization taken over the hypotheses $H_0, \ldots, H_{M-1}$, or equivalently over the indices $i = 0, \ldots, M - 1$. Note that what we have in this case is a zero-sum game (see Appendix A.2), where the decision maker is the minimizer and the "adversary" who picks the hypothesis is the maximizer. In what follows, we focus on binary hypothesis testing. Let $r(\pi_0, \delta)$ denote the Bayes risk when the decision rule δ is used and the prior probabilities of H_0 and H_1 are π_0 and $\pi_1 = 1 - \pi_0$, respectively. Then, the problem the decision maker has to solve can be stated as

$$\min_{\delta} \max_{0 \leq \pi_0 \leq 1} r(\pi_0, \delta). \tag{10.5}$$

This gives advantage to the "adversary," who picks the maximizing π_0 after the minimizer picks δ, and hence is known as the *upper value* of the game. An alternative (for the game) would be first for the adversary to pick π_0 and then the decision maker to pick δ which would give an advantage to the decision maker, leading to what is known as the *lower value* of the game:

$$\max_{0 \leq \pi_0 \leq 1} \min_{\delta} r(\pi_0, \delta), \tag{10.6}$$

where the adversary maximizes the minimum risk obtained by the decision maker.

The saddle point, if it exists, is a pair (π^*, δ^*) that satisfies

$$r(\pi_0, \delta^*) \leq r(\pi^*, \delta^*) \leq (\pi^*, \delta), \quad \forall \pi_0 \in [0,1], \forall \delta, \tag{10.7}$$

in which case the upper and lower values are equal.

Now coming back to the upper value (10.2), note that for each δ, the Bayes risk $r(\pi_0, \delta)$ is affine in π_0 (see (10.2) along with(10.3)), say

$$r(\pi_0, \delta) = a_\delta \pi_0 + b_\delta,$$

where a_δ and b_δ are parameters that depend on δ.[3] If $a_\delta > 0$, then the maximizing π_0 will be $\pi_0 = 1$, whereas if $a_\delta < 0$, the unique choice will be $\pi_0 = 0$. If $a_\delta = 0$, however, $r(\pi_0, \delta)$ would be independent of π_0, and hence any choice out of the interval $[0,1]$ would be a maximizing one for the adversary.

Now looking at the lower value (10.6), the minimizing decision rule will depend on the choices of π_0, denoted as δ_{π_0}, and this is the Bayes decision rule since it minimizes the Bayes risk $r(\pi_0, \delta)$. If we have a π_0 against which a particular δ is Bayes, and if these choices are consistent with the conclusion arrived at in the discussion of the upper value, then we have a saddle point as in (10.7), where π_0^* is known as the *least favorable prior* for hypothesis H_0. This discussion also covers the possibility that δ^* could be a *randomized* policy.

To explore the path outlined in the two preceding paragraphs, let $V(\pi_0) := r(\pi_0, \delta_{\pi_0})$, where δ_{π_0} is the Bayes rule associated with the prior π_0. Thus, $V(\pi_0)$ is the minimum Bayes risk for the prior π_0. It can be shown that $V(\pi_0)$ is a continuous concave function of π_0, with $\pi_0 \in [0,1]$. From references [138] and [120], the minimax decision rule is the Bayes rule at the so-called *least-favorable prior* π_L, where π_L could be either 0, or 1, or an interior point that maximizes $V(\pi_0)$. As shown in Figure 10.1, the line $r(\pi_0, \delta_{\pi_L})$ is tangent to the curve $V(\pi_0)$ at the point $(\pi_0, V(\pi_0))$ if $V(\pi_0)$ is differentiable at this point. If $V(\pi_0)$ is not differentiable at π_L (which, in this case, has to be an interior point), the decision will be randomized at π_L. Let $\tilde{\delta}(y)$ denote the probability of saying H_1 given that the observation $Y = y$, that is $\tilde{\delta}(y) \equiv Pr(\delta(y) = 1|y)$. Then, we have the minimax decision rule as follows

---

[3] The precise expressions for these parameters are: $a_\delta = R_0(\delta) - R_1(\delta)$ and $b_\delta = R_1(\delta)$.

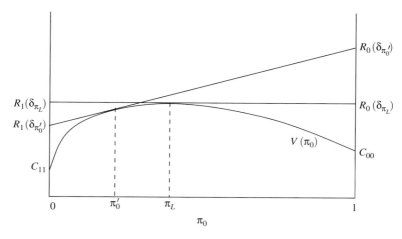

Figure 10.1 Minimax decision rule when π_L is an interior point and $V(\pi_0)$ is differentiable at π_L. (Figure adapted from reference [138].)

$$
\tilde{\delta}(y) = \begin{cases} 1, & \text{if } \dfrac{P_1(y)}{P_0(y)} > \tau \\[2mm] \beta, & \text{if } \dfrac{P_1(y)}{P_0(y)} = \tau \\[2mm] 0, & \text{if } \dfrac{P_1(y)}{P_0(y)} < \tau, \end{cases} \tag{10.8}
$$

where

$$
\tau = \frac{\pi_L(C_{10} - C_{00})}{(1 - \pi_L)(C_{01} - C_{11})},
$$

$$
\beta = \frac{V'\left(\pi_L^+\right)}{V'\left(\pi_L^+\right) - V'\left(\pi_L^-\right)}.
$$

Here, $V'\left(\pi_L^+\right)$ and $V'\left(\pi_L^-\right)$ are respectively the left-hand and right-hand derivatives of $V(\pi_0)$ at π_L. It should, however, be noted that if $Pr(P_1(y)/P_0(y) = \tau) = 0$ under H_0 and H_1, the value β becomes immaterial. The line $r(\pi_0, \delta_{\pi_L})$ of the minimax decision rule in this case is illustrated in Figure 10.2.

10.2.3 Neyman–Pearson hypothesis testing

The Neyman–Pearson problem is one of maximizing the detection probability (P_D) (or minimizing the miss probability $P_M = 1 - P_D$,) subject to an upper bound on the false alarm probability (P_F). Note that

$$
P_F(\delta) \equiv P_0\left(\delta(y) = 1\right), \tag{10.9}
$$
$$
P_D(\delta) \equiv P_1\left(\delta(y) = 1\right). \tag{10.10}
$$

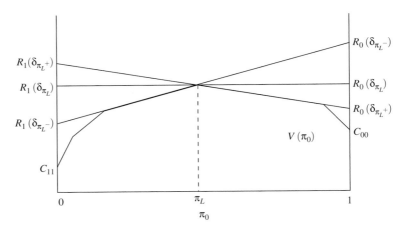

Figure 10.2 Minimax decision rule when $V(\pi_0)$ is not differentiable at π_L. (Figure adapted from reference [138].)

Then, mathematically, the Neyman–Pearson HT problem is

$$\text{maximize } P_D(\delta) \text{ subject to } P_F(\delta) \leq \alpha, \, 0 < \alpha < 1. \qquad (10.11)$$

The decision rule that solves this problem is again of the form as in (10.8) but with different threshold τ and $0 \leq \beta \leq 1$ values. Letting $P_1(y)/P_0(y) =: L$, the likelihood ratio, the false alarm probability and the detection probability resulting from this decision rule can be written as

$$P_D = P_1(\delta(y) = 1) = P_1(L > \tau) + \beta P_1(L = \tau), \qquad (10.12)$$
$$P_F = P_0(\delta(y) = 1) = P_0(L > \tau) + \beta P_0(L = \tau). \qquad (10.13)$$

The values of β and τ can then be computed as follows. First, we construct a function of the threshold t: $f(t) := P_0(L > t)$, and pick τ to be the smallest number such that $P_0(L > t) \leq \alpha$. Now, if $P_0(L > \tau) = \alpha$, which means $P_0(L = \tau) = 0$, then β can be anything in $[0, 1]$. Otherwise, if $P_0(L > \tau) < \alpha$, which means that $f(t)$ is not continuous at τ (in which case $P_0(L = \tau) > 0$), β will be chosen to satisfy $P_F = \alpha$, where P_F is given in (10.13). Specifically,

$$\beta = \frac{\alpha - P_0(L > \tau)}{P_0(L = \tau)}. \qquad (10.14)$$

10.2.4   Other hypothesis testing schemes

We briefly discuss here five other hypothesis testing schemes that can be used for network security: Centralized hypothesis testing with vector measurements, sequential hypothesis testing, and composite, nonparametric, and robust hypothesis testing.

Centralized hypothesis testing with vector measurements

In the previous sections, the hypothesis-testing problem with only one piece of observation (Y was taken as a scalar random variable) was examined. The solutions, however,

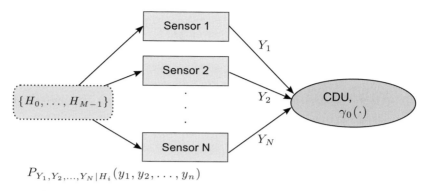

$$P_{Y_1, Y_2, \ldots, Y_N \mid H_i}(y_1, y_2, \ldots, y_n)$$

Figure 10.3 Centralized detection, where the CDU has full access to the observations of the sensors.

can easily be extended to the case where the observation is instead a random vector. In this case, the likelihood ratio can be calculated from the conditional joint densities of all the random variables that constitute the random vector. Consider the configuration given in Figure 10.3. This is a parallel configuration with a finite number of sensors and a central decision unit (CDU). The sensors all observe the M hypotheses ($M \geq 2$), $H_0, H_1, \ldots, H_{M-1}$. The observations of the sensors are $Y_1, Y_2, \ldots, Y_N$. Given hypothesis H_i, the joint density of the observations is $P_i(y_1, \ldots, y_N)$, where $i = 0, 1, \ldots, M - 1$. In this model, it is assumed that the CDU has full access to the observations of the sensors. It then uses all the data to decide finally which hypothesis is true. This is in fact a simple extension of the problems presented in the previous sections, where the observation of the CDU is now a vector instead of a scalar value. The Bayesian decision rule at the CDU for the case of binary hypotheses ($M = 2$) can thus be stated as follows:

$$\gamma_0(y_1, y_2, \ldots y_N) = \begin{cases} 1, & \text{if } \dfrac{P_1(y_1, y_2, \ldots y_N)}{P_0(y_1, y_2, \ldots y_N)} \geq \dfrac{\pi_0(C_{10} - C_{00})}{\pi_1(C_{01} - C_{11})} \\ 0, & \text{otherwise,} \end{cases} \tag{10.15}$$

where γ_0 is the decision rule at the CDU. For the centralized minimax and Neyman–Pearson problems, the decision rules of the CDU can be extended from those in the previous sections in a straightforward manner.

Sequential hypothesis testing

As shown above, once the sensors are up and running, centralized HT problems deal with a fixed sample size, i.e. fixed number of observations: the CDU makes the decision when it has received all the values of the observation vector. In network security, however, it is desirable to have a detector that is not only optimal in terms of detection probability and false alarm probability, but can also make a decision as quickly as possible. These requirements are addressed through the design of a sequential detector, where the expected number of samples is minimized given upper bounds on miss probability and false-alarm probability.

Sequential hypothesis testing (also known as *sequential analysis*, or *sequential probability ratio test* – SPRT) was developed by Abraham Wald during World War II [189]. A sequential detector entails two distinct rules: a *stopping rule* and a *terminal decision rule*. Suppose that the observation samples arrive as a sequence. Then, the stopping rule tells us when the detector has collected enough samples to make the final decision. When the detector has stopped sampling (or receiving measurements), then the terminal decision rule is used to make the final decision on the hypotheses.

Specifically, suppose having a sequence of observations in the form of continuous random variables with given probability density functions, $y_1, \ldots, y_k, \ldots$, involving the hypotheses H_0 and H_1. As before, let P_0 and P_1 denote the conditional pdfs of the observations under H_0 and H_1, respectively. At stage k, the logarithm of the likelihood ratio of observations $y_1, \ldots, y_k$ is

$$S_k = \log\left(\frac{P_1(y_1, \ldots, y_k)}{P_0(y_1, \ldots, y_k)}\right). \tag{10.16}$$

If the observations are conditionally independent, then S_k can be written as

$$S_k = \sum_{j=1}^{k} \log\left(\frac{P_1(y_j)}{P_0(y_j)}\right). \tag{10.17}$$

The decision possibilities at stage k are to *stop and declare H_1*, to *stop and declare H_0*, or to *continue*. More precisely,

$$\begin{aligned} S_k \geq b &\Rightarrow \delta(y_1, \ldots, y_k) = 1, \\ S_k < a &\Rightarrow \delta(y_1, \ldots, y_k) = 0, \\ a \leq S_k < b &\Rightarrow \text{take another observation}, \end{aligned} \tag{10.18}$$

where a and b ($a < b$) are the parameters to be determined, and will depend on P_F and P_M. The solution process is governed by the Wald–Wolfowitz theorem, which says that given upper bounds on P_F and P_M, the SPRT is the detector with the smallest expected value of sample size [138]. The values of a and b can be approximated using Wald's approximations

$$a \approx \ln \frac{P_M}{1 - P_F}, \quad b \approx \ln \frac{1 - P_M}{P_F}.$$

For more details on sequential hypothesis testing, we refer to the original book by Wald [189], the textbook by Poor [138], and the survey by Lai [92].

Composite, nonparametric, and robust hypothesis testing

A standard assumption in HT which may not be valid in network security is availability of the conditional density of the observations. If, however, each of the conditional densities can be considered to belong to a class of densities (also called an *uncertainty class*), one may use composite hypothesis testing, nonparametric hypothesis testing, or robust hypothesis testing, depending on how much information is given about these classes and how diverse they are. In the following, these hypothesis-testing schemes are briefly discussed.

To be specific, consider a centralized binary hypothesis testing problem, where the observations Y_j, $k = 1, 2, \ldots, N$, are independent and identically distributed given each hypothesis, and where we have

$$H_0 : Y_j \sim P \in \mathcal{P}_0,$$
$$H_1 : Y_j \sim P \in \mathcal{P}_1,$$

where $\mathcal{P}_0$ and $\mathcal{P}_1$ are two uncertainty classes given H_0 and H_1, respectively. If $\mathcal{P}_0$ and $\mathcal{P}_1$ can be parameterized by a finite-dimensional parameter, this problem is said to be *parametric*. If these classes are too broad to be parameterized by such a parameter, it is called a *nonparametric* hypothesis-testing problem.

Let us first consider the case where the uncertainty classes are given as

$$\mathcal{P}_0 := \{P_\Theta, \ \Theta \in \Lambda_0\},$$
$$\mathcal{P}_1 := \{P_\Theta, \ \Theta \in \Lambda_1\},$$

where Θ is some parameter of the probability densities in each class. In the example given in Section 10.2 where a digital signal is corrupted by additive Gaussian noise, we may have, for example, Θ as the mean of the noise, and $\Lambda_0 = (-\infty, 0)$ and $\Lambda_1 = [0, \infty)$. This class of problems is called *composite hypothesis* testing, as opposed to the problems presented in Sections 10.2.1 to 10.2.3 that now can be referred to as *simple hypothesis* testing problems. It has been shown that the Bayesian decision rule for the composite hypothesis test is in a similar form to the simple Bayesian HT problem in Section 10.2.1, that is

$$\delta(y) = \begin{cases} 1, & \text{if} \quad \dfrac{P(y|\Theta \in \Lambda_1)}{P(y|\Theta \in \Lambda_0)} \geq \dfrac{\pi_0(C_{10} - C_{00})}{\pi_1(C_{01} - C_{11})} \\ 0, & \text{otherwise.} \end{cases} \tag{10.19}$$

The minimax and the Neyman–Pearson formulations of the composite hypothesis testing problem can also be extended from those in Section 10.2.2 and Section 10.2.3.

When the uncertainty classes $\mathcal{P}_0$ and $\mathcal{P}_1$ cannot be parameterized, a possible approach is to use nonparametric hypothesis testing. An illustrative example demonstrating this can be found in reference [138] where $\mathcal{P}_0$ is the class of all densities on the reals that have $P(Y > 0) = 1/2$, and $\mathcal{P}_1$ is the class of all densities on the reals that have $P(Y > 0) \in (1/2, 1)$. Neither of these two classes of densities can be related to a finite-dimensional parameter through a one-to-one correspondence, and therefore the test of H_0 against H_1 is a nonparametric one. The Neyman–Pearson test for this problem is based on the number of samples y_j's that are positive, and is thus called a *sign test* [138] .

Consider now the cases where we have more information on the two uncertainty classes $\mathcal{P}_0$ and $\mathcal{P}_1$, say each of them consists of some well-defined distributions which depend on a parameter. A common approach in these situations is to optimize the worst case performance of the detector (over the distributions in each uncertainty class). Such a problem is called a robust hypothesis testing problem. Specifically, the Bayes, minimax, and Neyman–Pearson formulations of the robust test of H_0 against H_1 are given as

$$\min_{\delta} \left[\pi_0 \sup_{P \in \mathcal{P}_0} P_F(\delta, P) + \pi_1 \sup_{P \in \mathcal{P}_1} P_M(\delta, P) \right] \quad \text{(Bayes)},$$

$$\min_{\delta} \left[\max \left\{ \sup_{P \in \mathcal{P}_0} P_F(\delta, P), \sup_{P \in \mathcal{P}_1} P_M(\delta, P) \right\} \right] \quad \text{(Minimax), and}$$

$$\min_{\delta} \left[\sup_{P \in \mathcal{P}_1} P_M(\delta, P) \right] \text{ subject to } \left(\sup_{P \in \mathcal{P}_0} P_F(\delta, P) \le \alpha \right) \quad \text{(Neyman–Pearson)}.$$

Robust tests may be perceived as too conservative if the uncertainty classes $\mathcal{P}_0$ and $\mathcal{P}_1$ happen to be too large. If these classes instead contain slightly deviated versions of two nominal distributions corresponding to H_0 and H_1, robust HT schemes are good choices for maintaining good performance against the uncertainty of the conditional distributions. The interested reader is referred to [138] for a more complete treatment of these topics.

Example: misbehavior detection in wireless networks

An example application of hypothesis testing in network security is in reference [139], where robust sequential hypothesis testing is used to detect misbehaving nodes at the MAC layer in wireless networks. Recall that the random access protocol in IEEE 802.11 wireless LAN (Wi-Fi) is called Carrier Sense Multiple Access with Collision Avoidance (CSMA/CA). In CSMA/CA, if a node has data to transmit but the channel is sensed to be busy, it will pick a random back-off value c uniformly from the set $\{0, 1, \ldots, W - 1\}$, where W is the size of the contention window. The back-off counter is counted down at each time slot when the channel is sensed as idle, and the node starts transmitting after c idle slots. If, however, this transmission fails due to collision, the contention window size will be doubled; otherwise, if the transmission is successful, it will be reset to the minimum value of W. A malicious node can thus choose statistically smaller back-off values to gain advantage in transmission. In reference [139], it is assumed that an immediate neighbor (one within the transmission radius) of a node in question can measure its back-off times. The goal is to detect, as quickly as possible, whether this node is a regular one or a malicious one.

This problem can thus be addressed within the framework of sequential hypothesis testing presented earlier in this section under the following assumptions. First, although the contention window size changes along the test $(2^i W, i = 1, 2, \ldots)$, the back-off value (of a regular node) can be considered to be uniformly distributed in the set $\{0, 1, \ldots, W - 1\}$, due to the scaling property. Second, *pdf*s are used instead of probability mass functions (pmfs) for the sake of mathematical treatment.

Let H_0 be the hypothesis that the node in question is a regular node with the corresponding pdf f_0; H_1 be the hypothesis that the node in question is a malicious node with unknown pdf f. The pdf of the regular node is then $f_0(y) = 1/W$, and the conditional expectation of Y given H_0 is $\mathbb{E}_0[Y] = W/2$. The malicious node wants to ensure that $\mathbb{E}_1[Y] < W/2$. Thus, the uncertainty class under H_1 is defined as

$$\mathcal{F}_{\varepsilon} = \left\{ f(x) : \int_0^W x f(x)\, dx \leq \frac{W}{2} - \varepsilon \right\}, \tag{10.20}$$

where $0 < \varepsilon < W/2$.

This is a minimax problem, where the detector tries to minimize the number of observation samples against all the distributions $f \in \mathcal{F}_{\varepsilon}$ with given P_F and P_M. The malicious node, on the other hand, tries to maximize the number of samples against all the decision rules of the detector. The problem finally can be reduced to

$$\min_f \int_0^W f(x) \ln \frac{f(x)}{f_0(x)}\, dx \tag{10.21}$$

subject to the constraints

$$\int_0^W f(x)\, dx = 1 \quad \text{and} \quad \int_0^W x f(x)\, dx \leq \frac{W}{2} - \varepsilon. \tag{10.22}$$

At this point the saddle point (f^*, d^*) can be computed, where $f(x)$ has the exponential form, and d^* is a sequential probability ratio test, as described above.

For more details on the HT schemes mentioned in this section, the interested reader is referred to [80, 92, 138, 139, 186] and other references therein.

10.3 Decentralized hypothesis testing with correlated observations

The previous section has briefly discussed the centralized HT problem, where the peripheral sensors first send their observations to a CDU, which then makes the decision on the hypotheses. This is essentially no different from the central unit receiving all the measurements directly. In some practical situations, however, the peripheral sensors have to send out summaries of their observations instead. One typical reason is that the peripheral sensors may be located at distances far from the CDU and connected with the CDU through band-limited communication channels. If the channel bandwidth is not sufficient for frequent update of the observation, a peripheral sensor will have to compress the data it dispatches to the CDU. Another reason is that real-world IDSs may consist of multiple heterogeneous decision units, both signature- and anomaly-based, each making its own (local) decision. Such situations will give rise to a decentralized HT problem where, in place of the CDU, a fusion center is now needed to fuse all the summaries (or local decisions) to make the final decision on the hypotheses.

Accordingly, this section formulates the problem of decentralized HT with correlated observations. In Section 10.3.2, details on the fusion rule and the average probability of error at the fusion center are discussed.

10.3.1 Decentralized hypothesis testing

Consider a decentralized Bayesian HT problem with a parallel configuration. Each sensor uses a decision rule, which is a map $\gamma_j : \mathcal{Y}_j \mapsto \{0, 1, \ldots, D-1\}$, and then

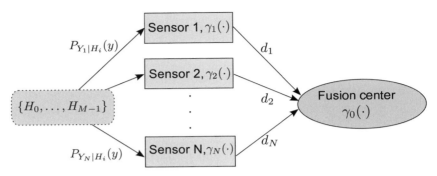

Figure 10.4 Decentralized hypothesis testing with N sensors and a fusion center.

sends the resulting message, which is an integer $d_j \in \{0, 1, \dots, D-1\}$, to the fusion center. The communication channels between the sensors and the fusion center are assumed to be perfect (no communication errors). At the fusion center, a fusion rule $\gamma_0 : \{0, 1, \dots, D-1\}^N \mapsto \{0, 1 \dots, M-1\}$ is employed to finally decide which hypothesis is true. The configuration of the N sensors and the fusion center is shown in Figure 10.4.

Naturally, given the same a priori probabilities of the hypotheses and conditional joint distributions of the observations, the decentralized configuration will yield an average probability of error that is no smaller than the centralized configuration. The reason is that there is generally information loss (and definitely no information gain) after the quantization at the sensors [175]. Putting it another way, given the observations of the sensors and assuming the use of a likelihood ratio test at the fusion center in the centralized configuration, the test in (10.15) will yield the minimum probability of error. The decentralized configuration, however, can always be considered as a special setup of the fusion center in the centralized case, where the observations from the sensors are quantized before being fused together.

For this decentralized detection problem, under the assumption that the observations are conditionally independent, there exists an optimal solution for the local sensors, which is a deterministic (likelihood ratio) threshold strategy [175]. When the observations are conditionally dependent, however, the threshold rule is no longer necessarily optimal [175]. In this case, obtaining the overall optimal non-threshold rule is a very challenging problem. In view of this, we restrict the analysis and discussion here to threshold-type rules at the local sensors (which are suboptimal) and seek optimality within that restricted class. The optimal fusion rule, as shown next, will also be a likelihood ratio test.

Taking a realization of the random variable Y_j and sending out a message in $\{0, 1, \dots, D-1\}$, each sensor can be considered as a quantizer. As mentioned earlier, reference [176] characterizes these quantizers based on the set of marginal distributions of the messages given each hypothesis. Following reference [176], let

$$q_d^j(\gamma_j | H_i) = Pr(\gamma_j(Y_j) = d | H_i),$$

for each $i = 0, \ldots, M-1$, $j = 1, \ldots, N$, and $d = 0, \ldots, D-1$. Here, $Pr(\gamma_j(Y_j) = d|H_i)$ denotes the probability of the event $\{Y_j \in \mathcal{Y}_j : \gamma_j(Y_j) = d\}$ conditioned on the occurrence of hypothesis H_i. For any $\gamma_j \in \Gamma_j$, where Γ_j is the set of all deterministic quantizers for sensor j, let

$$q^j(\gamma_j|H_i) = (q_0(\gamma_j|H_i), \ldots, q_{D-1}(\gamma_j|H_i)). \tag{10.23}$$

Define the vector $q^j(\gamma_j) \in R^{MD}$, for any $\gamma_j \in \Gamma_j$, as

$$q^j(\gamma_j) = (q^j(\gamma_j|H_0), \ldots, q^j(\gamma_j|H_{M-1})). \tag{10.24}$$

Now a quantizer can be represented by the vector $q(\gamma)$ for the purpose of detecting the hypotheses,

$$Q_j = \{q^j(\gamma_j) : \gamma_j \in \Gamma_j\}. \tag{10.25}$$

For a parallel configuration with N sensors, define

$$q(\gamma_1, \gamma_2, \ldots, \gamma_N) = (q^1(\gamma_1), q^2(\gamma_2), \ldots, q^N(\gamma_N)). \tag{10.26}$$

This yields $q(\gamma_1, \gamma_2, \ldots, \gamma_N) \in Q_a$, where Q_a is the Cartesian product of all $Q_j, j = 1, \ldots, N$, that is $Q_a = \times_{j=1}^N Q_j$.

As previously mentioned, Q_j is a compact set [176], and thus any cost function that is a continuous function on Q_j will attain a minimum (see Appendix A.1.3). In a parallel configuration with multiple sensors and a fusion center, if the sensor observations are independent given each hypothesis, then there exists an optimal solution over the set Q_a [176].

10.3.2 Decision rules at the sensors and at the fusion center

First, define two classes of decision rule at each sensor and the fusion center.[4] A *general rule* is one in which the observation space is partitioned into M regions, $R_i, i = 0, 1, \ldots, M-1$, and the sensor will pick H_i if $Y \in R_i$. Consistent with the earlier discussion, define the *threshold rule* for the case of binary hypotheses ($M = 2$) as one where

$$R_1 = \left\{ y \in \mathcal{Y} : \frac{P_1(y)}{P_0(y)} \geq \tau \right\} \tag{10.27}$$

$$R_0 = \left\{ y \in \mathcal{Y} : \frac{P_1(y)}{P_0(y)} < \tau \right\}. \tag{10.28}$$

Here, $\mathcal{Y}$ is the observation space of the sensor, and $P_0(y)$ and $P_1(y)$ are the conditional pdfs or the conditional pmfs (as the case may be) of the observation $Y = y$ given H_0 and H_1, respectively. In the case of a specific sensor j, these quantities are indexed by j; τ_j denotes the threshold, and $R_i^{(j)}$ the region i for sensor j.

---

[4] A fusion center can also be viewed as a sensor; thus we will occasionally use the term "sensor" to refer to both in this section.

Assuming uniform costs, the Bayes risk will become the average probability of error [138]. As mentioned earlier, the fusion center can be considered as a sensor with the observation being $(d_1, \ldots, d_N)$. Note that in order to minimize the Bayes risk, a joint optimization of the decision rules at the (local) sensors and the fusion rule at the fusion center is needed. However, if the decision rules at the (local) sensors have already been optimized over, or have been fixed according to some rule, then the fusion rule at the fusion center must be the solution to the centralized detection problem to minimize the Bayes risk. From (10.15), the fusion rule for binary hypotheses can be written as a likelihood ratio test:

$$
\gamma_0(d_1, d_2, \ldots, d_N) = \begin{cases} 1, & \text{if} \quad \dfrac{P_1(d_1, \ldots, d_N)}{P_0(d_1, \ldots, d_N)} \geq \dfrac{\pi_0}{\pi_1} \\ 0, & \text{otherwise,} \end{cases} \tag{10.29}
$$

and the corresponding average probability of error at the fusion center is

$$
\begin{aligned}
P_e &= \pi_0 P_0 \left(L_a \geq \frac{\pi_0}{\pi_1} \right) + \pi_1 P_1 \left(L_a < \frac{\pi_0}{\pi_1} \right) \\
&= \pi_0 \sum_{(d_1, \ldots, d_N): L_a \geq \frac{\pi_0}{\pi_1}} P_0(d_1, \ldots, d_N) \\
&\quad + \pi_1 \sum_{(d_1, \ldots, d_N): L_a < \frac{\pi_0}{\pi_1}} P_1(d_1, \ldots, d_N)
\end{aligned}
$$

$$
\text{where} \quad L_a = \frac{P_1(d_1, \ldots, d_N)}{P_0(d_1, \ldots, d_N)}. \tag{10.30}
$$

In the above, $P_i(d_1, d_2, \ldots, d_N)$, $i = 0, 1$, are the conditional probabilities (given H_i) of the decisions $d_1, d_2, \ldots, d_N$, which are observations to the fusion center, and by a slight abuse of notation $P_0(A)$ and $P_1(B)$ denote the probabilities of the events A and B conditioned on H_0 and H_1, respectively. If the observation sets of the peripheral sensors are infinite and the joint probabilities are given as pdfs, the conditional probabilities of the collection of decisions can be computed as

$$
P_i(d_1, d_2, \ldots, d_N) = \int_{R_{d_N}^{(N)}} \cdots \int_{R_{d_1}^{(1)}} P_i(y_1, \ldots, y_N) dy_1, \ldots, dy_N
$$

where $d_j = 0, 1, \ldots, D - 1$ and $R_{d_j}^{(j)}$ is the region where sensor j decides to send message d_j, $j = 1, \ldots, N$. Thus, it can be seen that in the optimal solution (which achieves the minimum P_e) the fusion rule is always a likelihood ratio test, but the decision rules at the local sensors can be general rules.

When the sensor observations are independent given each hypothesis, the optimal solution can be achieved with the decision rule at each sensor being also a threshold rule [175]. However, when the sensor observations are conditionally dependent, the global minimum of P_e cannot necessarily be achieved within the class of threshold rules at the local sensors [175]. It is also worth noting that, in general, the minimum P_e at the fusion center only depends on the decision rules at the sensors. If we restrict the sensors to threshold rules, however, the minimum P_e will only depend on the thresholds at the sensors, $\{\tau_1, \tau_2, \ldots, \tau_N\}$.

10.3.3   Decentralized Bayesian hypothesis testing

In this section, we first prove that when the observations are conditionally dependent, P_e can no longer be expressed as a function of the marginal distributions of the messages from the sensors. We then characterize P_e based on the set of joint distributions of the sensor messages. We show that this set is compact and that there exists an optimal solution (that minimizes P_e) when general rules are used at the sensors. We further show that there also exists an optimal solution when the sensors are restricted to threshold rules. The first two propositions below (Propositions 10.1 and 10.2) are stated for $D = 2$ and $M = 2$, which is sufficient for our purposes since they are negative results (on optimality of threshold rules); naturally, their statements can be extended to $M > 2$ and $D > 2$.

Proposition 10.1 *Let $f_0(y_1, y_2)$ and $f_1(y_1, y_2)$ be two nonidentical joint pdfs, where $f_i(y_1, y_2)$, $i = 0, 1$, are continuous on $\mathbb{R}^2$ and nonzero for $-\infty < y_1, y_2 < \infty$. Let $\Phi_i(y_1, y_2)$, $i = 0, 1$, denote the corresponding cumulative distribution functions. Let*

$$\alpha_0 = \Phi_0\left(y_1^*, y_2^*\right) = \int_{-\infty}^{y_1^*} \int_{-\infty}^{y_2^*} f_0(y_1, y_2)\, dy_2\, dy_1, \qquad (10.31)$$

$$\alpha_1 = \Phi_1\left(y_1^*, y_2^*\right) = \int_{-\infty}^{y_1^*} \int_{-\infty}^{y_2^*} f_1(y_1, y_2)\, dy_2\, dy_1, \qquad (10.32)$$

where (y_1^*, y_2^*) is an arbitrary point in $\mathbb{R}^2$. Then, specifying a value for $\alpha_0 \in (0, 1)$ does not uniquely determine the value of α_1, and vice versa.

Proof Let $g_i(y_1)$ and $h_i(y_2)$ be the marginal densities of y_1 and y_2 given H_i, where $i = 0, 1$. For each $0 < \alpha_0 < 1$, we can pick $\gamma_0 > 0$ such that $\alpha_0 + \gamma_0 < 1$. As the conditional marginal density $g_0(y_1)$ is continuous, we can always uniquely pick y_1^* such that $\int_{-\infty}^{y_1^*} g_0(y_1) dy_1 = \alpha_0 + \gamma_0$. Once y_1^* is specified, we can also choose y_2^* such that $\int_{-\infty}^{y_1^*} \int_{-\infty}^{y_2^*} f_0(y_1, y_2) dy_2 dy_1 = \alpha_0$. Thus, for each fixed value of γ_0, we have a unique pair (y_1^*, y_2^*). It can be seen that there are infinitely many values of γ_0 satisfying $\alpha_0 + \gamma_0 < 1$, each of which yields a different pair (y_1^*, y_2^*). Therefore, specifying a value for $\alpha_0 \in (0, 1)$ does not uniquely determine the value of α_1, and vice versa, unless $f_0(y_1, y_2)$ and $f_1(y_1, y_2)$ are identically equal. $\qquad \square$

Proposition 10.2 *Consider a parallel structure as in Figure 10.4 with the number of sensors $N \geq 2$, the number of messages $D = 2$, and the number of hypotheses $M = 2$. When the observations of the sensors are conditionally dependent, there exists a fusion rule γ_0 in which the minimum average probability of error P_e given in (10.30) cannot be expressed solely as a function of $q(\gamma_1, \ldots, \gamma_N)$ (given in (10.26)).*

Proof We first prove this proposition for the two-sensor case and then use induction to extend the result to $N > 2$. As before, let d_1 and d_2 denote the messages that sensor 1 and sensor 2 send to the fusion center. For notational simplicity, let $P_i(l_1, l_2)$ denote

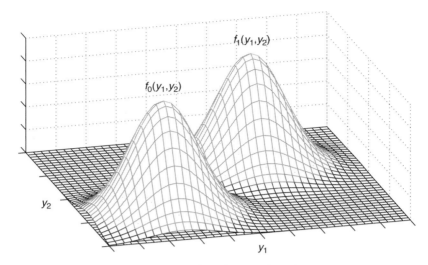

Figure 10.5 α_0 and α_1 are integrations of $f_0(y_1,y_2)$ and $f_1(y_1,y_2)$ over the same region.

$P(d_1 = l_1, d_2 = l_2 | H_i)$ where $l_1, l_2 \in \{0,1\}$. We have the following linear system of equations with $P_i(0,0), P_i(0,1), P_i(1,0)$, and $P_i(1,1)$ as the unknowns:

$$P_i(0,0) + P_i(0,1) = P_i(l_1 = 0)$$
$$P_i(1,0) + P_i(1,1) = P_i(l_1 = 1) = 1 - P_i(l_1 = 0)$$
$$P_i(0,0) + P_i(1,0) = P_i(l_2 = 0)$$
$$P_i(0,1) + P_i(1,1) = P_i(l_2 = 1) = 1 - P_i(l_2 = 0)$$

Note that the matrix of coefficients is singular. Solving this system yields

$$P_i(0,0) = \alpha_i$$
$$P_i(0,1) = P_i(l_1 = 0) - \alpha_i$$
$$P_i(1,0) = P_i(l_2 = 0) - \alpha_i$$
$$P_i(1,1) = 1 - P_i(l_1 = 0) - P_i(l_2 = 0) + \alpha_i$$

where $\alpha_i, i = 0,1$, corresponding to H_0, H_1 are real numbers in $(0,1)$.

Now we rewrite (10.30) for a fixed fusion rule γ_0:

$$P_e = \pi_0 \sum_{(d_1,d_2) \in R_1} P_0(d_1,d_2) + \pi_1 \sum_{(d_1,d_2) \in R_0} P_1(d_1,d_2) \tag{10.33}$$

where R_0 and R_1 are two partitions of the set of all possible values of (d_1, d_2) in which the fusion center decides, respectively, whether hypothesis H_0 or hypothesis H_1 is true. Now suppose that the fusion center uses the following fusion rule: it picks one if $(d_1, d_2) = (1,1)$ and picks zero for the remaining three cases. After some manipulation, expression (10.33) becomes

$$P_e = \pi_0(1 - P_0(d_1 = 0) - P_0(d_2 = 0) + \alpha_0) + \pi_1 (P_1(d_1 = 0) + P_1(d_2 = 0) - \alpha_1). \tag{10.34}$$

From Proposition 10.1, α_0 is not uniquely determined given α_1 and vice versa. Thus P_e in (10.33) cannot be expressed solely as a function of $q(\gamma_1, \gamma_2)$.

Now we prove the proposition for $N > 2$ by induction on N. Suppose that there exists a fusion rule $\gamma_0^{(N)}$ that results in $P_e^{(N)}$ which cannot be expressed solely as a function of $q(\gamma_1, \ldots, \gamma_N)$; we will then show that there exists a fusion rule $\gamma_0^{(N+1)}$ that yields $P_e^{(N+1)}$ which cannot be expressed solely as a function of $q(\gamma_1, \ldots, \gamma_{N+1})$.

Let $\tilde{R}_0^{(N)}$ and $\tilde{R}_1^{(N)}$ be the decision regions (for H_0 and H_1, respectively) at the fusion center when there are N sensors. Let $\tilde{R}_0^{(N+1)}$ and $\tilde{R}_1^{(N+1)}$ be those of the $(N+1)$-sensor case. Without loss of generality, we assume that the observation of sensor $(N+1)$ is independent of those of the first N sensors. Rewriting (10.30) for the N-sensor problem, we have:

$$P_e^{(N)} = \pi_0 \sum_{(l_1,\ldots,l_N) \in \tilde{R}_1^{(N)}} P_0(l_1,\ldots,l_N) + \pi_1 \sum_{(l_1,\ldots,l_N) \in \tilde{R}_0^{(N)}} P_1(l_1,\ldots,l_N). \qquad (10.35)$$

Now we construct $\tilde{R}_0^{(N+1)}$ and $\tilde{R}_1^{(N+1)}$ based on $\tilde{R}_0^{(N)}$ and $\tilde{R}_1^{(N)}$ as follows. $\tilde{R}_0^{(N+1)}$ consists of combinations of the forms $(l_1,\ldots,l_N,0)$ and $(l_1,\ldots,l_N,1)$ where $(l_1,\ldots,l_N) \in \tilde{R}_0^{(N)}$; $\tilde{R}_1^{(N+1)}$ consists of combinations of the forms $(l_1,\ldots,l_N,0)$ and $(l_1,\ldots,l_N,1)$ where $(l_1,\ldots,l_N) \in \tilde{R}_1^{(N)}$. Note that, for $i = 0, 1$,

$$P_i(l_1,\ldots,l_N,0) + P_i(l_1,\ldots,l_N,1) = P_i(l_1,\ldots,l_N).$$

Thus, P_e for the $(N+1)$-sensor case can be written as

$$
\begin{aligned}
P_e^{(N+1)} &= \pi_0 \sum_{(l_1,\ldots,l_N,l_{N+1}) \in \tilde{R}_1^{(N+1)}} P_0(l_1,\ldots,l_N,l_{N+1}) \\
&\quad + \pi_1 \sum_{(l_1,\ldots,l_N,l_{N+1}) \in \tilde{R}_0^{(N+1)}} P_1(l_1,\ldots,l_{N+1}) \\
&= \pi_0 \sum_{(l_1,\ldots,l_N) \in \tilde{R}_1^{(N)}} P_0(l_1,\ldots,l_N) + \pi_1 \sum_{(l_1,\ldots,l_N) \in \tilde{R}_0^{(N)}} P_1(l_1,\ldots,l_N) \\
&= P_e^{(N)}.
\end{aligned}
$$

But $P_e^{(N)}$ cannot be expressed solely as a function of $q(\gamma_1, \ldots, \gamma_N)$ and $q(\gamma_{N+1})$ due to the induction hypothesis and the independence assumption of sensor $(N+1)$'s observation. Thus, $P_e^{(N+1)}$ cannot be expressed solely as a function of $q(\gamma_1, \ldots, \gamma_{N+1})$. $\qquad \square$

Thus, for the case of conditionally dependent observations, instead of using conditional marginal distributions, we relate the Bayesian probability of error to the joint distribution of the decisions of the sensors. In what follows, we use γ as before collectively to denote $(\gamma_1, \gamma_2, \ldots, \gamma_N)$ and Γ to denote the Cartesian product of $\Gamma_1, \Gamma_2, \ldots, \Gamma_N$, where Γ_j is the set of all deterministic decision rules (quantizers) of sensor j, $j = 1, \ldots, N$. Also, we define

$$s_{d_1,\ldots,d_N}(\gamma|H_i) = Pr(\gamma_1 = d_1, \ldots, \gamma_N = d_N|H_i). \qquad (10.36)$$

Then, the D^N-tuple $s(\gamma|H_i)$ is defined as:

$$s(\gamma|H_i) := (s_{0,0,\ldots,0}(\gamma|H_i), s_{0,0,\ldots,1}(\gamma|H_i), \ldots, s_{D-1,D-1,\ldots,D-1}(\gamma|H_i)). \quad (10.37)$$

Finally, we define the $M \times D^N$-tuple $s(\gamma)$:

$$s(\gamma) := (s(\gamma|H_0), s(\gamma|H_1), \ldots, s(\gamma|H_{M-1})). \quad (10.38)$$

From (10.30), it can be seen that P_e is a continuous function on $s(\gamma)$ for a fixed fusion rule. We now prove that the set $S = \{s(\gamma) : \gamma_1 \in \Gamma_1, \ldots, \gamma_N \in \Gamma_N\}$ is compact, and therefore there exists an optimal solution for a fixed fusion rule. As the number of fusion rules is finite, we can then conclude that there exists an optimal solution for the whole system for each class of decision rule at the sensors.

Proposition 10.3 *The set S given by*

$$S := \{s(\gamma) : \gamma_1 \in \Gamma_1, \gamma_2 \in \Gamma_2, \ldots, \gamma_N \in \Gamma_N\} \quad (10.39)$$

is compact.

Proof [5] The proof is an extension of the proof of a similar result in reference [176] for the single sensor case. Let $\Phi = (\Phi_0 + \cdots + \Phi_{M-1})/M$, where $\Phi_0, \ldots, \Phi_{M-1}$ are the conditional distributions of the observations given $H_0, \ldots, H_{M-1}$, respectively, and we let $\mathcal{P}$ denote the corresponding probability measure. We use G to denote the set of all measurable functions from the observation space, $\mathcal{Y} = \mathcal{Y}_1 \times \mathcal{Y}_2 \times \cdots \times \mathcal{Y}_N$, into $\{0, 1\}$. Let $G^{(D^N)}$ denote the Cartesian product of D^N replicas of G. Let

$$F = \left\{ (f_{00\ldots0}, \ldots, f_{(D-1)(D-1)\cdots(D-1)}) \in G^{(D^N)} \right.$$
$$\left. \left| \mathcal{P}\left(\sum_{d_1,\ldots,d_N=0}^{D-1} f_{d_1,\ldots,d_N}(Y) = 1 \right) = 1 \right\}$$

For any $\gamma \in \Gamma$ and $d_1, \ldots, d_N \in \{0, \ldots, D-1\}$, we define $f_{d_1,\ldots,d_N}$ such that $f_{d_1,\ldots,d_N}(y) = 1$ if, and only if, $\gamma(y) = (d_1, \ldots, d_N)$, and $f_{d_1,\ldots,d_N}(y) = 0$ otherwise. Then, $f_{d_1,\ldots,d_N}$ will be the indicator function of the set $\gamma^{-1}(d_1, \ldots, d_N)$. It can be seen that $(f_{00\ldots0}, \ldots, f_{(D-1)(D-1)\cdots(D-1)}) \in F$. Also, we have

$$s_{d_1,\ldots,d_N}(\gamma|H_i) = Pr(\gamma(y) = (d_1, \ldots, d_N)|H_i) = \int f_{d_1,\ldots,d_N}(y)d\Phi_i(y).$$

Conversely, for any $f = (f_{00\ldots0}, \ldots, f_{(D-1)(D-1)\cdots(D-1)}) \in F$, define $\gamma \in \Gamma$ as follows.

- If $\sum_{d_1,\ldots,d_N=0}^{D-1} f_{d_1,\ldots,d_N}(y) = 1$, then

$$\gamma(y) = (d_1, \ldots, d_N) \text{ such that } f_{d_1,\ldots,d_N}(y) = 1.$$

- If $\sum_{d_1,\ldots,d_N=1}^{D} f_{d_1,\ldots,d_N}(y) \neq 1$, then

$$\gamma(y) = (1, 1, \ldots, 1).$$

---

[5] The proof uses some topological notions not introduced in the book, and it can be skipped in a first reading. Details of the proof are not needed to follow the main result of the theorem stated next.

As $\mathcal{P}\left(\sum_{d_1,\ldots,d_N=1}^{D} f_{d_1,\ldots,d_N}(Y) \neq 1\right) = 0$, (10.40) still holds. Now we define a mapping $h : F \to \mathfrak{R}^{MD^N}$ such that

$$h_{i,d_1,\ldots,d_N}(f) = \int f_{d_1,\ldots,d_N} d\Phi_i(y). \tag{10.40}$$

It can be seen that $S = h(F)$. If we can find a topology on G in which F is compact and h is continuous, S will be a compact set.

Let $L_1(\mathcal{Y};\mathcal{P})$ denote the set of all measurable functions $f : \mathcal{Y} \to \mathbb{R}$ that satisfy $\int |f(y)| d\Phi(y) < \infty$, $L_\infty(\mathcal{Y};\mathcal{P})$ denote the set of all measurable functions $f : \mathcal{Y} \to \mathbb{R}$ such that f is bounded after removing the set $\mathcal{Y}_z \subset \mathcal{Y}$ that has $\mathcal{P}(\mathcal{Y}_z) = 0$. Then G is a subset of $L_\infty(\mathcal{Y};\mathcal{P})$. It is known that $L_\infty(\mathcal{Y};\mathcal{P})$ is the dual of $L_1(\mathcal{Y};\mathcal{P})$ [101]. Consider the weak* topology on $L_\infty(\mathcal{Y};\mathcal{P})$, which is the weakest topology where the mapping

$$f \to \int f(y)g(y)d\Phi(y) \tag{10.41}$$

is continuous for every $g \in L_1(\mathcal{Y};\mathcal{P})$. Using Alaoglu's theorem [101], we have that the unit ball in $L_\infty(\mathcal{Y};\mathcal{P})$ is weak*-compact. Thus G is compact. Then $G^{(D^N)}$, which is a Cartesian product of D^N compact sets, is also compact. Now, from (10.40), every point $\left(f_{00\cdots0},\ldots,f_{(D-1)(D-1)\cdots(D-1)}\right) \in F$ satisfies

$$\int_A \sum_{d_1,\ldots,d_N=0}^{D-1} f_{d_1,\ldots,d_N}(y)d\Phi(y) = \mathcal{P}(A), \tag{10.42}$$

where A is any measurable subset of $\mathcal{Y}$. If we let X_A denote the indicator function of A, it follows that

$$\int \sum_{d_1,\ldots,d_N=0}^{D-1} f_{d_1,\ldots,d_N}(y)X_A(y)d\Phi(y) = \mathcal{P}(A). \tag{10.43}$$

As $X_A \in L_1(\mathcal{Y};\mathcal{P})$ and the mapping in (10.41) is continuous for every $g \in L_1(\mathcal{Y};\mathcal{P})$, we have that the map $f \to \mathcal{P}(A)$ is also continuous. Furthermore, F is a subset of the compact set $G^{(D^N)}$, and thus F is also compact.

Let $g_i, i = 0,\ldots,M-1$ denote the Radon–Nikodym derivative of Φ_i with respect to $\mathcal{P}$, $g_i(y) = d\Phi_i(y)/d\Phi(y)$. Then we have $g_i \in L_1(\mathcal{Y};\mathcal{P})$ [176]. Also, we have that

$$\int f_{d_1,\ldots,d_N}(y)d\Phi_i(y) = \int f_{d_1,\ldots,d_N}(y)g_i(y)d\Phi(y), \; \forall i,d_1,\ldots,d_N.$$

From (10.41), (10.44), and the fact that $g_i \in L_1(\mathcal{Y};\mathcal{P})$, it follows that the mapping $f \to \int f_{d_1,\ldots,d_N}(y)d\Phi_i(y)$ is continuous. Therefore, the mapping h given in (10.40) is continuous. As $S = h(F)$, we finally have that S is compact. □

We next state the counterpart of Proposition 10.3 when sensors are restricted to threshold rules (again in the multiple hypothesis case), whose proof is similar to that of Proposition 10.3. First a definition to pave the way.

Definition 10.4 *[176] (a) A threshold set T is a vector $\underline{t} = (t_1,\ldots,t_{D-1}) \in [0,\infty]^{D-1}$, satisfying $0 \leq t_1 \leq t_2 \leq \ldots \leq t_{D-1}$, and the intervals associated with this threshold set are defined by $I_1 = [0,t_1], I_1 = [t_1,t_2],\ldots,I_D = [t_{D-1},\infty]$.*

(b) A decision rule γ is a monotone likelihood ratio quantizer (LRQ) with threshold vector $\underline{t} \in T$ if

$$P(\gamma(Y) = d \text{ and } L(Y) \notin I_d | H_i) = 0, \ \forall d, i. \tag{10.44}$$

(c) A decision rule is an LRQ if there exists some permutation function π : $\{1, \ldots, D\} \mapsto \{1, \ldots, D\}$ such that $\pi(\gamma(Y))$ is a monotone LRQ.

Proposition 10.5 *Let $\Gamma_j^{(t)}$ denote the set of all monotone LRQs for the sensor j, $j = 1, \ldots, N$. The set $S^{(t)}$ given by*

$$S^{(t)} = \left\{ s(\gamma) : \gamma_1 \in \Gamma_1^{(t)}, \gamma_2 \in \Gamma_2^{(t)}, \ldots, \gamma_N \in \Gamma_N^{(t)} \right\} \tag{10.45}$$

is compact.

We can now state and prove the following theorem.

Theorem 10.6 *There exists an optimal solution for the general rules at the sensors, and there also exists an optimal solution for the special case where the sensors are restricted to threshold rules on likelihood ratios.*

Proof For each fixed fusion rule γ_0 at the fusion center, the probability of error P_e given in (10.30) is a continuous function on the compact set S. Thus, by the Weierstrass theorem [101], there exists an optimal solution that minimizes P_e for each γ_0. Furthermore, there is a finite number of fusion rules γ_0 at the fusion center (in particular, this is the number of ways to partition the set $\{d_1, d_2, \ldots, d_N\}$ into two subsets, which is 2^N). Therefore, there exists an optimal solution over all the fusion rules at the fusion center. Note that the use of the general rule or the threshold rule will result in different fusion rules, but will not affect the reasoning in this proof. The optimal solutions in each case, however, will be different in general. More specifically, the set of all the decision rules (of the sensors) based on the threshold rule will be a subset of the set of all decision rules (of the sensors), thus the optimal solution in the former case will be no better (actually worse) than that of the latter. $\square$

10.3.4 Decentralized Neyman–Pearson hypothesis testing

We now examine the decentralized Neyman–Pearson problem, but for the case $M = 2$, i.e. the case of only two hypotheses. Consider a finite sequence of deterministic strategies $\{\gamma^{(k)} \in \Gamma, \ k = 1, \ldots, K\}$, where $\gamma^{(k)} \in \Gamma$ stands for $\left\{ \gamma_1^{(k)} \in \Gamma_1, \gamma_2^{(k)} \in \Gamma_2, \ldots, \gamma_N^{(k)} \in \Gamma_N \right\}$. Suppose that each deterministic strategy $\gamma^{(k)}$ is used with probability $0 \le p_k \le 1$, where $\sum_{k=1}^{K} p_k = 1$. Let $\bar{\Gamma}$ denote the set of all such randomized strategies. For $\bar{\gamma} \in \bar{\Gamma}$, we have that

$$s(\overline{\gamma}) = \sum_{k=1}^{K} p_k s(\gamma^{(k)}). \tag{10.46}$$

Note that the set of strategies resulting from this randomization scheme includes (as a subset) those generated by the "independent randomization" scheme, where the strategies of each peripheral sensor are randomized independently. From (10.46), it can be seen that the set $\overline{S}$ of all such $s(\overline{\gamma})$ is the convex hull of S defined in (10.3), $\overline{S} = co(S)$. As shown in reference [127], S is a finite-dimensional space and is clearly bounded. Thus, $\overline{S}$ is also finite-dimensional and bounded. Furthermore, it is shown in reference [127] that S is a closed set. As $\overline{S}$ is the convex hull of S, it is also a closed set. Thus, we can state the following result.

Proposition 10.7 *The set $\overline{S}$ defined by $\overline{S} := \{s(\overline{\gamma}) : \overline{\gamma} \in \overline{\Gamma}\}$ is compact.*

Note that for the Bayesian formulation, the extension to randomized rules will not improve the optimal solution, as stated in the following proposition.

Proposition 10.8 *Consider the problem of minimizing the Bayes risk P_e on the set of randomized rules $\overline{\Gamma}$. There exists an optimal solution that entails deterministic rules at peripheral sensors.*

Proof Consider a fixed fusion rule, where the Bayes risk is given by

$$P_e = \pi_0 \sum_{(d_1,\ldots,d_N):(d_1,\ldots,d_N)\in R_1} P_0(d_1,\ldots,d_N)$$
$$+ \pi_1 \sum_{(d_1,\ldots,d_N):(d_1,\ldots,d_N)\in R_0} P_1(d_1,\ldots,d_N),$$

where R_0 and R_1 are the regions in which the fusion center decides H_0 and H_1, respectively. If randomized rules are used at peripheral sensors, the Bayes risk can be written as

$$P_e = \pi_0 \sum_{(d_1,\ldots,d_N):(d_1,\ldots,d_N)\in R_1} s_{d_1,\ldots,d_N}(\overline{\gamma}|H_0)$$
$$+ \pi_1 \sum_{(d_1,\ldots,d_N):(d_1,\ldots,d_N)\in R_0} s_{d_1,\ldots,d_N}(\overline{\gamma}|H_1),$$

where $s(\gamma)$ is given by

$$s(\overline{\gamma}) = \sum_{k=1}^{K} p_k s(\gamma^{(k)}). \tag{10.47}$$

In particular, we have

$$s_{d_1,\ldots,d_N}(\overline{\gamma}|H_j) = \sum_{k=1}^{K} p_k s_{d_1,\ldots,d_N}(\gamma^{(k)}|H_j), \quad j = 0, 1. \tag{10.48}$$

Thus, the Bayes risk is now minimized over the convex hull of K points $(s(\gamma^{(1)}),\ldots,s(\gamma^{(K)}))$. Using the fundamental theorem of linear programming (see, for example, reference [100]), if there is an optimal solution, there is an optimal solution that is an extreme point of the convex hull, which corresponds to deterministic rules at the peripheral sensors.   □

The extension from deterministic strategies to randomized strategies helps the Neyman–Pearson test to be accommodated at peripheral sensors. Now, similarly to decentralized Bayesian, the fusion center can be considered as a sensor with the observation being $\bar{d} \equiv (d_1, d_2, \ldots, d_N)$. We seek a joint optimization of the decision rules at the peripheral sensors and the fusion rules at the fusion center to solve the Neyman–Pearson problem at the fusion center. The decentralized Neyman–Pearson problem at the fusion center can be stated as follows:

$$\text{maximize } P_D(\gamma) \text{ subject to } P_F(\gamma) \leq \alpha, \, 0 < \alpha < 1, \tag{10.49}$$

where the false alarm probability (P_F) and the detection probability (P_D) are given by

$$P_F \equiv P_0 \left(\gamma_0(\bar{d}) = 1 \right), \tag{10.50}$$

$$P_D \equiv P_1 \left(\gamma_0(\bar{d}) = 1 \right). \tag{10.51}$$

Here we have used γ_0 to denote the fusion rule at the fusion center. Note that when the decision rules at the peripheral sensors have already been optimized on, or picked according to some predetermined rule, the fusion rule at the fusion center must be the solution to the centralized Neyman–Pearson detection problem.

Let $\tilde{\gamma}_0(\bar{d}) \equiv Pr(\gamma_0(\bar{d}) = 1 | \bar{d})$. From a natural extension of (10.8) to vector valued observations (see reference [138]), the fusion rule can be written as a likelihood ratio test:

$$\tilde{\gamma}_0(\bar{d}) = \begin{cases} 1, & \text{if } \dfrac{P_1(\bar{d})}{P_0(\bar{d})} > \tau \\[2mm] \beta, & \text{if } \dfrac{P_1(\bar{d})}{P_0(\bar{d})} = \tau \\[2mm] 0, & \text{if } \dfrac{P_1(\bar{d})}{P_0(\bar{d})} < \tau, \end{cases} \tag{10.52}$$

where τ is the threshold and $0 \leq \beta \leq 1$. Letting $L_a \equiv \frac{P_1(\bar{d})}{P_0(\bar{d})}$, the false-alarm probability and the detection probability resulting from this fusion rule can be written as

$$\begin{aligned} P_F &= P_0 \left(\gamma_0(\bar{d}) = 1 \right) = P_0 \left(L_a > \tau \right) + \beta P_0 \left(L_a = \tau \right) \\ &= \sum_{(d_1, \ldots, d_N): L_a > \tau} P_0(\bar{d}) + \beta \sum_{(d_1, \ldots, d_N): L_a = \tau} P_0(\bar{d}), \end{aligned} \tag{10.53}$$

$$\begin{aligned} P_D &= P_1 \left(\gamma_0(\bar{d}) = 1 \right) = P_1 \left(L_a > \tau \right) + \beta P_1 \left(L_a = \tau \right) \\ &= \sum_{(d_1, \ldots, d_N): L_a > \tau} P_1(\bar{d}) + \beta \sum_{(d_1, \ldots, d_N): L_a = \tau} P_1(\bar{d}). \end{aligned} \tag{10.54}$$

Here, $P_i(\overline{d}) \equiv P_i(d_1, d_2, \ldots, d_N)$, $i = 0, 1$, are the conditional probabilities (given H_i) of the decision $\overline{d}$, which can be computed as follows

$$P_i(d_1, \ldots, d_N) = \sum_{k=1}^{K} p_k P_i^{(k)}(d_1, \ldots, d_N), \tag{10.55}$$

$$P_i^{(k)}(d_1, \ldots, d_N) = \int_{R_{d_N}^{(N)}} \cdots \int_{R_{d_1}^{(1)}} P_i(y_1, \ldots, y_N) dy_1, \ldots, dy_N, \tag{10.56}$$

where $d_j = 0, 1, \ldots, D-1$, and $R_{d_j}^{(j)}$ is the region where sensor j decides to send message d_j, $j = 1, \ldots, N$ for the deterministic decision profile $\gamma^{(k)}$. (Note that the partitions of sensor observation spaces on the right-hand side of (10.56) are of a specific deterministic strategy k; however, we have omitted the superscript k to simplify the formula; note also that P_i's are probability densities.) Thus, it can be seen that in the optimal solution, the fusion rule is always a likelihood ratio test (10.52), but the decision rules at the peripheral sensors can be general rules. We now formally state this in the following theorem.

Theorem 10.9 *There exists an optimal solution for the decentralized configuration in Figure 10.4 with the Neyman–Pearson criterion, where the decision rules at peripheral sensors lie in $\overline{\Gamma}$, and the fusion rule at the fusion center is a standard Neyman–Pearson likelihood ratio test.*

Proof For each fixed fusion rule γ_0 at the fusion center, the false alarm probability P_F given in (10.53) and the detection probability P_D given in (10.54) are both continuous functions on the compact set $\overline{S}$. Hence, the set $\overline{\Gamma}^0 \equiv \{\gamma \in \overline{\Gamma} : P_F(\gamma) \leq \alpha\}$ is also closed and bounded. Also, recall that $\overline{\Gamma}$ is a finite-dimensional space. Thus $\overline{\Gamma}^0$ is a compact set. Therefore, by the Weierstrass theorem [101], there exists an optimal solution that maximizes P_D given that $P_F \leq \alpha$ for each γ_0. Furthermore, there is a finite number of fusion rules γ_0 at the fusion center (in particular, this is upper-bounded by the number of ways to partition the set $\{d_1, d_2, \ldots, d_N\}$ into three subsets with $L_a > \tau$, $L_a = \tau$, and $L_a < \tau$, which is 3^N). Note that once this partition is fixed, τ and β can be calculated accordingly. Therefore, there exists an optimal solution over all the fusion rules at the fusion center. $\qquad\square$

In what follows, we introduce a special case where we can further characterize the optimal solution. First, we present the following definition [175].

Definition 10.10 *A likelihood ratio $L_j(y_j)$ is said to have no point mass if*

$$Pr(L_j(y_j) = x|H_i) = 0, \ \forall x \in [0, \infty], \ i = 1, 2. \tag{10.57}$$

It can be seen that this property holds when $P_i(y_j)$, $i = 1, 2$, are both continuous.

Proposition 10.11 *If all peripheral sensors are restricted to threshold rules on likelihood ratios, and $L_j(y_j)$, $j = 1, \ldots, N$, have no point mass, there exists an optimal solution that is a deterministic rule at peripheral sensors, that is, $\gamma \in \Gamma$.*

Proof When $L_j(y_j)$, $j = 1, \ldots, N$ have no point mass, $Pr(L_j(y_j) = \tau_d) = 0$, thus what each sensor does at the boundary of decision regions is immaterial.   □

10.4   The majority vote versus the likelihood ratio test

In the rest of this chapter, we consider the discrete case where the distributions of the observations are given as pmfs. In this section, we first show that if the observations of the sensors are conditionally independent, given the set of thresholds at the local sensors, any sensor switching from decision 0 to decision 1 will increase the likelihood ratio at the fusion center. Furthermore, if the observations are conditionally independent and identically distributed and the sensors all use the same threshold for the likelihood ratio test, the likelihood ratio test at the fusion center becomes equivalent to a majority vote. In the general case, where the observations are not independent and identically distributed, this property no longer holds; we provide towards the end of the section an example where the likelihood ratio test and the majority vote yield different results.

Recall that the fusion rule at the fusion center is given by (10.29). If the observations of the sensors are conditionally independent, the likelihood ratio at the fusion center becomes:

$$\frac{P_1(d_1, d_2, \ldots, d_N)}{P_0(d_1, d_2, \ldots, d_N)} = \frac{\prod_{n=1}^N P_1(d_n)}{\prod_{n=1}^N P_0(d_n)} = \prod_{n=1}^N \frac{P_1(d_n)}{P_0(d_n)}.$$

Let us denote by $\mathcal{N}$ the set of all local sensors (represented by their indices). We divide $\mathcal{N}$ into two partitions: $\mathcal{N}_0$, the set of local sensors that send 0 to the fusion center, and $\mathcal{N}_1$, the set of local sensors that send 1 to the fusion center. Then, we have $\mathcal{N}_0 \cup \mathcal{N}_1 = \mathcal{N}$ and $\mathcal{N}_0 \cap \mathcal{N}_1 = \emptyset$. Note that, given the conditional joint probabilities of the observations, $\mathcal{N}_0$ and $\mathcal{N}_1$ are set-valued functions of the thresholds $\{\tau_1, \tau_2, \ldots, \tau_N\}$. Let N_0 and N_1 denote the cardinalities of $\mathcal{N}_0$ and $\mathcal{N}_1$, respectively. Obviously, $N_0, N_1 \in \mathcal{Z}$ (where $\mathcal{Z}$ is the set of all integers), $0 \leq N_0, N_1 \leq N$, and $N_0 + N_1 = N$. Now the likelihood ratio can be written as:

$$\frac{P_1(d_1, d_2, \ldots, d_N)}{P_0(d_1, d_2, \ldots, d_N)} = \prod_{n \in \mathcal{N}_0} \frac{P_1(d_n = 0)}{P_0(d_n = 0)} \prod_{m \in \mathcal{N}_1} \frac{P_1(d_m = 1)}{P_0(d_m = 1)}. \tag{10.58}$$

From the definitions of the decision regions in (10.64) and (10.65), it follows that

$$P_1(d_n = 1) = \sum_{Y_n : L_{Y_n} \geq \tau_n} P_1(Y_n) \text{ and } P_0(d_n = 1) = \sum_{Y_n : L_{Y_n} \geq \tau_n} P_0(Y_n).$$

Consider the region where sensor n decides 1 (defined in (10.64)), $\{R_{n1} : Y_n \in \mathcal{Y}_n : L_{Y_n} = P_1(Y_n)/P_0(Y_n) \geq \tau_n\}$. In this region, we have that

$$P_1(d_n = 1) = \sum_{Y_n : L_{Y_n} \geq \tau_n} P_1(Y_n) \geq \tau_n \sum_{Y_n : L_{Y_n} \geq \tau_n} P_0(Y_n) \geq \tau_n P_0(d_n = 1),$$

or

$$\frac{P_1(d_n = 1)}{P_0(d_n = 1)} \geq \tau_n. \tag{10.59}$$

Similarly, summing over the region where sensor n decides 0 (defined in (10.65)), $\{R_{n0} : Y_n \in \mathcal{Y}_n : L_{Y_n} = P_1(Y_n)/P_0(Y_n) < \tau_n\}$, we have that

$$\frac{P_1(d_n = 0)}{P_0(d_n = 0)} < \tau_n. \tag{10.60}$$

From (10.58), (10.59), and (10.60), we can see that any sensor switching from decision 0 to decision 1 will increase the likelihood ratio at the fusion center.

Now, if the observations are conditionally independent and identically distributed and all the sensors use the same threshold, then

$$P_i(d_n = 1) = \sum_{Y_n : L_{Y_n} \geq \tau} P_i(Y_n) = P_i(d_m = 1)$$

where $i = 0, 1$; $0 \leq m, n \leq N$. Thus we can write (10.58) as follows:

$$\frac{P_1(d_1, d_2, \ldots, d_N)}{P_0(d_1, d_2, \ldots, d_N)} = \left(\frac{P_1(d = 0)}{P_0(d = 0)}\right)^{N - N_1} \left(\frac{P_1(d = 1)}{P_0(d = 1)}\right)^{N_1}. \tag{10.61}$$

The fusion rule compares the likelihood ratio in (10.61) with the ratio π_0/π_1. Again, using (10.59) and (10.60), it can be seen that the likelihood ratio is a non-decreasing function of N_1. Therefore, the likelihood ratio test becomes equivalent to a majority vote rule in this case.

In what follows, we present an example where $L(001) > L(110)$ for the case of three sensors. The observations are taken to be conditionally independent but not conditionally identically distributed. If we use the majority vote, the fusion center will output H_1 if it receives $(1, 1, 0)$ and H_0 if it receives $(0, 0, 1)$. On the contrary, we will show that, if the likelihood ratio test is used, the fusion center will pick $(0, 0, 1)$ against $(1, 1, 0)$ for H_1. Under the independence assumption, the likelihood ratios are:

$$L(110) = \frac{P_1(110)}{P_0(110)} = \frac{P_1(d_1 = 1)}{P_0(d_1 = 1)} \frac{P_1(d_2 = 1)}{P_0(d_2 = 1)} \frac{P_1(d_3 = 0)}{P_0(d_3 = 0)},$$

$$L(001) = \frac{P_1(001)}{P_0(001)} = \frac{P_1(d_1 = 0)}{P_0(d_1 = 0)} \frac{P_1(d_2 = 0)}{P_0(d_2 = 0)} \frac{P_1(d_3 = 1)}{P_0(d_3 = 1)}.$$

Consider the ratio

$$\frac{L(001)}{L(110)} = \frac{P_1(d_1 = 0)P_0(d_1 = 1)}{P_1(d_1 = 1)P_0(d_1 = 0)} \frac{P_1(d_2 = 0)P_0(d_2 = 1)}{P_1(d_2 = 1)P_0(d_2 = 0)} \frac{P_1(d_3 = 1)P_0(d_3 = 0)}{P_1(d_3 = 0)P_0(d_3 = 1)}$$

$$= \frac{[1 - P_1(d_1 = 1)][1 - P_0(d_1 = 0)]}{P_1(d_1 = 1)P_0(d_1 = 0)} \frac{[1 - P_1(d_2 = 1)][1 - P_0(d_2 = 0)]}{P_1(d_2 = 1)P_0(d_2 = 0)}$$

$$\frac{P_1(d_3 = 1)P_0(d_3 = 0)}{[1 - P_1(d_3 = 1)][1 - P_0(d_3 = 0)]}. \tag{10.62}$$

As d_1, d_2, and d_3 are conditionally independent given each hypothesis, we can choose their conditional probabilities such that the ratio in (10.62) is larger than 1. For example, we can choose the conditional probabilities as follows:

$$P_1(d_1 = 1) = P_0(d_1 = 0) = P_1(d_2 = 1) = P_0(d_2 = 0) = 0.6,$$
$$P_1(d_3 = 1) = P_0(d_3 = 0) = 0.9.$$

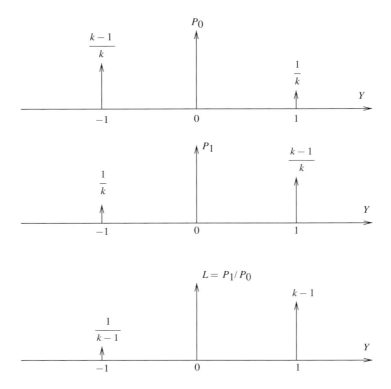

Figure 10.6 The majority vote versus the likelihood ratio test: If P_0 and P_1 of each sensor is as shown, the thresholds for all three quantizers satisfy $1/(k-1) < \tau < k-1$ with $k = 2.5$ for sensor 1 and sensor 2 and $k = 10$ for sensor 3, then $L(001) > L(110)$. A majority vote will output H_1 if it receives $(1,1,0)$ and H_0 if it receives $(0,0,1)$, while the likelihood ratio test favors $(0,0,1)$ for H_1.

Such conditional probabilities can be obtained if we choose P_0 and P_1 as in Figure 10.6 with $k = 2.5$ for sensor 1 and sensor 2, and $k = 10$ for sensor 3; and the thresholds for all three quantizers satisfy $1/(k-1) < \tau < k-1$.

10.5   An algorithm to compute the optimal thresholds

The binary decentralized detection problem with two sensors, binary messages, and the fusion rule fixed a priori is known to be NP-complete [177]. In the light of this, we study in this section a brute-force search algorithm (as a starting point) to solve the underlying optimization problem, which is suitable for small sensor networks.[6]

For each combination of the thresholds at the sensors, $\{\tau_1, \tau_2, \ldots, \tau_N\}$, the fusion rule ($\gamma_0$) is determined based on the likelihood ratio test at the fusion center given in (10.52).

---

[6] For a discussion on the complexity of this kind of algorithm, see references [175] and [177].

Recall that the average probability of error at the fusion center is then given by (10.30). As we are considering the discrete case, where the conditional pdfs of observations are replaced by pmfs, the conditional joint pmfs of the local decisions can be written as:

$$P_i(d_1, d_2, \ldots, d_N) = \sum_{Y_N \in R_{Ni_N}} \cdots \sum_{Y_1 \in R_{1i_1}} P_i(Y_1, Y_2, \ldots, Y_N), \qquad (10.63)$$

where $i_n = 0, 1$, and R_{ni_n} is the region where sensor n decides to send bit i_n, $n = 1, \ldots, N$. These regions are given by

$$R_{n1} = \left\{ Y_n \in \mathcal{Y}_n : L_{Y_n} = \frac{P_1(Y_n)}{P_0(Y_n)} \geq \tau_n \right\} \qquad (10.64)$$

$$R_{n0} = \left\{ Y_n \in \mathcal{Y}_n : L_{Y_n} = \frac{P_1(Y_n)}{P_0(Y_n)} < \tau_n \right\}, \qquad (10.65)$$

where $L_{Y_n} = P_1(Y_n)/P_0(Y_n)$ is the likelihood ratio at sensor n.

Our goal is to find the combination $\{\tau_1, \tau_2, \ldots, \tau_N\}$ that yields the minimum probability of error at the fusion center. If the number of threshold candidates for every sensor is finite, the number of combinations of thresholds will also be finite. Then there is an optimal solution, i.e. a combination of thresholds $\{\tau_1, \tau_2, \ldots, \tau_N\}$ that yields the minimum probability of error. In the algorithms to follow, we address the question of how to pick the threshold candidates for each sensor.

Suppose that we are given a training dataset each record of which has been labeled with either "Normal" or "Attack." Suppose further that each record consists of N parameters, each of which takes values in a finite set. We do not assume that the observations of the sensors (the parameters) are conditionally independent nor identically distributed. The a priori probabilities and the conditional joint pmfs given each hypothesis then can be learnt from the training dataset. Once the optimal thresholds for the sensors have been computed (offline), we can use Algorithm 10.12 to detect attacks in the system. Finally, if we have a labeled dataset where each record has been marked as "Normal" or "Attack," we can compute the error probabilities using Algorithm 10.13. The underlying search algorithm for the optimal thresholds is presented in Algorithm 10.14.

Algorithm 10.12 Using the optimal thresholds for attack detection.

1: Given R records of connection.
2: **for** $r = 1$ to R **do**
3:   Each local sensor quantizes the corresponding parameter into a single bit of information (indicating whether or not there is an attack).
4:   The fusion center collects all the bits from the local sensors and computes the likelihood ratio using (10.63) (the joint conditional pmfs are drawn from the training data).
5:   The fusion center makes the final decision using (10.52).
6: **end for**

Algorithm 10.13 Computing the probabilities of error.

1: Given R records of connection.
2: Compute the actual a priori probabilities ($\overline{\pi}_0$ and $\overline{\pi}_1$), the false-alarm probability ($P_F = P_0(\gamma_0(.) = 1)$, and the misdetection probability ($P_M = P_1(\gamma_0(.) = 0)$).
3: Compute the average probability of error using the equation

$$P_e = \overline{\pi}_0 \times P_F + \overline{\pi}_1 \times P_M. \tag{10.66}$$

Algorithm 10.14 An algorithm to compute the optimal thresholds at the sensors.

1: Given hypotheses H_0 ("Normal") and H_1 ("Attack") and N parameters $\{1, 2, \ldots, N\}$.
2: **for** $j = 1$ to N **do**
3:     Group all possible values of parameter j into b_j equally spaced bins. {In general, b_j's do not have to be equal.}
4: **end for**
5: **for** $i = 0$ to 1 **do**
6:     Compute the a priori probability π_i of hypothesis H_i.
7:     Compute the conditional joint pmfs $P_i(d_1, \ldots, d_n)$ and the conditional marginal pmfs $P_i(d_j)$ of the parameters for hypothesis i.
8: **end for**
9: **for** $j = 1$ to N **do**
10:     Compute the likelihood ratios for parameter j: $\tau_n^1, \tau_n^2, \ldots, \tau_n^{b_n}$ $\{0 \leq \tau_n^1 \leq \tau_n^2 \ldots \leq \tau_n^{b_n} \leq \infty.\}$
11: **end for**
12: **for** $j = 1$ to N **do**
13:     Remove threshold duplications in the likelihood ratios computed from Step 10.14 to have the candidates for the local likelihood ratio test of parameter j:

$$\tau_j^0 = 0 < \tau_j^1 < \tau_j^2 \ldots < \tau_j^{b_j'} < \tau_j^{b_j'+1} = \infty. \tag{10.67}$$

$\{\tau_j^1, \tau_j^2, \ldots, \tau_j^{b_j'}$ are the b_j' values of likelihood ratio of parameter j $\left(b_j' \leq b_j\right).\}$
14: **end for**
15: **for** $j_1 = 0$ to $b_1' + 1$ **do**
16:     **for** $\ldots$ **do**
17:         **for** $j_n = 0$ to $b_n' + 1$ **do**
18:             For each combination $\left\{\tau_1^{j_1}, \tau_2^{j_2}, \ldots, \tau_n^{j_n}\right\}$, determine the fusion rule (γ_0) based on the likelihood ratio test at the fusion center given in (10.52).
19:             For each combination $\left\{\tau_1^{j_1}, \tau_2^{j_2}, \ldots, \tau_n^{j_n}\right\}$, evaluate the average probability of error P_e using (10.30) and (10.63).
20:         **end for**
21:     **end for** $\ldots$
22: **end for**
23: Choose a combination that minimizes P_e.

10.6   Discussion and further reading

While Section 10.2 provides a summary from classical sources as cited in the text and discussed next, Sections 10.3, 10.4, and 10.5 are mainly based on references [126] and [127].

Centralized HT has been examined in many papers and texts (see, for example, reference [138]). Tenney and Sandell [172] were the first to study HT within a decentralized setting, where each of two sensors locally selected its threshold for the likelihood ratio test to minimize a common cost function. Sadjadi [152] later extended this work to accommodate arbitrary numbers of sensors and hypotheses, without, however, considering a fusion center: the cost was a function of the sensor decisions and the true hypothesis.

A comprehensive survey of decentralized detection can be found in reference [175], which examined different decentralized detection structures with both conditionally independent and correlated sensor observations. The complexity of decentralized detection problems was also studied in reference [177]. In reference [77], Hoballah and Varshney proposed a person-by-person optimization (PBPO) scheme to optimize a distributed detection system using the Bayesian criterion. The decentralized detection problem with quantized observations was addressed in reference [53], where the authors also introduced a joint power constraint on the sensors. An extension to reference [53] was given in reference [87], where the constraint was placed on the average cost of the system. References [184] and [185] have studied decentralized sequential detection where either the fusion center or the sensors perform the sequential tests.

For a single sensor, it has been shown in reference [176] that the set of conditional distributions of sensor messages is a compact set, and thus any cost function that is a continuous function on this set will attain a minimum, which corresponds to an optimal quantizer. In a parallel configuration with multiple sensors and a fusion center, if the sensor observations are independent given each hypothesis, it has been shown in reference [176] that there exists an optimal solution over the Cartesian product of the sets of conditional marginal probabilities of sensor messages. However, in several applications of HT such as sensor networks and attack/anomaly detection, it is generally seen that the observations from different sensors may be correlated (see, for example, references [54], [126], [190], [178]). This chapter provides a leading way toward such detection problems.

A Optimization, game theory, and optimal and robust control

This appendix provides some background material on those aspects of optimization, game theory, and optimal and robust control theory which are frequently used in the text. It also serves to introduce the reader to our notation and terminology. For more detailed expositions on the topics covered here, standard references are [31, 136] for game theory, [35, 36] for optimal control, and [29] for robust (H$^\infty$) control.

A.1   Introduction to optimization

We discuss in this section elements of and some key results from optimization in finite-dimensional spaces, including nonlinear, convex and linear programming, and distributed computation. Before we do this, however, it will be useful to introduce the notions of sets, spaces, norms, and functionals, which are building blocks of a theory of optimization.

A.1.1   Sets, spaces, and norms

A set S is a collection of elements. If s is a member (element) of S, we write $s \in S$; if s does not belong to S, we write $s \notin S$. If S contains a finite number of elements, it is called a *finite set*; otherwise it is called an *infinite set*. If the number of elements of an infinite set is countable (i.e. if there is a one-to-one correspondence between its elements and positive integers), we say that it is a *denumerable* (countable) set, otherwise it is a *nondenumerable* (uncountable) set.

A set S with some specific structure attached to it is called a *space*, and it is called a linear *(vector) space* if this specific structure is of an algebraic nature with certain well-known properties, such as it being closed under addition and scalar multiplication, having a unique *zero* element, and satisfying associative and distributive laws. If S is a vector space, a subset of S which is also a vector space is called a *subspace*. An example of a vector space is the *n-dimensional Euclidean space* (denoted by $\mathbb{R}^n$), each element of which is determined by n real numbers. A vector $x \in \mathbb{R}^n$ can be written either as a *row vector* $x = (x_1, \ldots, x_n)$ where $x_1, \ldots, x_n$ are real numbers and denote the components of x, or as a *column vector* which is the "transpose" of $(x_1, \ldots, x_n)$ (written as $x = (x_1, \ldots, x_n)^T$).

Given a finite set of vectors $s_1, \ldots, s_n$ in a vector space S, we say that this set of vectors is *linearly independent* if the equation $\sum_{i=1}^n \alpha_i s_i = 0$ implies that $\alpha_i = 0 \; \forall i = 1, \ldots, n$. Furthermore, if every element of S can be written as a linear combination of these vectors, we say that this set of vectors *generates* S. Now, if S is generated by such a linearly independent finite set (say, X), it is said to be *finite dimensional* with its unique "dimension" being equal to the number of elements of X; otherwise, S is *infinite dimensional*.

A linear vector space S is called a *normed linear vector space* if there is a real-valued function defined on it, which maps each element $u \in S$ into a real number $\|u\|$; such a function is called the *norm* of u. Norm satisfies the following three axioms:

1. $\|u\| \geq 0 \; \forall u \in S$; $\|u\| = 0$ if, and only if, $u = 0$.
2. $\|u + v\| \leq \|u\| + \|v\|$ for each $u, v \in S$.
3. $\|\alpha u\| = |\alpha| \cdot \|u\| \; \forall \alpha \in \mathbb{R}$ and for each $u \in S$.

An infinite sequence of vectors $\{s_1, s_2, \ldots, s_i \ldots\}$ in a normed vector space S is said to *converge* to a vector s if, given an $\varepsilon > 0$, there exists an N such that $\|s - s_i\| < \varepsilon$ for all $i \geq N$. In this case, we write $s_i \rightarrow s$, or $\lim_{i \to \infty} s_i = s$, and call s the *limit point* of the sequence $\{s_i\}$. More generally, a point s is said to be a *limit point* of an infinite sequence $\{s_i\}$ if it has an infinite subsequence $\{s_{i_k}\}$ that converges to s.

An infinite sequence $\{s_i\}$ in a normed vector space is said to be a *Cauchy sequence* if, given an $\varepsilon > 0$, there exists an N such that $\|s_n - s_m\| < \varepsilon$ for all $n, m \geq N$. A normed vector space S is said to be *complete*, or a *Banach space*, if every Cauchy sequence in S is convergent to an element of S. Let S be a normed vector space. Given an $s \in S$ and an $\varepsilon > 0$, the set $N_\varepsilon(s) = \{x \in S : \|x - s\| < \varepsilon\}$ is said to be an *ε-neighborhood* of s. A subset X of S is *open* if, for every $x \in X$, there exists an $\varepsilon > 0$ such that $N_\varepsilon(x) \subset X$. A subset X of S is *closed* if its complement in S is open; equivalently, X is closed if every convergent sequence in X has its limit point in X. Given a set $X \subset S$, the largest subset of X which is open is called the *interior* of X.

A subset X of a normed vector space S is said to be *compact* if every infinite sequence in X has a convergent subsequence whose limit point is in X. If X is finite dimensional, compactness is equivalent to being closed and bounded.

A.1.2   Functionals, continuity, and convexity

A mapping f of a vector space S into a vector space $\mathcal{T}$ is called a *transformation* or a *function*, and is written symbolically as $f : S \rightarrow \mathcal{T}$ or $y = f(x)$, for $x \in S$, $y \in \mathcal{T}$. A transformation f is said to be a *functional* if $\mathcal{T} = \mathbb{R}$.

Let $f : S \rightarrow \mathcal{T}$ where S and $\mathcal{T}$ are normed linear spaces. f is said to be *continuous* at $x_0 \in S$ if, for every $\varepsilon > 0$, there exists a $\delta > 0$ such that $f(x) \in N_\varepsilon(f(x_0))$ for every $x \in N_\delta(x_0)$. If f is continuous at every point of S it is said to be *continuous everywhere* or, simply, *continuous*.

A subset C of a vector space S is said to be *convex* if for every $u, v \in C$ and every $\alpha \in [0, 1]$, we have $\alpha u + (1 - \alpha)v \in C$. A functional $f : C \rightarrow \mathbb{R}$ defined over a convex subset C of a vector space S is said to be *convex* if, for every $u, v \in C$ and every scalar

$\alpha \in [0,1]$, we have $f(\alpha u + (1-\alpha)v) \leq \alpha f(u) + (1-\alpha)f(v)$. If this is a strict inequality for every $\alpha \in (0,1)$, then f is said to be *strictly convex*. A functional f is said to be *concave* if $(-f)$ is convex, and *strictly concave* if $(-f)$ is strictly convex.

A functional $f : \mathbb{R}^n \to \mathbb{R}$ is said to be differentiable if, with $x = (x_1, \ldots, x_n)^T \in \mathbb{R}^n$, the partial derivatives of f with respect to the components of x exist, in which case we write

$$\nabla f(x) = \left[\frac{\partial f(x)}{\partial x_1}, \ldots, \frac{\partial f(x)}{\partial x_n} \right].$$

$\nabla f(x)$ is called the *gradient* of f at x and is a row vector. We shall also use the notation $f_x(x)$ or $df(x)/dx$ to denote the same quantity. If we partition x into two vectors y and z of dimensions n_1 and $n - n_1$, respectively, and are interested only in the partial derivatives of f with respect to the components of y, then we use the notation $\nabla_y f(y,z)$ or $\partial f(y,z)/\partial y$ to denote this partial gradient.

Let $g : \mathbb{R}^n \to \mathbb{R}^m$ be a vector-valued function whose components are differentiable with respect to the components of $x \in \mathbb{R}^n$. Then, we say that $g(x)$ is differentiable, with the derivative $dg(x)/dx$ being an $m \times n$ matrix whose ij-th element is $\partial g_i(x)/\partial x_j$. (Here g_i denotes the i-th component of g.) The gradient $\nabla f(x)$ being a vector, its derivative (which is the second derivative of $f : \mathbb{R}^n \to \mathbb{R}$) will thus be an $n \times n$ matrix, assuming that $f(x)$ is twice continuously differentiable in terms of the components of x. This matrix, denoted by $\nabla^2 f(x)$, is symmetric, and is called the *Hessian matrix* of f at x. This Hessian matrix is non-negative definite for all $x \in \mathbb{R}^n$ if, and only if, f is convex. In general, an $n \times n$ real symmetric matrix M is positive (negative) definite, if $z^T M z > 0$ ($z^T M z < 0$) for all nonzero vectors $z \in \mathbb{R}^n$. Likewise, M is non-negative definite or positive-semidefinite if $z^T M z \geq 0$ for all nonzero $z \in \mathbb{R}^n$.

A.1.3   Optimization of functionals

Given a functional $f : \mathcal{S} \to \mathcal{R}$, where $\mathcal{S}$ is a vector space, and given a subset $\mathcal{X} \subseteq \mathcal{S}$, by the optimization problem

$$\text{minimize } f(x) \text{ subject to } x \in \mathcal{X}$$

we mean the problem of finding an element $x^* \in \mathcal{X}$ (called a *minimizing element* or an *optimal solution*) such that

$$f(x^*) \leq f(x) \qquad \forall x \in \mathcal{X}.$$

This is sometimes also referred to as a *globally minimizing solution* in order to differentiate it from the other alternative – a *locally minimizing solution*. An element $x^\circ \in \mathcal{X}$ is called a locally minimizing solution if we can find an $\varepsilon > 0$ such that

$$f(x^\circ) \leq f(x) \qquad \forall x \in N_\varepsilon(x^\circ) \cap \mathcal{X},$$

i.e. we compare $f(x^\circ)$ with values of $f(x)$ in that part of a certain ε-neighborhood of x°, N_ε, which lies in $\mathcal{X}$.

For a given optimization problem, it is not necessarily true that an optimal (minimizing) solution exists; an optimal solution will exist if the set of real numbers $\{f(x) : x \in \mathcal{X}\}$ is bounded below and there exists an $x^* \in \mathcal{X}$ such that $\inf\{f(x) : x \in \mathcal{X}\} = f(x^*)$, in which case we write

$$f(x^*) = \inf_{x \in \mathcal{X}} f(x) = \min_{x \in \mathcal{X}} f(x).$$

If such an x^* cannot be found, even though $\inf\{f(x) : x \in \mathcal{X}\}$ is finite, we simply say that an optimal solution does not exist; but we declare the quantity

$$\inf\{f(x) : x \in \mathcal{X}\} \text{ or } \inf_{x \in \mathcal{X}} f(x)$$

as the *optimal value* (or the *infimum*) of the optimization problem. If $\{f(x) : x \in \mathcal{X}\}$ is not bounded below, i.e. $\inf_{x \in \mathcal{X}} f(x) = -\infty$, then neither an optimal solution nor an optimal value exists.

An optimization problem which involves maximization instead of minimization may be converted into a minimization problem by simply replacing f by $-f$. Any optimal solution of this minimization problem is also an optimal solution for the initial maximization problem, and the optimal value of the latter, denoted $\sup_{x \in \mathcal{X}} f(x)$, is equal to minus the optimal value of the former. When a *maximizing element* $x^* \in \mathcal{X}$ exists, then $\sup_{x \in \mathcal{X}} f(x) = \max_{x \in \mathcal{X}} f(x) = f(x^*)$.

Existence of optimal solutions

In the minimization problem formulated above, an optimal solution exists if $\mathcal{X}$ is a finite set, since then there is only a finite number of comparisons to make. When $\mathcal{X}$ is not finite, however, the existence of an optimal solution is not always guaranteed; it is guaranteed if f is continuous and $\mathcal{X}$ is compact – a result known as the *Weierstrass theorem*. For the special case when $\mathcal{X}$ is finite dimensional, we should recall that compactness is equivalent to being closed and bounded.

Necessary and sufficient conditions for optimality in the absence of constraints

Let $\mathcal{S} = \mathbb{R}^n$, and $f : \mathbb{R}^n \to \mathbb{R}$ be a differentiable function. If $\mathcal{X}$ is an open set, a first-order necessary condition for an optimal solution to satisfy is

$$\nabla f(x^*) = 0.$$

If, in addition, f is twice continuously differentiable on $\mathbb{R}^n$, a second-order necessary condition is

$$\nabla^2 f(x^*) \geq 0,$$

by which we mean that all eigenvalues of the symmetric matrix $\nabla^2 f(x^*)$ are non-negative. The pair of conditions $\{\nabla f(x^*) = 0, \nabla^2 f(x^*) > 0\}$ (that is, stationarity together with all eigenvalues of $\nabla^2 f(x^*)$ being positive) is sufficient for $x^* \in \mathcal{X}$ to be a locally minimizing solution. These conditions are also sufficient for global optimality if, in addition, $\mathcal{X}$ is a convex set and f is a convex functional on $\mathcal{X}$.

Necessary conditions for optimality in the presence of constraints

If X is not an open set, then there is the possibility for an optimum to be on the boundary of X. To capture these cases, we now bring some structure to the characterization of the constraint set X. Again working with the finite-dimensional case, that is, with $S = \mathbb{R}^n$, let $g_j : \mathbb{R}^n \to \mathbb{R}$ be a continuously differentiable function, for $j = 1, \ldots, m$, where $m < n$. Let $X = \{g_j(x) = 0, \; j = 1, \ldots, m\}$, and consider again the minimization of f over X. This is known as an *optimization problem with equality constraints*. We can write these constraints also using the compact notation $g(x) = 0$, where $g := (g_1, \ldots, g_m)^T$. Let $x^* \in X$ be a locally minimizing solution for this optimization problem, and x^* be a *regular point* of the constraints, meaning that the $m \times n$ matrix $dg(x^*)/dx$ is of full rank m, that is, the *Jacobian* of g is full rank. Then, there exist m scalars, $\{\lambda_j, j = 1, \ldots, m\}$ (called *Lagrange multipliers*), such that the *Lagrangian*

$$L(x;\lambda) = f(x) + \sum_{j=1}^{m} \lambda_j g_j(x) =: f(x) + \lambda^T g(x)$$

has a stationary point at $x = x^*$, that is, $\nabla_x L(x^*;\lambda) = 0$, along with the condition $g(x^*) = 0$. This is, of course, a necessary condition (under the regularity assumption) also for global minima.

We now modify the constraint set by including inequality constraints. Let $h_k : \mathbb{R}^n \to \mathbb{R}$ be a continuously differentiable function, for $k = 1, \ldots, p$, and $X = \{x \in \mathbb{R}^n : g_j(x) = 0, h_k(x) \leq 0, \; j = 1, \ldots, m; k = 1, \ldots, p\}$. Consider again the minimization of f over X. This is now known as an *optimization problem with equality and inequality constraints*. We can write the inequality constraints again using a compact notation as $h(x) \leq 0$, where $h := (h_1, \ldots, h_p)^T$. Let $x^* \in X$ be a locally minimizing solution for this optimization problem. We say that an inequality constraint corresponding to an index k is *active* at $x = x^*$ if $h_k(x^*) = 0$; otherwise an inequality constraint is *inactive* at $x = x^*$. Let $\mathcal{K}^*$ be the set of all indices corresponding to active inequality constraints at $x = x^*$. Then the counterpart of the regularity condition in this case (or rather extension of it) is the so-called *Karush–Kuhn–Tucker constraint qualification condition*, which requires that the vectors

$$\frac{\partial g_j(x^*)}{\partial x}, j = 1, \ldots, m; \quad \frac{\partial h_k(x^*)}{\partial x}, k \in \mathcal{K}^*$$

be linearly independent. Then, there exist multipliers, $\lambda_j, j = 1, \ldots, m$; $\mu_k \geq 0, k = 1, \ldots, p$, such that the *Lagrangian*

$$L(x;\lambda,\mu) = f(x) + \sum_{j=1}^{m} \lambda_j g_j(x) + \sum_{k=1}^{p} \mu_k h_k(x) =: f(x) + \lambda^T g(x) + \mu^T h(x)$$

has a stationary point at $x = x^*$, that is

$$\nabla_x L(x^*;\lambda,\mu) = 0,$$

along with the conditions

$$g(x^*) = 0, \; \mu' h(x^*) = 0, \; h(x^*) \leq 0, \; \mu \geq 0.$$

Note that the condition $\mu^T h(x^*) = 0$ above along with the non-negativity condition on the μ's force $\mu_k = 0$ for indices k not in $\mathcal{K}^*$.

Duality and convex programming

The optimization problem above is known as the *nonlinear programming problem*. Let λ^* and μ^* be the Lagrange multiplier vectors associated with this problem, corresponding to a locally optimal solution x^*. Assume that the Lagrangian $L(x; \lambda^*, \mu^*)$ is twice continuously differentiable at $x = x^*$, and the associated matrix of second partials (that is, the Hessian matrix $\nabla_x^2 L(x; \lambda^*, \mu^*)$) evaluated at $x = x^*$ is positive definite. Consider the unconstrained local minimization of $L(x; \lambda, \mu)$ in an open neighborhood of $x = x^*$ for each λ and $\mu \geq 0$. Denote this local minimum value by $\phi(\lambda, \mu)$, that is

$$\phi(\lambda, \mu) = \min_x L(x; \lambda, \mu),$$

which is known as the *dual function* associated with the nonlinear programming problem. One of the fundamental results in nonlinear programming is that λ^* and μ^* introduced earlier locally maximize the dual function $\phi(\lambda, \mu)$ when the components of μ are restricted to be non-negative. This maximization problem is known as the *dual problem*.

If, in addition to the other assumptions, f and h_k, $k = 0, \ldots, p$ are convex, and g_j, $j = 0, \ldots, m$ are affine (linear plus constant), then we have a *convex programming problem*, and any local minimum is also a global minimum, and thus x^* is globally minimizing. Furthermore, the Lagrangian $L(x; \lambda, \mu)$ is convex for each $\lambda, \mu \geq 0$, and thus the dual function $\phi(\lambda, \mu)$ is obtained as a result of global minimization. Moreover, ϕ is concave in its arguments, and hence λ^* and μ^* are obtained as a result of global maximization of $\phi(\lambda, \mu)$, with λ unconstrained and μ constrained to be non-negative.

What we see above for the convex programming problem is that x^* is the result of an unconstrained **minimization** problem involving the Lagrangian L, when the multipliers are held at their optimum values, whereas the optimum values of the multipliers are obtained by maximizing the dual function, which (one can show) is equivalent to global **maximization** of the Lagrangian L with respect to the multipliers λ and $\mu \geq 0$, when x is held at its optimum value x^*, namely

$$L(x^*; \lambda^*, \mu^*) = \min_{x \in \mathbb{R}^n} L(x; \lambda^*, \mu^*) = \max_{(\lambda, \mu \geq 0)} L(x^*; \lambda, \mu).$$

The preceding equation says that x^* is in *saddle-point equilibrium* with (λ^*, μ^*) with respect to the function L.

The result above in fact holds with somewhat more tightening on x, restricting it to a convex subset, say Ω, of $\mathbb{R}^n$, in other words $\mathcal{X}$ can now be taken as $\mathcal{X} = \{x \in \Omega : g_j(x) = 0, h_k(x) \leq 0, \ j = 1, \ldots, m; k = 1, \ldots, p\}$, assuming that it is not empty, where as before g_j's are affine, h_k's are convex, and the constraint qualification condition holds. $\mathcal{X}$ is then convex, and L is also convex on Ω as a function of x. Saddle-point property again holds, with simply the minimization now being over Ω, that is

$$L(x^*; \lambda^*, \mu^*) = \min_{x \in \Omega} L(x; \lambda^*, \mu^*) = \max_{\lambda, \mu \geq 0} L(x^*; \lambda, \mu).$$

Linear programming

A special case of the convex programming problem that is of wide applicability is the *linear program* (LP), where f is linear, written as $c^T x$, where c is an n-vector, and $h_k(x) = -x_k, k = 1, \ldots, n$. Further, we can write the equality constraint as (since g is affine) $Ax = b$, where A is an $m \times n$ matrix and b is an m-vector. Hence, the constraint set in this case is $X = \{x \in \mathbb{R}^n : Ax = b, x \geq 0\}$, which we assume to be nonempty. This is a convex polytope, which need not be bounded. If it is bounded, however, then X is a convex polyhedron, and the minimum of the linear function $f(x) = c^T x$ has to appear on the boundary of X. Hence, we have either $f(x)$ unbounded (from below) on X, in which case the LP does not admit a solution, or $f(x)$ is bounded from below on X, in which case the LP does admit a globally optimal solution, which is on the boundary.

To obtain the dual function associated with the LP, it is convenient to view the non-negativity constraint on x being captured by the earlier defined convex set Ω, that is $\Omega = \{x \in \mathbb{R}^n : x \geq 0\}$, and forgo the inequality constraint. Then the dual function would have only λ as its argument:

$$\phi(\lambda) = \min_{x \geq 0}[c^T x + \lambda^T(b - Ax)].$$

The dual function would be well defined only if the coefficient of x is non-negative (that is, $c - A^T \lambda \geq 0$), in which case its minimum value is $\lambda^T b$. Hence, the dual problem associated with the LP is

$$\max_{\{\lambda \in \mathbb{R}^m : A^T \lambda \leq c\}} b^T \lambda,$$

which is another LP. Let the solution of the dual problem (if it exists) be λ^*, and the solution of the original LP (known as *primal problem*) be x^*. Then, a fundamental result in linear programming, known as the *dual theorem of linear programming*, is that if either problem has a finite optimal solution, then so does the other. In other words, the existence of x^* implies existence of λ^*, and vice versa. In this case, the corresponding values are equal, $c^T x^* = b^T \lambda^*$. Furthermore, if either problem has an unbounded objective (on the corresponding constraint set), then the other problem has no feasible solution.

Lagrangian decomposition and distributed computation

Consider the original general nonlinear programming problem, with f, g_j's, and h_k's having the following additive structures:

$$f(x) = \sum_{i=1}^{n} f_i(x_i), \quad g_j(x) = \sum_{i=1}^{n} g_{ji}(x_i), \quad h_k(x) = \sum_{i=1}^{n} h_{ki}(x_i).$$

Let $g^i := (g_{1i}, \ldots, g_{mi})^T$ and $h^i := (h_{1i}, \ldots, h_{pi})^T$. Then the Lagrangian can be written as

$$L(x; \lambda, \mu) = \sum_{i=1}^{n} f_i(x_i) + \lambda^T \sum_{i=1}^{n} g^i(x_i) + \mu^T \sum_{i=1}^{n} h^i(x_i) \equiv \sum_{i=1}^{n} L_i(x_i; \lambda, \mu),$$

where

$$L_i(x_i; \lambda, \mu) := f_i(x_i) + \lambda^T g^i(x_i) + \mu h^i(x_i).$$

It can be interpreted as the Lagrangian associated with the i-th component of x. Now note that if λ^* and μ^* are the optimal values of the multipliers for the nonlinear programming problem (local or global), then the optimal (again local or global) value of x is obtained from

$$L(x^*;\lambda^*,\mu^*) = \min_x L(x;\lambda^*,\mu^*) \equiv \sum_{i=1}^{n} \min_{x_i} L_i(x_i;\lambda^*,\mu^*) = \sum_{i=1}^{n} L_i(x_i^*;\lambda^*,\mu^*) \,,$$

that is, instead of solving a single optimization problem in an n-dimensional space, one solves n one-dimensional optimization problems. This leads to savings in complexity in computation (particularly if n is large) in addition to opening the door for a *distributed computation* of the solution, with each computational unit solving only a scalar problem. Such a computation would of course be possible if there is a way of obtaining the optimal values of the multipliers centrally, without the need for the values of x. This can actually be done using the dual function introduced earlier, by maximizing it with respect to the multipliers. One way of doing this would be to use a gradient ascent algorithm, assuming that the dual function is differentiable; for more on this topic, see reference [38].

A.2   Introduction to noncooperative game theory

Game theory deals with strategic interactions among multiple decision makers, called *players*. Each player's preference ordering among multiple alternatives is captured in an objective function for that player, which s/he tries to either maximize (in which case the objective function is a *utility* function or *benefit* function) or minimize (in which case we refer to the objective function as a *cost* function or a *loss* function).

For a nontrivial game, the objective function of a player depends on the choices (*actions*, or equivalently *decision variables*) of at least one other player, and generally of all the players. Hence, a player cannot simply optimize his/her own objective function independent of the choices of the other players. This results in a coupling between the actions of the players, and binds them together in decision making even in a noncooperative environment.

If the players were able to enter into a cooperative agreement so that the selection of actions or decisions is done collectively and with full trust, so that all players would benefit to the extent possible, and no inefficiency would arise, then we would be in the realm of *cooperative game theory*, with issues of bargaining, coalition formation, excess utility distribution, etc. of relevance there. Cooperative game theory will not be covered in this overview; see for example references [67, 136, 188].

If no cooperation is allowed or possible among the players, then we are in the realm of *noncooperative game theory*, where one has to introduce first a satisfactory solution concept. Leaving aside for the moment the issue of how the players can reach a satisfactory solution point, let us address the following question: if the players are at such a satisfactory solution point, what are the minimum features one would expect to see there? To first order, such a solution point should have the property that if all players but one stay put, then the player who has the option of moving away from the solution point should not have any incentive to do so because s/he cannot improve his/her payoff. Note that we cannot allow two or more players to move collectively from the solution point, because such a collective move requires cooperation, which is not allowed in a noncooperative game. Such a solution point where none of the players can improve his/her payoff by a unilateral move is known as a *noncooperative equilibrium* or *Nash equilibrium* (*NE*), named after John Nash, who introduced it and proved that it exists in finite games (that is, games where each player has only a finite number of alternatives), some 60 years ago [124, 125]. This is what we discuss below, following some terminology, a classification of noncooperative games according to various attributes, and a mathematical formulation.

We say that a noncooperative game is *nonzero-sum* if the sum of the players' objective functions cannot be made zero even after appropriate positive scaling and/or translation that do not depend on the players' decision variables. We say that a two-player game is *zero-sum* if the sum of the objective functions of the two players is *zero* or can be made zero by appropriate positive scaling and translation that do not depend on the decision variables of the players. If the two players' objective functions add up to a constant (without scaling or translation), then the game is sometimes called *constant sum*, but according to our convention such games are also zero-sum (since it can be converted to one).

A game is a *finite game* if each player has only a finite number of alternatives, that is the players pick their actions out of finite sets (action sets); otherwise the game is an *infinite game*; finite games are also known as *matrix games*. An infinite game is said to be a *continuous-kernel game* if the action sets of the players are continua (continuums), and the players' objective functions are continuous with respect to action variables of all players. A game is said to be *deterministic* if the players' actions uniquely determine the outcome, as captured in the objective functions, whereas if the objective function of at least one player depends on an additional variable (state of nature) with a known probability distribution, then we have a *stochastic game*.

A game is a *complete information* game if the description of the game (that is, the players, the objective functions, and the underlying probability distributions – if stochastic) is common information to all players; otherwise we have an *incomplete information* game. Finally, we say that a game is *static* if each player acts only once, and none of the players has access to information on the actions of any of the other players; otherwise what we have is a *dynamic game*. A dynamic game is said to be a *differential game* if the evolution of the decision process (controlled by the players over time) takes place in continuous time, and generally involves a differential equation. In this section, we will be covering only static, deterministic, complete information noncooperative games.

A.2.1   General formulation for noncooperative games and equilibrium solutions

We consider an N-player game, with $\mathcal{N} := \{1, \ldots, N\}$ denoting the Players set. The decision or action variable of Player i is denoted by $x_i \in \mathcal{X}_i$, where $\mathcal{X}_i$ is the action set of Player i. Let x denote the N-tuple of action variables of all players, $x := (x_1, \ldots, x_N)$. Allowing for possibly coupled constraints, let $\Omega \subset X$ be the constraint set for the game, where X is the N-product of $X_1, \ldots, X_N$. Hence, for an N-tuple of action variables to be feasible, we need $x \in \Omega$. The players are minimizers, with the objective function (loss function or cost function) of Player i denoted by $L_i(x_i, x_{-i})$, where x_{-i} stands for the action variables of all players except the i-th one.

Now, an N-tuple of action variables $x^* \in \Omega$ is a Nash equilibrium (or, *noncooperative equilibrium*) if, for all $i \in \mathcal{N}$,

$$L_i\left(x_i^*, x_{-i}^*\right) \leq L_i\left(x_i, x_{-i}^*\right), \quad \forall x_i \in X_i, \text{ such that } \left(x_i, x_{-i}^*\right) \in \Omega. \tag{A.1}$$

If $N = 2$, and $L_1 \equiv -L_2 =: L$, then we have a two-player zero-sum game, with Player 1 minimizing L and Player 2 maximizing the same quantity. In this case, the Nash equilibrium becomes the *saddle-point equilibrium*, which is formally defined as follows: a pair of actions $(x_1^*, x_2^*) \in X$ is in *saddle-point equilibrium* for a game with cost function L, if

$$L\left(x_1^*, x_2\right) \leq L\left(x_1^*, x_2^*\right) \leq L\left(x_1, x_2^*\right), \quad \forall (x_1, x_2) \in X. \tag{A.2}$$

Here, we leave out the coupling constraint Ω or simply assume it to be equal to the product set $X := X_1 \times X_2$. This definition also implies that the order in which minimization and maximization are carried out is inconsequential, that is

$$\min_{x_1 \in X_1} \max_{x_2 \in X_2} L(x_1, x_2) = \max_{x_2 \in X_2} \min_{x_1 \in X_1} L(x_1, x_2) = L(x_1^*, x_2^*) =: L^*,$$

where the first expression on the left is known as the *upper value* of the game, the second expression is the *lower value* of the game, and L^* is known as the value of the game.[1]

Note that the value of a game, whenever it exists (which it certainly does if there exists a saddle point), is *unique*. Hence, if there exists another saddle-point solution, say $(\hat{x}_1, \hat{x}_2)$, then $L(\hat{x}_1, \hat{x}_2) = L^*$. Moreover, these multiple saddle points are *orderly interchangeable*, that is, the pairs $(x_1^*, \hat{x}_2)$ and $(\hat{x}_1, x_2^*)$ are also in saddle-point equilibrium. This property that saddle-point equilibria enjoy does not extend to multiple Nash equilibria (for nonzero-sum games): multiple Nash equilibria are generally not interchangeable, and further, they do not lead to the same values for the Players' cost functions, the implication being that when players switch from one equilibrium to another, some players may benefit from that (in terms of reduction in cost) while others may see an increase in their costs. Further, if the players pick randomly (for their actions) from the multiple Nash equilibria of the game, then the resulting N-tuple of actions may not be in Nash equilibrium.

Now coming back to the zero-sum game, a saddle point does not exist if there is no value, which essentially means that the upper and lower values are not equal, i.e. the former is strictly higher than the latter:

$$\min_{x_1 \in X_1} \max_{x_2 \in X_2} L(x_1, x_2) > \max_{x_2 \in X_2} \min_{x_1 \in X_1} L(x_1, x_2).$$

---

[1] Upper and lower values are defined in more general terms using infimum (inf) and supremum (sup) replacing minimum and maximum, respectively, to account for the fact that minima and maxima may not exist. When the action sets are finite, however, the latter always exist.

We then say in this case that *the zero-sum game does not have a saddle point in pure strategies*. This opens the door for looking for a *mixed-strategy* equilibrium.

A *mixed strategy* is for each player a probability distribution over his action set, which we denote by p_i for Player i. This argument also extends to the general N-player game, which may not have a Nash equilibrium in pure strategies (actions, in this case). In search of a mixed-strategy equilibrium, L_i is replaced by its expected value taken with respect to the mixed strategy choices of the players, which we denote for Player i by $J_i(p_1, \ldots, p_N)$. Nash equilibrium over mixed strategies is then introduced as before, with just J_i's replacing L_i's, and p_i's replacing x_i's, and $p_i \in \mathcal{P}_i$, where $\mathcal{P}_i$ is the set of all probability distributions on $\mathcal{X}_i$. We do not bring Ω into the picture here, assuming that the constraint sets are rectangular. If $\mathcal{X}_i$ is finite, then p_i will be a probability vector, taking values in the probability simplex determined by $\mathcal{X}_i$. Then, the N-tuple $(p_1^*, \ldots, p_N^*)$ is in (mixed-strategy) Nash equilibrium if

$$J_i\left(p_i^*, p_{-i}^*\right) \leq J_i\left(p_i, p_{-i}^*\right), \quad \forall\, p_i \in \mathcal{P}_i. \tag{A.3}$$

This readily leads, in the case of zero-sum games, as a special case, to the following definition of a saddle point in mixed strategies: a pair (p_1^*, p_2^*) constitutes a *saddle point in mixed strategies* (or a *mixed-strategy saddle-point equilibrium*), if

$$J\left(p_1^*, p_2\right) \leq J\left(p_1^*, p_2^*\right) \leq J\left(p_1, p_2^*\right), \quad \forall\, (p_1, p_2) \in \mathcal{P},$$

where $J(p_1, p_2) = E_{p_1, p_2}[L(x_1, x_2)]$, and $\mathcal{P} := \mathcal{P}_1 \times \mathcal{P}_2$. Here, $J^* = J(p_1^*, p_2^*)$ is the value of the zero-sum game in mixed strategies.

A.2.2   Existence of Nash and saddle-point equilibria in finite games

Let us first consider zero-sum finite games, or equivalently matrix games. For any such game we have to specify the cardinality of action sets $\mathcal{X}_1$ and $\mathcal{X}_2$ (card $(\mathcal{X}_1)$ and card $(\mathcal{X}_2)$), and the objective function $L(x_1, x_2)$ defined on the product of these finite sets. As per our earlier convention, Player 1 is the minimizer and Player 2 the maximizer. Let card $(\mathcal{X}_1) = m$ and card $(\mathcal{X}_2) = n$, that is, the minimizer has m choices and the maximizer has n choices, and let the elements of $\mathcal{X}_1$ and $\mathcal{X}_2$ be ordered according to some (could be arbitrary) convention. We can equivalently associate an $m \times n$ matrix A with this game, whose entries are the values of $L(x_1, x_2)$, following the same ordering as that of the elements of the action sets, that is ij-th entry of A is the value of $L(x_1, x_2)$ when x_1 is the i-th element of $\mathcal{X}_1$ and x_2 is the j-th element of $\mathcal{X}_2$. Player 1's choices are then the rows of the matrix A and Player 2's are its columns.

It is easy to come up with example matrix games where a saddle point does not exist in pure strategies, with perhaps the simplest one being the game known as *Matching Pennies*, where

$$A = \begin{pmatrix} 1 & -1 \\ -1 & 1 \end{pmatrix}. \tag{A.4}$$

Here, there is no row-column combination at which the players would not have an incentive to deviate and improve their returns.

The next question is whether there exists a saddle point in mixed strategies, in terms of the matrix A, and the probability vectors p_1 and p_2 (both column vectors) we had introduced earlier. Note that p_1 is of dimension m and p_2 is of dimension n, and components of each are non-negative and add up to 1. We can in this case rewrite the expected cost function as

$$J(p_1, p_2) = p_1^T A p_2.$$

By the *minimax theorem*, due to John von Neumann [179], J indeed admits a saddle point, which means that the matrix game A has a saddle point in mixed strategies, that is, there exists a pair (p_1^*, p_2^*) such that for all other probability vectors p_1 and p_2, of dimensions m and n, respectively, the following pair of saddle-point inequalities holds:

$$p_1^{*T} A p_2 \leq p_1^{*T} A p_2^* \leq p_1^T A p_2^*.$$

The quantity $p_1^{*T} A p_2^*$ is the *value* of the game in mixed strategies. This result is now captured in the following *Minimax theorem*.

Theorem A.1 *Every finite two-person zero-sum game has a saddle point in mixed strategies.*

Extension of this result to N-player finite games was provided by John Nash [125], as captured in the following theorem.

Theorem A.2 *Every finite N-player nonzero-sum game has an NE in mixed strategies.*

A standard proof for this result uses Brouwer's fixed point theorem [31].[2] Note that clearly the minimax theorem follows from this one since zero-sum games are special cases of nonzero-sum games. The main difference between the two, however, is that in zero-sum games the *value* is unique (even though there may be multiple saddle-point solutions), whereas in genuine nonzero-sum games the expected cost N-tuple to the players under multiple Nash equilibria need not be the same. In zero-sum games, multiple equilibria have the ordered interchangeability property, whereas in nonzero-sum games they do not.

A.2.3 Existence and uniqueness of Nash and saddle-point equilibria in continuous-kernel (infinite) games

We now go back to the general class of N-player games introduced through (A.1), with $\mathcal{X}_i$ being a finite-dimensional space (for example, m_i-dimensional Euclidean space, $\mathbb{R}^{m_i}$), for $i \in \mathcal{N}$; L_i a continuous function on the product space $\mathcal{X}$, which of course is also finite-dimensional (for example, if $\mathcal{X}_i = \mathbb{R}^{m_i}$, $\mathcal{X}$ can be viewed as $\mathbb{R}^m$, where

---

[2] Brouwer's theorem says that a continuous mapping, f, of a closed, bounded, convex subset (S) of a finite-dimensional space into itself (that is, S) has a fixed point, that is, a $p \in S$ such that $f(p) = p$.

$m := \sum_{i \in \mathcal{N}} m_i$); and the constraint set Ω a subset of $\mathcal{X}$. This class of games is known as *continuous-kernel (or infinite) games with coupled constraints*, and of course if the constraints are not coupled, for example with each player having a separate constraint set $\Omega_i \subset \mathcal{X}_i$, this would also be covered as a special case. Now, further assume that Ω is closed, bounded, and convex, and for each $i \in \mathcal{N}$, $L_i(x_i, x_{-i})$ is convex in $x_i \in \mathcal{X}_i$ for every $x_{-i} \in \times_{j \neq i} \mathcal{X}_j$. Then the basic result for such games is that they admit a Nash equilibrium in pure strategies (but the equilibria need not be unique), as stated in the theorem below; for a proof, see references [31] and [147].

Theorem A.3 *For the N-player nonzero-sum continuous-kernel game formulated above, with the constraint set Ω a closed, bounded, and convex subset of $\mathbb{R}^m$, and with $L_i(x_i, x_{-i})$ convex in x_i for each x_{-i}, and each $i \in \mathcal{N}$, there exists a Nash equilibrium in pure strategies.*

For the special class of similarly structured two-person zero-sum games, the same result clearly holds, implying the existence of a saddle-point solution (in pure strategies). Note that in this case the single objective function ($L \equiv L_1 \equiv -L_2$) to be minimized by Player 1 and maximized by Player 2, is convex in x_1 and concave in x_2, in view of which such zero-sum games are known as *convex–concave games*. Even though convex–concave games could admit multiple saddle-point solutions, they are ordered interchangeable, and the values of the games are unique (which is not the case for multiple Nash equilibria in genuine nonzero-sum games). Now, if the convexity–concavity is replaced by strict convexity–concavity, then the result can be sharpened as below, which, however, has no counterpart for Nash equilibrium in genuine nonzero-sum games.

Theorem A.4 *For a two-person zero-sum game on closed, bounded, and convex finite-dimensional action sets $\Omega_1 \times \Omega_2$, defined by the continuous kernel $L(x_1, x_2)$, let $L(x_1, x_2)$ be strictly convex in x_1 for each $x_2 \in \Omega_2$ and strictly concave in x_2 for each $x_1 \in \Omega_1$. Then, the game admits a unique pure-strategy saddle-point equilibrium.*

If the structural assumptions of Theorem A.3 do not hold, then a pure-strategy Nash equilibrium may not exist, but there may exist one in mixed strategies. In this case, mixed strategy for a player (say, Player i) is a probability distribution on that player's action set, which we take to be a closed and bounded subset, Ω_i, of $\mathcal{X}_i = \mathbb{R}^{m_i}$. We denote a mixed strategy for Player i by p_i, and the set of all probability distribution on Ω_i by $\mathcal{S}_i$. Nash equilibrium, then, is defined using the expected values of L_i's given mixed strategies of all the players, which we denote by J_i as in Section A.2.1. Nash equilibrium is then defined by the N-tuple of inequalities (A.3). The following theorem now states the basic result on existence of mixed-strategy Nash equilibrium in continuous-kernel games.

Theorem A.5 *In the N-player nonzero-sum continuous-kernel game formulated above, let the constrained action set Ω_i for Player i be a closed and bounded subset of $\mathbb{R}^{m_i}$, and $L_i(x_i, x_{-i})$ be continuous on $\Omega = \Omega_1 \times \cdots \times \Omega_N$, for each $i \in \mathcal{N}$. There exists a mixed-strategy Nash equilibrium, $(p_1^*, \ldots, p_N^*)$, in this game satisfying (A.3).*

The existence of a pure-strategy Nash equilibrium does not preclude the existence also of a genuine mixed-strategy Nash equilibrium,[3] and all such (multiple) Nash equilibria are generally non-interchangeable, unless the game is (or is strategically equivalent to) a zero-sum game.

A.2.4   Online computation of Nash equilibrium policies

We discuss here some online recursive schemes for the computation of Nash equilibrium in continuous-kernel games. We consider games for which the structural assumptions of Theorem A.3 hold, and further, the action sets of the players are not jointly coupled, that is, $\Omega = \Omega_1 \times \cdots \times \Omega_N$. For simplicity in exposition, let us further assume that for each $i \in \mathcal{N}$, $L_i(x_i, x_{-i})$ is strictly convex in x_i, and not only convex. We already know that for such games there exists a pure-strategy NE.

Consider now the following update mechanism, known as *parallel update*:

$$x_i^{(n+1)} = \arg\min_{\xi \in \Omega_i} L\left(\xi, x_{-i}^{(n)}\right) =: \nu_i\left(x_{-i}^{(n)}\right), \quad n = 0, 1, \ldots; \ i \in \mathcal{N},$$

where the minimum exists and is unique because Ω_i is a closed, bounded, convex subset of a finite-dimensional space, and L_i is strictly convex in x_i. Note that in this update algorithm, each player is responding to the action choices of the other players at the previous step of the iteration, that is, the sequence is generated using the optimum response (ν_i) of each player, and the players update *in parallel*. The algorithm generally starts with arbitrary selections out of the players' action sets, at time $t = 0$, and clearly if the sequence generated converges, that is, $x_i^n \to x_i^*$, $i \in \mathcal{N}$, then the x_i^*'s constitute an NE, and if, furthermore, the limit is independent of the starting choices, that is, of $\left\{x_i^{(0)}, i \in \mathcal{N}\right\}$, then the NE is unique.

Parallel update is not the only possible update scheme in the online computation of NE. One could have *round robin*, where the players take turns in their responses, with only one player being active at each step of the iteration, following a strict order; or *random polling*, where at each point of time one player is picked at random to be the active player. Another alternative is the stochastic *asynchronous update* or *random update*, where at each step of the iteration a subset of $\mathcal{N}$ is picked at random, and only those players belonging to that subset become active players, generating their next actions as in parallel update. For convergence in this latter case (as well as in random polling), it is necessary for each player to be active infinitely many times, i.e. no player can be left out of the update process for extended periods of time. It is worth noting that in all these online computational algorithms, the players do not have to know the objective functions of the other players; the only information each player needs is the most recent actions of other players and his/her own reaction function, which is unique under the assumption of strict convexity.

---

[3] The qualifier *genuine* is used here to stress the point that mixed strategies in this statement are not pure strategies (even though pure strategies are indeed special types of mixed strategy, with all probability weight concentrated on one point).

All four of these update mechanisms can be written in a unifying compact form: for all $i \in \mathcal{N}$, and for $n = 0, 1, \ldots,$

$$x_i^{(n+1)} = \mathsf{v}_i \left(x_{-i}^{(n)} \right), \quad \text{if } i \in K_n$$
$$= x_i^{(n)}, \quad \text{else}$$

where K_n is the set of active players at step n. It is $\mathcal{N}$ for parallel update; $K_n = \{(n + k) \bmod N + 1\}$ with k arbitrary, in round robin; it is an independent process on $\mathcal{N}$ in random polling; and it is an independent set-valued process on all subsets of $\mathcal{N}$ in stochastic asynchronous update. In each case, we have a mapping, say ρ, of Ω into itself, and NE is a *fixed point* of ρ, that is, $x^* = \rho(x^*)$. If ρ is a contraction on Ω,[4] then the sequence it generates converges, and the NE is unique. This is in fact one of the most effective ways of showing the uniqueness of NE. For a demonstration of this for different classes of nonzero-sum games, see references [27, 30, 97], and for applications in communication systems and networks, see, e.g., references [7, 10, 21].

If a noncooperative game does not admit a pure-strategy NE, but a mixed-strategy one, then the iterations above will have to be modified. For example, in the two-person zero-sum matrix game of (A.4), a parallel update or round robin will lead to indefinite cycling and would never converge. In fact for a mixed-strategy equilibrium,[5] the players will have to build empirical frequencies of the other players' actions over the course of the game, and pick their actions at each stage so as to optimize their expected costs under these empirical probabilities. If the sequence of empirical probabilities thus generated converge, then they converge to the mixed-strategy NE (assuming that it is unique). Such a scheme was first introduced in the context of zero-sum matrix games by Robinson [145], and is known as *fictitious play*. This is still an active area of research, and is also discussed in Chapter 5 of this book, where further references can be found.

A.3   Introduction to optimal and robust control theory

Optimal control theory deals with the selection of best inputs over time for a dynamic system – *best* in the sense of optimization of a performance index. *Robust control theory* deals with the selection of inputs that are *robust* to the presence of unmodeled disturbances. The time interval over which the inputs are picked could be continuous or discrete, with the underlying problems respectively called *continuous-time* or *discrete-time*.

In this section, we discuss some key elements of optimal and robust control problems, and present some of the main results. The section comprises four subsections. The first two deal with *dynamic programming* applied to discrete-time and

---

[4] *Contraction* means that there exists a $\lambda \in (0, 1)$, such that, using the Euclidean norm, $\| \cdot \|$, $\| \rho(x) - \rho(\xi) \| \leq \lambda \| x - \xi \|$, $\forall x, \xi \in \Omega$. If $\rho : \Omega \to \Omega$, and Ω is a compact subset of a complete vector space, then ρ has a fixed point.

[5] In the zero-sum matrix game of (A.4), the unique mixed-strategy saddle point is for each player to mix between his/her two choices with equal probability, $1/2$.

continuous-time optimal control problems, the third is on the alternative theory provided by the *minimum principle*, and the fourth and last section deals with robust optimal control (more precisely, H^∞ control) for only linear-quadratic systems.

A.3.1 Dynamic programming for discrete-time systems

The method of dynamic programming (DP) is based on *the principle of optimality*, which states that an optimal strategy (or decision) has the property that, whatever the initial state and time are, all remaining decisions from that particular initial state and particular initial time onwards must also constitute an optimal strategy. To exploit this principle, one works in retrograde time, starting at all possible final states with the corresponding final times.

Consider the dynamic system whose n-dimensional state x evolves according to the difference equation:

$$x_{k+1} = f_k(x_k, u_k), \quad u_k \in \mathcal{U}_k, \ k = 1, \ldots, K,$$

where u_k is the r-dimensional control input at the discrete time instant k, chosen from the constraint set $\mathcal{U}_k$, and f_k is the function characterizing the system dynamics. The performance index is given by the *cost functional* (to be minimized)

$$J(u) = \sum_{k=1}^{K} g_k(x_{k+1}, u_k, x_k),$$

where $u := \{u_k, k = 1, \ldots, K\}$, and $g_k, k = 1, \ldots, K$, is the so-called stagewise cost.

The control is allowed to depend on the current value of the state, $u_k = \mu_k(x_k)$, where $\mu_k(\cdot)$ denotes a permissible (control) strategy at stage k. Such a control function is known as *feedback control*, and dynamic programming makes it possible to obtain the optimal feedback control, as opposed to the open-loop control which depends only on the initial state, x_1, for all time k. Now, in order to determine the minimizing feedback control strategy, we will need the expression for the *minimum* cost from any starting point (state) at any starting time. This is also called the *value function*, and is defined as

$$V(k,x) = \min_{\mu_k, \ldots, \mu_K} \left[\sum_{i=k}^{K} g_i(x_{i+1}, u_i, x_i) \right]$$

with $u_i = \mu_i(x_i) \in \mathcal{U}_i$ and $x_k = x$. A direct application of the principle of optimality leads to the recursive relation (DP equation)

$$V(k,x) = \min_{u_k \in U_k} [g_k(f_k(x, u_k), u_k, x) + V(k+1, f_k(x, u_k))].$$

If the optimal control problem admits a solution $u^* = \{u_k^*, k = 1, \ldots, K\}$, then the solution $V(1, x_1)$ of the recursive equation above should equal $J(u^*)$, and furthermore each u_k^* should be determined as an argument of the right-hand-side of the DP equation.

Affine-quadratic problems

An important special case is the so-called *affine-quadratic (or linear-quadratic) discrete-time optimal control* problem, which is described by the state equation

$$x_{k+1} = A_k x_k + B_k u_k + c_k,$$

and cost functional

$$J(u) = \frac{1}{2} \sum_{k=1}^{K} \left(x_{k+1}^T Q_{k+1} x_{k+1} + u_k^T R_k u_k \right),$$

where, $c_k \in \mathbb{R}^n$, $R_k > 0$ is an $r \times r$ matrix, $Q_{k+1} \geq 0$ is an $n \times n$ matrix for all $k = 1, \ldots, K$, and $A_{(\cdot)}$, $B_{(\cdot)}$ are matrices of appropriate dimensions. Here, the corresponding expression for f_k is obvious, but the one for g_k is not uniquely defined; however, it is more convenient to take it as

$$g_k(u_k, x_k) = \begin{cases} u_1^T R_1 u_1 & , k = 1 \\ u_k^T R_k u_k + x_k^T Q_k x_k & , k \neq 1, K+1 \\ x_{K+1}^T Q_{K+1} x_{K+1} & , k = K+1. \end{cases}$$

It follows by inspection that $V(k, x)$ is a general quadratic function of x for all $k = 1, \ldots, K$, and that $V(K+1, x_{K+1}) = \frac{1}{2} x_{K+1}^T Q_{K+1} x_{K+1}$. This leads to the structural form $V(k, x) = \frac{1}{2} x^T S_k x + x^T s_k + q_k$. Substitution of this in the DP equation yields the unique solution of the affine-quadratic optimal control problem to be:

$$u_k^* = \mu_k^*(x_k) = -P_k S_{k+1} A_k x_k - P_k(s_{k+1} + S_{k+1} c_k), \ k = 1, \ldots, K,$$

where

$$P_k = \left[R_k + B_k^T S_{k+1} B_k \right]^{-1} B_k^T$$
$$S_k = Q_k + A_k^T S_{k+1}[I - B_k P_k S_{k+1}] A_k; \quad S_{k+1} = Q_{K+1}$$
$$s_k = A_k^T [I - B_k P_k S_{k+1}]^T [s_{k+1} + S_{k+1} c_k]; \quad s_{K+1} = 0.$$

The equation generating the $\{S_k\}$ above is known as the *discrete-time Riccati equation*. Furthermore, the minimum value of $J(u)$ is

$$J(u^*) = \frac{1}{2} x_1^T S_1 x_1 + x_1^T s_1 + q_1$$

where

$$q_1 = \frac{1}{2} \sum_{k=1}^{K} c_k^T S_{k+1} c_k - (s_{k+1} + S_{k+1} c_k)^T P_k^T B_k^T (s_{k+1} + S_{k+1} c_k) + 2 c_k^T s_{k+1}.$$

Infinite horizon linear-quadratic problems

We now consider the *time-invariant* version of the *linear-quadratic* optimal control problem (that is, with $c_k = 0$ for all k, and matrices $A_k, B_k, Q_k,$ and R_k independent of k, which we henceforth write without the index k) when the number of stages is infinite (that is, $K \to \infty$). This is known as the *discrete-time infinite-horizon linear-quadratic optimal control problem*, which is formulated as

$$\min_u \frac{1}{2} \sum_{k=1}^{\infty} \left(x_{k+1}^T Q x_{k+1} + u_k^T R u_k \right); \quad Q \geq 0, R > 0$$

subject to

$$x_{k+1} = A x_k + B u_k, \; k = 1, \ldots$$

For the problem to be well defined, one has to assume that there exists at least one control sequence under which the infinite-horizon cost is finite, and that the optimization process yields a control that results in a stable closed-loop system.

The conditions that guarantee this are stabilizability of the matrix pair (A, B), and detectability of the matrix pair (A, D), where D is a matrix such that $D^T D = Q$. These notions of stabilizability and detectability belong to the realm of the theory of linear systems; see, for instance, references [83] or [22]. The pair (A, B) is stabilizable if an $r \times n$ matrix F exists such that $(A + BF)$ is a stable matrix for a discrete-time linear system. In other words, all vector sequences generated by the system $x_{k+1} = (A + BF) x_k$ corresponding to all initial conditions x_1 converge asymptotically to *zero*, or equivalently all eigenvalues of $(A + BF)$ lie strictly within the unit circle. The pair (A, D) is detectable if its "dual pair," (A^T, D^T) is stabilizable. In terms of the cost function, this means that all unstable modes of the open-loop system (of the matrix A), which affect the term $x^T D D^T x$, have to be made stable through the optimization process such that the cost remains bounded. This is of course only possible if those modes are stabilizable, which is already guaranteed by the first condition.

Under the conditions above, and for the infinite-horizon optimal control problem posed, there exists a unique time-invariant optimal feedback control law,

$$u^*(t) = \mu^*(x(t)) = F^* x(t) \equiv -[R + B^T SB]^{-1} B^T SA x(t),$$

where S is the unique solution of the discrete-time algebraic Riccati equation (ARE)

$$S = Q + A^T S [I - B(R + B^T SB)^{-1} BS] A \qquad \text{(DT-ARE)}$$

within the class of non-negative definite matrices.[6,7] Furthermore, the (closed-loop) matrix $A - B(R + B^T SB)^{-1} BS$ is stable (that is, all its eigenvalues lie strictly within the unit circle), and the minimum value of the cost functional is $\frac{1}{2} x_1^T S x_1$.

A.3.2 Dynamic programming for continuous-time systems

We provide here the counterparts of the results of the previous section in the continuous time. The system dynamics and the cost function for a finite-horizon problem are given by

---

[6] This solution is the convergent limit of the sequence $\{S_k\}$ generated by the discrete-time Riccati equation, for any starting condition $S_{K+1} = Q_{K+1} \geq 0$ as $k \to -\infty$, or equivalently if $S_k^{(K)}$ denotes the solution of the finite-horizon discrete-time Riccati equation, then $S_k^{(K)} \to S$, the unique solution of (DT-ARE) in the class of non-negative definite matrices, for each finite k as $K \to \infty$.

[7] Under the further condition that (A, D) is *observable*, every non-negative-definite matrix solution of DT-ARE is positive definite.

$$\dot{x} = f(t, x, u(t)), \; x(0) = x_0, t \geq 0$$

and

$$J(u) = \int_0^T g(t, x, u) \, dt + q(x(T)),$$

where f, g, and q are assumed to be continuous in their arguments. Here the interval of interest is $[0, T]$, q denotes the terminal cost, and g the incremental cost. The control u is generated again using the feedback policy, $u(t) = \mu(t, x(t))$, where μ has to satisfy additional conditions such that when it is substituted for u in the state equation, the resulting differential equation admits a unique continuously differentiable solution for each initial condition $x(0) = x_0$. One such set of conditions is for $\mu(t, x)$ to be continuous in t and Lipschitz in x, in addition to f being continuous in t and Lipschitz in x and u. Such control laws are said to be *admissible*; we henceforth restrict the analysis to the class of admissible feedback strategies.

The minimum cost-to-go from any initial state (x) and any initial time (t) is described by the so-called *value function*, $V(t, x)$, which is defined by

$$V(t, x) = \min_{\{u(s), s \geq t\}} \left[\int_t^T g(s, x(s), u(s)) \, ds + q(x(T)) \right],$$

satisfying the boundary condition

$$V(T, x) = q(x).$$

Let us assume that V is continuously differentiable in both arguments. Then, a direct application of the *principle of optimality* on this problem leads to the HJB equation, which is a partial differential equation:

$$-\frac{\partial V(t, x)}{\partial t} = \min_u \left[\frac{\partial V(t, x)}{\partial x} f(t, x, u) + g(t, x, u) \right]; \; V(T, x) = q(x).$$

If a continuously differentiable solution to this HJB equation exists, then it generates the optimal control law through the static (pointwise) minimization problem defined by its right-hand side, leading to $u^*(t) = \mu^*(t, x)$. Note that this is a closed-loop state-feedback controller, which has the property that if some non-optimal control is applied initially in some subinterval $[0, s)$, then μ^* is still the optimal controller from that point on, that is, on the remaining interval $[s, T]$; such controllers are called *strongly time consistent* [31].

Affine-quadratic problems

The HJB equation does not generally admit a closed-form solution, but it does for *affine-quadratic* (or *linear-quadratic*) problems. In these problems (as the continuous-time counterpart of the earlier discrete-time problem), the system dynamics are described by

$$\dot{x} = A(t)x + B(t)u(t) + c(t); \quad x(0) = x_0,$$

and the cost functional to be minimized is given as

$$J(u) = \frac{1}{2} x^T(T) Q_f x(T) + \frac{1}{2} \int_0^T (x^T Q x + 2x^T p + u^T R u) \, dt,$$

where $A(\cdot)$, $B(\cdot)$, $Q(\cdot) \geq 0$, $R(\cdot) > 0$ are matrices of appropriate dimensions and with continuous entries on $[0, T]$. Q_f is a non-negative definite matrix, and $c(\cdot)$ and $p(\cdot)$ are continuous vector-valued functions, taking values in $\mathbb{R}^n$. Then, the value function is general quadratic, in the form

$$V(t, x) = \frac{1}{2} x^T S(t) x + k^T(t) x + m(t),$$

where $S(\cdot)$ is a symmetric $n \times n$ matrix with continuously differentiable entries, $k(\cdot)$ is a continuously differentiable n-vector, and $m(\cdot)$ is a continuously differentiable function. Substitution of this structure into the HJB equation leads to

$$-\frac{1}{2} x^T \dot{S} x - x^T \dot{k} - \dot{m}$$
$$= \min_{u \in \mathbb{R}^n} \left[(Sx + k)^T (Ax + Bu + c) + \frac{1}{2} x^T Q x + x^T p + \frac{1}{2} u^T R u \right].$$

Carrying out the minimization on the right-hand side yields

$$u^*(t) = \mu^*(t, x(t)) = -R^{-1} B^T [S(t) x(t) + k(t)],$$

substitution of which back into the HJB equation leads to an identity relation which is readily satisfied if

$$\dot{S} + SA + A^T S - SBR^{-1} B^T S + Q = 0, \quad S(T) = Q_f,$$
$$\dot{k} + (A - BR^{-1} B^T S)^T k + Sc + p = 0, \quad k(T) = 0,$$
$$\dot{m} + k^T c - \frac{1}{2} k^T BR^{-1} B^T k = 0, \quad m(T) = 0,$$

where the S equation is known as the *Riccati differential equation* (RDE), which admits a unique continuously differentiable solution which is non-negative definite. The other two equations (for k and m) being affine also admit unique continuously differentiable solutions. The end result is that μ^* above is indeed the unique optimal control law, and the minimum value of the cost functional is

$$J(\mu^*) = \frac{1}{2} x_0^T S(0) x_0 + k^T(0) x_0 + m(0).$$

Infinite-horizon linear-quadratic problems

Meaningful continuous-time linear-quadratic optimal control problems when $T \to \infty$ can also be formulated, paralleling the development in the discrete-time case discussed in the previous section. Toward that end we take $c(t) \equiv 0$, and the matrices $A(t), B(t), Q(t)$ and $R(t)$ to be independent of t. Then the problem becomes

$$\min_u \frac{1}{2} \int_0^\infty (x^T Q x + u^T R u) \, dt; \quad Q \geq 0, \ R > 0,$$

subject to

$$\dot{x} = Ax + Bu.$$

Conditions which ensure that this problem is well defined (with a finite minimum) are precisely those of the discrete-time analog of this problem, i.e. the matrix pair (A, B)

must be stabilizable and the matrix pair (D,A), where D is a matrix such that $D^T D = Q$, must be detectable; see, for instance, reference [83] for discussions on these notions in the continuous time. The pair (A,B) is stabilizable if an $r \times n$ matrix F exists such that $(A+BF)$ is stable in the sense of continuous-time systems, i.e. all its eigenvalues lie in the open left half of the complex plane (note the difference of this definition of stability from that for its discrete-time counterpart). Furthermore, the pair (D,A) is detectable if its "dual pair," (A^T, D^T) is stabilizable. Under these two conditions, the infinite-horizon LQ optimal control problem (also known as the LQR) admits a unique time-invariant optimal feedback control law,

$$u^*(t) = \mu^*(x(t)) = -R^{-1} B^T S x(t),$$

where S is the unique solution of the continuous-time ARE

$$SA + A^T S - SBR^{-1}B^T S + Q = 0 \qquad \text{(CT-ARE)}$$

within the class of non-negative definite matrices.[8,9] Furthermore, the (closed-loop) matrix $A - BR^{-1}B^T S$ is stable (i.e. all its eigenvalues lie in the open left half of the complex plane), and the minimum value of the cost functional is $\frac{1}{2}x_0^T S x_0$.

A.3.3   The minimum principle

The dynamic programming approach to optimal control, discussed in the previous section, yields a state-feedback control policy, μ^*, which is a function of time, t, and the current value of the state, x. A corresponding open-loop control policy, that is, one that depends only on time and the initial state, can be obtained by substituting the feedback control in the state equation, solving for the corresponding (optimal) state trajectory, and then evaluating the feedback control policy on that optimal trajectory. Hence, if $x^*(t), t \geq 0$, is the optimal state trajectory (in continuous time), then it will be generated by the differential equation

$$\dot{x}^* = f(t, x^*, \mu^*(t, x^*)), \quad x^*(0) = x_0,$$

and the optimal open-loop control will be given by

$$u^*(t) = \mu^*(t, x^*(t)),$$

where we have suppressed the dependence on the initial state x_0. This *control function* also minimizes the cost

$$J(u) = \int_0^T g(t, x, u)\, dt + q(x(T)),$$

---

[8] This solution S has the property that for fixed t, and for every $Q_f \geq 0$, the unique non-negative definite solution to the continuous-time RDE, $S^T(t)$, converges to S (uniformly in $t < T$) as $T \to \infty$.

[9] Under the stronger condition that (A,D) is observable (a sufficient condition for which is $Q > 0$), the solution to CT-ARE (in the class of non-negative definite matrices) is positive definite.

under the dynamic state constraint

$$\dot{x} = f(t, x, u(t)), \; x(0) = x_0, t \geq 0,$$

and hence solves the optimal control problem. But, it is not strongly time consistent, that is, if there is any deviation in the state trajectory at some time $t = s$ due to some non-optimal control input over the interval $(0, s)$, u^* will no longer be optimal from that point on. If there is no such deviation, however, then re-optimizing from time s on (with new initial condition $x(s) = x^*(s)$) will lead still to u^* as the optimal control over $[s, T]$; to capture this one says that u^* is *time consistent*, but not strongly (as in the case of the optimal state-feedback control).

Now the question is whether the optimal open-loop control can be derived directly, without resorting to dynamic programming and the solution of the HJB equation. The answer to this question lies in the *minimum principle*, developed by Pontryagin in the late 1950s [137]. Introduce the Hamiltonian

$$H(t, p, x, u) = g(t, x, u) + p^T(t) f(t, x, u),$$

where $p(\cdot)$ is the *co-state variable*. If $u^*(\cdot)$ is the open-loop controller that solves the optimal control problem, then it minimizes the Hamiltonian for each t, that is[10]

$$u^*(t) = \arg \min_u H(t, p(t), x^*(t), u),$$

where $x^*(\cdot)$ is the optimum state

$$\dot{x}^*(t) = \left(\frac{\partial H}{\partial p} \right)^T = f(t, x^*(t), u^*(t)), \; x^*(0) = x_0,$$

and the co-state p solves the retrograde differential equation

$$\dot{p}^T(t) = -\frac{\partial H(t, p(t), x^*(t), u^*(t))}{\partial x}, \; p^T(T) = \frac{\partial q(x^*(T))}{\partial x}.$$

The minimum principle applies to a broader class of optimal control problems where the solution need not be smooth (for example, the optimal open-loop control could be of the *bang-bang* type, switching back and forth between extreme values), but when the solution is smooth, and the HJB equation admits a continuously differentiable solution, the co-state variable is equal to the gradient of the optimal value function evaluated on the optimal trajectory, that is

$$p^T(t) = \partial V(t, x^*(t))/\partial x,$$

which is the connection between the HJB approach to optimal control and the minimum principle.

---

[10] If there are any pointwise constraints on u for each t, such as $u(t) \in U(t)$, where $U(t)$ is a given constraint set in $\mathbb{R}^r$, then the minimization below will be over this set $U(t)$.

A.3.4   H$^\infty$-optimal control

We discuss in this section a robust control approach to optimal control, first when the state dynamics are disturbed by an unknown input (so-called *disturbance input*), and then when the measurements are also perturbed by disturbance (or unknown noise). Hence, the earlier state dynamics are now replaced by

$$\dot{x} = f(t; x(t), u(t), w(t)), \quad x(0) = x_0, \ t \geq 0,$$

where $w(\cdot)$ is the unknown disturbance input, of dimension not necessarily the same as that of u or that of x. The cost functional is still given as before

$$L(u, w) = \int_0^T g(t, x(t), u(t)) \, dt + q(x(T)),$$

where now we have w also appearing as an argument because L depends on w through x. The objective is to find the controller (say, state-feedback controller) under which some *safe* value of the performance is obtained regardless of the choice of w. One criterion is to make the ratio of L to the square of the norm of w as small as possible for worst (maximizing) values of w. For mathematical convenience, one also lumps the norm of x_0 with that of w, by viewing x_0 as also unknown, leading to the ratio

$$L(u, w) / [\|w\|^2 + q_o(x_0)]$$

where

$$\|w\|^2 := \int_0^T w^T(t) w(t) \, dt,$$

and $q_o(x_0)$ is a positive cost on x_0, which is *zero* at zero, such as the square of the Euclidean norm. We want to find the state-feedback control law μ which minimizes the maximum of the ratio over all (unrestricted) w's and x_o's. If that *min max*, or more precisely the *sup inf* value is denoted by γ^*, even though it may not be achieved, one would be looking at feedback controls that achieve (for the ratio) some $\gamma > \gamma^*$, known as *disturbance attenuation*. In the linear-quadratic case (that is, when the dynamics are linear and L is quadratic, supremum (over all w's) of this ratio (or of its square root) is known as the H$^\infty$ norm, and hence the controller sought would be minimizing the H$^\infty$ norm — the reason why this class of problems is known as *H$^\infty$-optimal control* or simply *H$^\infty$-control*. An application of H$^\infty$-optimal control to network security is presented in Chapter 7.

It can be shown that minimization of this H$^\infty$ norm is equivalent to solving a game (a zero-sum differential game), with dynamics as given above, and with a parameterized objective function

$$L_\gamma(u, w) = L(u, w) - \gamma^2 \|w\|^2 - \gamma^2 q_0(x_0),$$

where the scalar $\gamma > \gamma^*$, and we take x_0 at this point to be fixed and known (instead of being controlled by an adversary). This objective function will be maximized by the disturbance w (which we can call the *maximizing player*) and minimized by the controller, $u(t) = \mu(t, x(t))$ (which we can call the *minimizing player*), and the maximizing disturbance can also be allowed to depend on the state, that is, $w(t) = v(t, x(t))$. If this game admits a saddle-point solution, $(\mu_\gamma^*, v_\gamma^*)$ over the policy spaces of the two players,[11] that is

$$\min_\mu L_\gamma(\mu, v_\gamma^*) = \max_v L_\gamma(\mu_\gamma^*, v) = L_\gamma(\mu_\gamma^*, v_\gamma^*) =: L_\gamma^*,$$

then L_γ^* is known as the *value* of the game. If we consider a version of the same game over a (truncated) shorter interval, say $[s, T]$, with the value of the state at time $t = s$ being $x(s) = x$, then we can introduce the *value* of that truncated (in time) game as a function of s and x, which we denote by $V_\gamma(s, x)$. If V is continuously differentiable in the two arguments, then it satisfies the *Isaacs* equation (or the *Hamilton–Jacobi–Isaacs* (*HJI*) equation),[12] which is the counterpart (in fact, generalization) of the HJB equation that arises in optimal control, which was introduced in the previous section:

$$\frac{\partial V(t, x)}{\partial t} = \inf_u \sup_w \left[\frac{\partial V(t, x)}{\partial x} f(t, x, u, w) + g(t, x, u) - \gamma^2 |w|^2 \right]$$

$$= \frac{\partial V(t, x)}{\partial x} f(t, x, \mu^*(t, x), v^*(t, x))$$

$$+ g(t, x, \mu^*(t, x)) - \gamma^2 |v^*(t, x)|^2$$

$$V(T, x) = q(x),$$

where we have suppressed the dependence of the saddle-point policies on the parameter γ. Then the value of the game over the original horizon, $[0, T]$, will be $V(0, x_0) - \gamma^2 q_o(x_0)$, and if x_0 is also viewed as being controlled by the disturbance (adversary), its worst value will be the one that maximizes this function, that is

$$x_0^* = \arg \max_{x \in R^n} \left[V(0, x) - \gamma^2 q_o(x) \right],$$

leading to $V_\gamma(0, x_0^*) - \gamma^2 q_o(x_0^*)$ as the value of the game for each $\gamma > \gamma^*$.

Of course, γ^* is not known a priori, and has to be computed a posteriori. We now discuss this further for the case when the state dynamics are linear and the objective function is quadratic, and the initial state is $x_0 = 0$, which is the original H^∞-control problem, when $T \to \infty$. But we first discuss the finite-horizon time-varying problem, and subsequently the infinite-horizon one.

LQ systems with state-feedback controller: finite-horizon case

Consider now the following structural choices for f, g, and q, and take the initial state to be $x_0 = 0$:

$$f(t, x, u, w) = A(t) x + B(t) u(t) + D(t) w(t),$$

$$g(t, x, u) = \frac{1}{2} \left[x^T Q(t) x + u^T u \right], \quad q(x) = \frac{1}{2} x^T Q_f x,$$

---

[11] Since the game is parameterized in γ, the saddle-point solution will depend on this parameter, which is why we have γ appearing as a subscript in the two policies.

[12] This was first introduced by Rufus Isaacs while he was working at RAND Corporation in the early 1950s; his book, *Differential Games*, later appeared in 1965 [81].

where $A(\cdot), B(\cdot), D(\cdot), Q(\cdot) \geq 0$ are matrices of appropriate dimensions with continuous entries, and $Q_f \geq 0$. Using this structure in the HJI equation leads to the solution $V_\gamma(t, x) = \frac{1}{2} x^T Z_\gamma(t) x$, where $Z_\gamma(\cdot) \geq 0$ solves the following generalized Riccati differential equation:

GRDE-1 :

$$\dot{Z} + A^T Z + ZA - Z(BB^T - \gamma^{-2} DD^T)Z + Q = 0; \ Z(T) = Q_f,$$

assuming of course that such a solution exists. Under this condition, the saddle-point solution of the underlying differential game is

$$\mu_\gamma^*(t, x) = -B(t)^T Z_\gamma(t) x, \quad v_\gamma^*(t, x) = \gamma^{-2} D(t)^T Z_\gamma(t) x.$$

Note that the GRDE-1 becomes identical to the RDE encountered in the context of the LQ optimal control problem when $\gamma^{-2} = 0$, that is, when there is infinite cost on the disturbance, in which case we have already noted that there exists a unique solution. In view of this, let us introduce the following:

$$\hat{\gamma} := \inf\{\gamma > 0 : \text{GRDE-1 admits a solution on } [0, T]\}.$$

Then, $\gamma^* = \hat{\gamma}$, that is the optimum disturbance attenuation level, γ^*, can be computed offline in connection with the solution of the GRDE-1. For each $\gamma > \gamma^*$, the saddle-point property of the state-feedback control policy, μ_γ^*, introduced above, implies that it achieves the performance bound γ for the linear system, that is

$$\sup_w \left[L\left(\mu_\gamma^*, w\right) / \|w\|^2 \right] \leq \gamma^2,$$

and γ^* is the smallest such γ for any state-feedback controller. The controller

$$\mu_\gamma^*(t, x) = -B(t)^T Z_\gamma(t) x, \quad \text{for } \gamma < \gamma^*,$$

is known as the H^∞ controller.

LQ systems with state-feedback controller: infinite-horizon case

For the infinite-horizon case, as in the case of the LQR problem, we take all matrices to be time invariant, $Q_f = 0$, and let $T \to \infty$. Then, the GRDE-1 is replaced with the generalized algebraic Riccati equation (GARE):

GARE-1 : $\quad A^T Z + ZA - Z(BB^T - \gamma^{-2} DD^T)Z + Q = 0$

for which we seek a non-negative-definite solution. Note again that for $\gamma^{-2} = 0$, this equation becomes identical to CT-ARE introduced for the LQR problem earlier. The counterpart of $\hat{\gamma}$ in this case is

$$\hat{\gamma}_\infty := \inf\{\gamma > 0 : \text{GARE-1 admits a non-negative-definite solution}\}.$$

As in the case of LQR, we require the pair (A, B) to be stabilizable, and the pair (H, A) to be detectable, where H is any matrix with the property $H^T H = Q$. Then, we have the following results:

1. γ_∞^* is finite, and $\gamma_\infty^* = \hat{\gamma}_\infty$.

2. For each $\gamma > \gamma_\infty^*$, GARE-1 admits a unique *minimal* non-negative-definite solution, $\bar{Z}_\gamma^+$, in the sense that if there is some other non-negative-definite solution, $\hat{Z} \neq \bar{Z}^+$, then $\hat{Z} - \bar{Z}^+ \geq 0$.

3. For each $\gamma > \gamma_\infty^*$, there exists a time-invariant state-feedback controller that achieves the performance bound γ, which is given by

$$\mu_\gamma^*(x) = -B^T \bar{Z}_\gamma^+ x.$$

   Equivalently, if $L_2[0, \infty)$ denotes the space of square-integrable functions on $[0, \infty)$ of dimension equal to that of w,

$$L\left(\mu_\gamma^*, w\right) \leq \gamma^2 \|w\|, \quad \forall w \in L_2[0, \infty).$$

4. The state-feedback controller above leads to a bounded input bounded state (BIBS) stable system, that is, the matrix $A - BB^T \bar{Z}_\gamma^+$ has all its eigenvalues in the open left half plane (that is, it is *Hurwitz*).

LQ systems with imperfect state measurements

We now discuss the solution of the H$^\infty$-control problem when the controller has access to only a noise-corrupted version of the state, with the noise being again an unknown disturbance. We cover only the infinite-horizon (time-invariant) case. The finite-horizon case and details of the derivation of the solution below for the infinite-horizon case can be found in reference [29].

The m-dimensional measured output available to the controller is

$$y(t) = Cx(t) + E v(t), \quad EE^T =: N, \quad N > 0,$$

where C is an $m \times n$ matrix, v is an additional disturbance of dimension no lower than m, and the positive definiteness of EE^T implies that all components of y are corrupted by noise. The control has access to the current as well as past values of y; we call such control policies *closed-loop measurement feedback*. The performance index (PI) now also has the norm of v entering alongside w:

$$\left(\|x\|_Q^2 + \|u\|^2\right) / \left(\|w\|^2 + \|v\|^2\right)$$

where

$$\|x\|_Q^2 := \int_{-\infty}^{\infty} x(t)^T Q x(t) \, dt$$

with similar definitions applying to other norms. The H$^\infty$-control problem is one of finding a closed-loop measurement feedback controller which yields the lowest possible value to PI when both w and v are maximizers. Denote this lowest possible value by $\gamma_{f\infty}^*$, which is the optimum disturbance attenuation for the measurement output feedback problem.

In addition to GARE-1 introduced earlier, let us introduce a second GARE:

GARE-2 : $A\Sigma + \Sigma A^T - \Sigma(C^T N^{-1} C - \gamma^{-2} H^T H)\Sigma + DD^T = 0$

Let $\hat{\gamma}_{I\infty}$ be the infimum of the set of all $\gamma > 0$ such that GARE-1 and GARE-2 admit non-negative-definite solutions $\bar{Z}_\gamma$ and $\bar{\Sigma}_\gamma$, with the property that $\rho(\bar{\Sigma}_\gamma \bar{Z}_\gamma) < \gamma^2$, where $\rho(\cdot)$ is the spectral radius.

Assume that (A,B) and (A,D) are stabilizable, and (H,A) and (C,A) are detectable. Then, we have the following:

1. $\gamma^*_{I\infty}$ is finite, and $\gamma^*_{I\infty} = \hat{\gamma}_{I\infty}$.

2. For each $\gamma > \gamma^*_{I\infty}$, GARE-1 admits a unique minimal non-negative-definite solution, $\bar{Z}^+_\gamma$, and GARE-2 admits a unique minimal non-negative-definite solution, $\bar{\Sigma}^+_\gamma$, which further satisfy the spectral radius condition $\rho\left(\bar{\Sigma}^+_\gamma \bar{Z}^+_\gamma\right) < \gamma^2$.

3. For each $\gamma > \gamma^*_{I\infty}$, there exists a controller that achieves the performance bound γ, which is given by

$$\mu^*_\gamma(y) = -B^T \bar{Z}^+_\gamma \hat{x}(t),$$

   where

$$\dot{\hat{x}} = \left[A - (BB^T - \gamma^{-2} DD^T)\bar{Z}^+_\gamma\right]\hat{x} + \left[I - \gamma^{-2}\bar{\Sigma}^+_\gamma \bar{Z}^+_\gamma\right]^{-1} \bar{\Sigma}^+_\gamma C^T N^{-1}(y - C\hat{x}).$$

   Equivalently, with $u(t) = \mu^*_\gamma(y) = -B^T \bar{Z}^+_\gamma \hat{x}(t)$,

$$\|x\|^2_Q + \|u\|^2 \leq \gamma^2 \left(\|w\|^2 + \|v\|^2\right), \quad \forall w \in L_2(-\infty, \infty) \text{ and } \forall v \in L_2(-\infty, \infty).$$

4. The controller above leads to a BIBS stable $2n$-dimensional system, that is, the matrices $A - BB^T \bar{Z}^+_\gamma$ and $A - \bar{\Sigma}^+_\gamma C^T N^{-1} C$ are Hurwitz.

It is instructive to consider the limiting case of this solution (as well as of the problem) when $\gamma^{-2} = 0$. As we have commented earlier, GARE-1 becomes the standard CT-ARE in this case:

$$A^T Z + ZA - ZBB^T Z + Q = 0.$$

GARE-2, on the other hand, also becomes a standard GARE:

$$A\Sigma + \Sigma A^T - \Sigma C^T N^{-1} C\Sigma + DD^T = 0,$$

which is that of *Kalman filtering*, where in the measurement equation, as well as in the system dynamics, v and w are taken as independent white noise processes. Under the given stabilizability and detectability conditions, there exist unique non-negative-definite solutions to these two equations, which we denote by $\bar{Z}$ and $\bar{\Sigma}$, respectively. The differential equation generating $\hat{x}$ now becomes

$$\dot{\hat{x}} = [A - BB^T \bar{Z}]\hat{x} + \bar{\Sigma} C^T N^{-1}(y - C\hat{x}),$$

which is the evolution of the conditional mean in the Kalman filter. The robust controller, μ_γ^*, then becomes

$$u^*(t) = \mu^*(\hat{x}) = -B^T \bar{Z} \hat{x}(t), \; t \geq 0,$$

which is the unique optimal control associated with the linear-quadratic-Gaussian (LQG) problem. Note that since this corresponds to $\gamma = \infty$, the optimum LQG controller does not guarantee any level of disturbance attenuation.

References

[1] A. Agah, S. K. Das, K. Basu, and M. Asadi, "Intrusion detection in sensor networks: A noncooperative game approach," in *Proc. 3rd Intl Symp. on Network Computing and Applications (NCA)*. Washington, DC, USA: IEEE Computer Society, 2004, pp. 343–346.

[2] S. Agarwal, "Ranking on graph data," in *Proc. Intl Conf. on Machine Learning, ICML*, 2006, pp. 25–32.

[3] B. H. Ahn, "Solution of nonsymmetric linear complementarity problems by iterative methods," *J. Optimization Theor. Applic.*, **33**, 185–197, 1981.

[4] T. Alpcan, "Noncooperative games for control of networked systems," PhD dissertation, University of Illinois at Urbana-Champaign, Urbana, IL, May 2006.

[5] T. Alpcan and T. Başar, "A game theoretic approach to decision and analysis in network intrusion detection," in *Proc. 42nd IEEE Conf. on Decision and Control (CDC)*, Maui, HI, Dec. 2003, pp. 2595–2600.

[6] ——, "A game theoretic analysis of intrusion detection in access control systems," in *Proc. 43rd IEEE Conf. on Decision and Control (CDC)*, Paradise Island, Bahamas, Dec. 2004, pp. 1568–1573.

[7] ——, "Distributed algorithms for Nash equilibria of flow control games," in *Advances in Dynamic Games: Applications to Economics, Finance, Optimization, and Stochastic Control*, ser. Annals of Dynamic Games. Boston, MA: Birkhauser, 2005, vol. 7, pp. 473–498.

[8] ——, "An intrusion detection game with limited observations," in *12th Intl Symp. on Dynamic Games and Applications*, Sophia Antipolis, France, July 2006.

[9] T. Alpcan and N. Bambos, "Modeling dependencies in security risk management," in *Proc. 4th Intl Conf. on Risks and Security of Internet and Systems (CRISIS)*, Toulouse, France, Oct. 2009.

[10] T. Alpcan, T. Başar, R. Srikant, and E. Altman, "CDMA uplink power control as a noncooperative game," *Wireless Networks*, **8**(6), 659–670, Nov. 2002.

[11] T. Alpcan and C. Bauckhage, "A discrete-time parallel update algorithm for distributed learning," in *Proc. 19th Intl Conf. on Pattern Recognition (ICPR)*, Tampa, FL, USA, Dec. 2008.

[12] ——, "A distributed machine learning framework," in *Proc. 48th IEEE Conf. on Decision and Control (CDC)*, Shanghai, China, Dec. 2009.

[13] T. Alpcan, C. Bauckhage, and A.-D. Schmidt, "A probabilistic diffusion scheme for anomaly detection on smartphones," in *Proc. Workshop on Information Security Theory and Practices (WISTP)*, Passau, Germany, April 2010.

[14] T. Alpcan and S. Buchegger, "Security games for vehicular networks," *IEEE Trans. Mob. Comput.* (to appear).

[15] T. Alpcan, X. Fan, T. Başar, M. Arcak, and J. T. Wen, "Power control for multicell CDMA wireless networks: A team optimization approach," *Wireless Networks*, **14**(5), 647–657, Oct. 2008.

[16] T. Alpcan and X. Liu, "A game theoretic recommendation system for security alert dissemination," in *Proc. IEEE/IFIP Intl Conf. on Network and Service Security (N2S 2009)*, Paris, France, June 2009.

[17] T. Alpcan, C. Orencik, A. Levi, and E. Savas, "A game theoretic model for digital identity and trust in online communities," in *Proc. 5th ACM Symp. on Information, Computer and Communications Security (ASIACCS)*, April 2010.

[18] T. Alpcan and L. Pavel, "Nash equilibrium design and optimization," in *Proc. Intl Conf. on Game Theory for Networks (GameNets)*, Istanbul, Turkey, May 2009.

[19] T. Alpcan, L. Pavel, and N. Stefanovic, "A control theoretic approach to noncooperative game design," in *Proc. 48th IEEE Conf. on Decision and Control (CDC)*, Shanghai, China, December 2009.

[20] T. Alpcan, N. Shimkin, and L. Wynter, "International Workshop on Game Theory in Communication Networks (GameComm)," October 2008. [Online]. Available: http://www.game-comm.org

[21] E. Altman and T. Başar, "Multiuser rate-based flow control," *IEEE Trans. Commun.*, **46**(7), 940–949, July 1998.

[22] B. D. O. Anderson and J. B. Moore, *Optimal Control: Linear Quadratic Methods*. Upper Saddle River, NJ, USA: Prentice-Hall, Inc., 1990.

[23] R. J. Anderson, *Security Engineering: A Guide to Building Dependable Distributed Systems*, 2nd edn. Indianapolis, IN, USA: Wiley Publishing, 2008.

[24] S. Axelsson, "The base-rate fallacy and its implications for the difficulty of intrusion detection," in *Proc. 6th ACM Conf. on Computer and Communications Security (CCS)*, Kent Ridge Digital Labs, Singapore, 1999, pp. 1–7.

[25] R. G. Bace, *Intrusion Detection*. Indianapolis, IN, USA: Macmillan Technical Publishing, 2000.

[26] R. G. Bace and P. Mell, "Intrusion detection systems," NIST Special Publication on Intrusion Detection Systems, Nov. 2001. [Online]. Available: http://www.snort.org/docs/nist-ids.pdf

[27] T. Başar, "Relaxation techniques and asynchronous algorithms for on-line computation of noncooperative equilibria," *J. Econ. Dyn. Control*, **11**(4), 531–549, Dec. 1987.

[28] ——, "International Conference on Game Theory for Networks (GameNets)," May 2009. [Online]. Available: http://www.gamenets.org

[29] T. Başar and P. Bernhard, *H^∞-Optimal Control and Related Minimax Design Problems: A Dynamic Game Approach*, 2nd edn. Boston, MA: Birkhäuser, 1995.

[30] T. Başar and S. Li, "Distributed computation of nash equilibria in linear-quadratic stochastic differential games," *SIAM J. Control Optim.*, **27**(3), 563–578, 1989.

[31] T. Başar and G. J. Olsder, *Dynamic Noncooperative Game Theory*, 2nd edn. Philadelphia, PA: SIAM, 1999.

[32] C. Bauckhage, "Image tagging using pagerank over bipartite graphs," in *LNCS: Pattern Recognition*, G. Rigoll, ed. Berlin / Heidelberg: Springer, 2008, vol. 5096/2008.

[33] A. Beresford and F. Stajano, "Mix zones: User privacy in location-aware services," in *Proc. 2nd IEEE Conference on Pervasive Computing and Communications Workshops, 2004*, March 2004, pp. 127–131.

[34] U. Berger, "Fictitious play in 2xn games," *J. Econ. Theor.*, **120**(2), 139–154, 2005.

[35] D. Bertsekas, *Nonlinear Programming*, 2nd edn. Belmont, MA: Athena Scientific, 1999.

[36] ——, *Dynamic Programming and Optimal Control*, 3rd edn. Belmont, MA: Athena Scientific, 2007, vol. 2.

[37] D. Bertsekas and R. Gallager, *Data Networks*. 2nd edn. Upper Saddle River, NJ: Prentice Hall, 1992.

[38] D. Bertsekas and J. N. Tsitsiklis, *Parallel and Distributed Computation: Numerical Methods*. Upper Saddle River, NJ: Prentice Hall, 1989.

[39] I. Bilogrevic, M. H. Manshaei, M. Raya, and J.-P. Hubaux, "Optimal revocations in ephemeral networks: A game-theoretic framework," in *Proc. of 8th Intl. Symp. on Modeling and Optimization in Mobile, Ad Hoc, and Wireless Networks (WiOpt)*, Avignon, France, May 2010.

[40] C. M. Bishop, *Pattern Recognition and Machine Learning (Information Science and Statistics)*. Secaucus, NJ, USA: Springer-Verlag New York, Inc., 2006.

[41] M. Bloem, T. Alpcan, and T. Başar, "Intrusion response as a resource allocation problem," in *Proc. 45th IEEE Conf. on Decision and Control (CDC)*, San Diego, CA, Dec. 2006, pp. 6283–6288.

[42] ——, "Optimal and robust epidemic response for multiple networks," in *Proc. 46th IEEE Conf. on Decision and Control (CDC)*, New Orleans, LA, Dec. 2007.

[43] ——, "An optimal control approach to malware filtering," in *Proc. 46th IEEE Conf. on Decision and Control (CDC)*, New Orleans, LA, Dec. 2007.

[44] ——, "Optimal and robust epidemic response for multiple networks," *IFAC Control Eng Practice* (to appear).

[45] M. Bloem, T. Alpcan, S. Schmidt, and T. Başar, "Malware filtering for network security using weighted optimality measures," in *Proc. IEEE Conf. on Control Applications (CCA)*, Singapore, Oct. 2007.

[46] S. Buchegger and T. Alpcan, "Security games for vehicular networks," in *Proc. Allerton Conf. on Communication, Control, and Computing*, Urbana-Champaign, IL, USA, Sept. 2008.

[47] S. Buchegger and J.-Y. Le Boudee, "Self-policing mobile ad hoc networks by reputation systems," *IEEE Commun. Mag.*, **43**(7), 101–107, July 2005.

[48] S. Buchegger, C. Tissieres, and J. Le Boudec, "A test-bed for misbehavior detection in mobile ad-hoc networks – how much can watchdogs really do?" in *Proc. 6th IEEE Workshop on Mobile Computing Systems and Applications (WMCSA)*, Dec. 2004, pp. 102–111.

[49] L. Buttyan, T. Holczer, and I. Vajda, "On the effectiveness of changing pseudonyms to provide location privacy in VANETs," in *Proc. 4th European Workshop on Security and Privacy in Ad Hoc and Sensor Networks (ESAS)*, 2007.

[50] L. Buttyan and J.-P. Hubaux, *Security and Cooperation in Wireless Networks*. Cambridge, UK: Cambridge University Press, 2008. [Online]. Available: http://secowinet.epfl.ch

[51] R. Bye, S. Schmidt, K. Luther, and S. Albayrak, "Application-level simulation for network security," in *Proc. 1st Intl Conf. on Simulation Tools and Techniques (Simutools)*. Marseille, France: ICST, March 2008, pp. 1–10.

[52] G. R. Cantieni, G. Iannaccone, C. Barakat, C. Diot, and P. Thiran, "Reformulating the monitor placement problem: Optimal network-wide sampling," in *Proc. 2nd Conf. on Future Networking Technologies (CoNeXT)*, Lisbon, Portugal, Dec. 2006.

[53] J.-F. Chamberland and V. V. Veeravalli, "Asymptotic results for decentralized detection in power constrained wireless sensor networks," *IEEE J. Sel. Areas in Commun.*, **22**(6), 1007–1015, Aug. 2004.

[54] ——, "How dense should a sensor network be for detection with correlated observations?" *IEEE Trans. Inf. Theor.*, **52**(11), 5099–5106, 2006.

[55] H. Chan, V. D. Gligor, A. Perrig, and G. Muralidharan, "On the distribution and revocation of cryptographic keys in sensor networks," *IEEE Trans. Dependable Secure Comput.*, **2**(3), 233–247, 2005.

[56] L. Chen, "On selfish and malicious behaviors in wireless networks – a noncooperative game theoretic approach," PhD dissertation, TELECOM ParisTech, 2008.

[57]   T. M. Chen and N. Jamil, "Effectiveness of quarantine in worm epidemics," in *Proc. IEEE Conf. on Communication (ICC)*, Istanbul, Turkey, June 2006, pp. 2142–2147.

[58]   L. Cranor and S. Garfinkel, *Security and Usability: Designing Secure Systems that People Can Use*. Sebastopol, CA: O'Reilly Media Inc., 2005.

[59]   A. A. Daoud, T. Alpcan, S. Agarwal, and M. Alanyali, "A stackelberg game for pricing uplink power in wide-band cognitive radio networks," in *Proc. 47th IEEE Conf. on Decision and Control (CDC)*, Cancun, Mexico, Dec. 2008.

[60]   N. Eagle and A. S. Pentland, "Reality mining: Sensing complex social systems," *Personal Ubiquitous Comput.*, **10**(4), 255–268, 2006.

[61]   R. Ellison, D. Fisher, R. Linger, H. Lipson, T. Longstaff, and N. Mead, "Survivable Network Systems: An Emerging Discipline," Software Engineering Institute, Carnegie Mellon University, Pittsburgh, PA, USA, Tech. Rep., Nov. 1997.

[62]   R. Ellison, R. Linger, H. Lipson, N. Mead, and A. Moore, "Foundations for survivable systems engineering," *J. Defense Software Eng.* pp. 10–15, July 2002.

[63]   J. C. Ely, D. Fudenberg, and D. K. Levine, "When is Reputation Bad?" *SSRN eLibrary*, 2004.

[64]   M. Felegyhazi, J.-P. Hubaux, and L. Buttyan, "Nash equilibria of packet forwarding strategies in wireless ad hoc networks," *IEEE Trans. Mob. Comput*, **5**(5), 463–476, 2006.

[65]   J. Freudiger, M. H. Manshaei, J. L. Boudec, and J. P. Hubaux, "On the age of pseudonyms in mobile ad hoc networks," in *Proc. 29th IEEE Conf. on Computer Communications (Infocom)*, San Diego, CA, USA, March 2010.

[66]   J. Freudiger, M. H. Manshaei, J.-P. Hubaux, and D. C. Parkes, "On noncooperative location privacy: A game-theoretic analysis," in *Proc. 16th ACM Conf. on Computer and Communications Security (CCS)*. New York, NY, USA: ACM, 2009, pp. 324–337.

[67]   D. Fudenberg and J. Tirole, *Game Theory*. Cambridge, MA, USA: MIT Press, 1991.

[68]   P. R. Garvey, *Analytical Methods for Risk Management: A Systems Engineering Perspective*, ser. Statistics: a Series of Textbooks and Monographs. Boca Raton, FL, USA: Chapman and Hall/CRC, 2009.

[69]   P. D. Groom, "The IT security model," *IEEE Potentials*, **22**(4), 6–8, Oct./Nov. 2003.

[70]   J. Grossklags and B. Johnson, "Uncertainty in the weakest-link security game," in *Proc. Intl Conf. on Game Theory for Networks (GameNets)*, May 2009, pp. 673–682.

[71]   J. Grossklags, N. Christin, and J. Chuang, "Secure or insure?: A game-theoretic analysis of information security games," in *Proc. 17th Intl Conf. on World Wide Web (WWW)*. New York, NY, USA: ACM, 2008, pp. 209–218.

[72]   J. Y. Halpern and R. Pass, "Iterated regret minimization: A more realistic solution concept," in *Proc. 21st Intl Joint Conf. on Artificial Intelligence (IJCAI)*, Pasadena, CA, USA, July 2009.

[73]   Z. Han, N. Marina, M. Debbah, and A. Hjørungnes, "Physical layer security game: How to date a girl with her boyfriend on the same table," in *Proc. Intl Conf. on Game Theory for Networks (GameNets)*, May 2009, pp. 287–294.

[74]   C. Herley, "So long, and no thanks for the externalities: The rational rejection of security advice by users," in *Proc. 2009 Workshop on New Security Paradigms (NSPW)*. New York, NY, USA: ACM, 2009, pp. 133–144.

[75]   H. W. Hethcote, "The mathematics of infectious diseases," *SIAM Rev.*, **42**(4), 599–653, 2000.

[76]   T. K. Ho and E. M. Kleinberg, "Building projectable classifiers of arbitrary complexity," in *Proc. 13th Intl Conf. on Pattern Recognition (ICPR)*. Washington, DC, USA: IEEE Computer Society, 1996, p. 880.

[77]   I. Y. Hoballah and P. K. Varshney, "Distributed Bayesian signal detection," *IEEE Trans. Inf. Theor.*, **35**(5), 995–1000, Sept. 1989.

[78]   R. Horn and C. Johnson, *Matrix Analysis*. New York, NY: Cambridge University Press, 1985.

[79]   L. Huang, K. Matsuura, H. Yamane, and K. Sezaki, "Enhancing wireless location privacy using silent period," in *Proc. IEEE Wireless Communications and Networking Conf. (WCNC)*, New Orleans, LA, USA, March 2005, vol. 2, pp. 1187–1192.

[80]   P. J. Huber, "A robust version of the probability ratio test," *Ann. Math. Statist.*, **36**, 1753–1758, 1965.

[81]   R. Isaacs, *Differential Games: A Mathematical Theory with Applications to Warfare and Pursuit, Control and Optimization*, ser. Dover Books on Mathematics. New York: Courier Dover Publications, 1999 (1965).

[82]   A. Josang, R. Ismail, and C. Boyd, "A survey of trust and reputation systems for online service provision," *Decision Support Systs*, **43**(2), 618–644, 2007.

[83]   T. Kailath, *Linear Systems*. Upper Saddle River, NJ, USA: Prentice Hall, 1980.

[84]   F. Kargl, P. Papadimitratos, L. Buttyan, M. Muter, E. Schoch, B. Wiedersheim, T. Ta-Vinh, G. Calandriello, A. Held, A. Kung, and J.-P. Hubaux, "Secure vehicular communication systems: implementation, performance, and research challenges," *IEEE Commun. Mag.*, **46**(11), 110–118, November 2008.

[85]   H. Kashima, K. Tsuda, and A. Inokuchi, "Kernels for graphs," in *Kernel Methods in Computational Biology*, B. Schölkopf, K. Tsuda, and J.-P. Vert, eds. Cambridge, MA, USA: MIT Press, 2004, pp. 155–170.

[86]   A. Kashyap, T. Başar, and R. Srikant, "Correlated jamming on MIMO Gaussian fading channels," *IEEE Trans. Inf. Theory*, **50**(9), 2119–2123, Sept. 2004.

[87]   ——, "Asymptotically optimal quantization for detection in power constrained decentralized networks," in *Proc. 2006 American Control Conf. (ACC)*, Minneapolis, MN, USA, June 2006, pp. 14–16.

[88]   M. Kodialam and T. V. Lakshman, "Detecting network intrusions via sampling: A game theoretic approach," in *Proc. 22nd IEEE Conf. on Computer Communications (Infocom)*, San Fransisco, CA, April 2003, vol. 3, pp. 1880–1889.

[89]   O. Kreidl and T. Frazier, "Feedback control applied to survivability: A host-based autonomic defense system," *IEEE Trans. Reliability*, **53**(1), 148–166, March 2004.

[90]   J. F. Kurose and K. W. Ross, *Computer Networking: A Top-Down Approach*, 5th ed. USA: Addison-Wesley, 2009.

[91]   M. G. Lagoudakis and R. Parr, "Value function approximation in zero sum Markov games," in *Proc. 18th Conf. on Uncertainty in Artificial Intelligence (UAI)*. Morgan Kaufmann, 2002, pp. 283–292.

[92]   T. L. Lai, "Sequential analysis: some classical problems and new challenges," *Statistica Sinica*, **11**, 303–408, 2001.

[93]   A. Langville and C. Meyer, "A survey of eigenvector methods for Web information retrieval," *SIAM Rev.*, **47**(1), 135–161, 2005.

[94]   S. Leyffer, "The return of the active set method," *Oberwolfach Reports*, **2**(1), 2005.

[95]   K. Leyton-Brown and Y. Shoham, *Essentials of Game Theory: A Concise, Multidisciplinary Introduction*, ser. Synthesis Lectures on Artificial Intelligence and Machine Learning. Morgan Claypool Publishers, 2008.

[96]   M. Li, K. Sampigethaya, L. Huang, and R. Poovendran, "Swing & swap: User centric approaches towards maximizing location privacy," in Workshop on Privacy in the Electronic Society *(WPES)*, 2006.

[97]   S. Li and T. Başar, "Distributed algorithms for the computation of noncooperative equilibria," *Automatica*, **23**(4), 523–533, 1987.

[98] M. L. Littman, "Markov games as a framework for multi-agent reinforcement learning," in *Proc. 11th Intl Conf. on Machine Learning (ICML)*, San Francisco, CA, 1994, pp. 157–163.

[99] Y. Liu, C. Comaniciu, and H. Man, "A Bayesian game approach for intrusion detection in wireless ad hoc networks," in *Proc. 2006 Workshop on Game Theory for Communications and Networks (GameNets'06)*. New York, NY, USA: ACM, Oct. 2006, p. 4.

[100] D. G. Luenberger, *Linear and Nonlinear Programming*, 2nd edn. Reading, MA: Addison-Wesley, 1984.

[101] ——, *Optimization by Vector Space Methods*, ser. Series in Decision and Control. New York, NY, USA: John Wiley & Sons, Inc., 1997.

[102] R. Mallik, R. Scholtz, and G. Papavassilopoulos, "Analysis of an on–off jamming situation as a dynamic game," *IEEE Trans. Commun.*, **48**(8), 1360–1373, Aug. 2000.

[103] M. A. Maloof, ed., *Machine Learning and Data Mining for Computer Security*, ser. Advanced Information and Knowledge Processing. London, UK: Springer London.

[104] N. Mandayam, G. Editor, S. Wicker, J. Walrand, T. Başar, J. Huang, and D. Palomar, "Game theory in communication systems," *IEEE J. Selected Areas in Commun. (JSAC)*, **26**(7), 1042–1046, Sept. 2008.

[105] W. Mao, X. Su, and X. Xu, "Comments on "correlated jamming on MIMO Gaussian fading channels"," *IEEE Trans. Inf. Theor.*, **52**(11), 5163–5165, Nov. 2006.

[106] I. Maros, "A general pricing scheme for the simplex method," *Ann. Oper. Res.*, **124**(1–4), 193–203, 2004.

[107] L. Mastroleon, "Scalable resource control in large-scale computing/networking infrastructures," PhD dissertation, Stanford University, Palo Alto, CA, USA, July 2009.

[108] M. Mavronicolas, V. G. Papadopoulou, A. Philippou, and P. G. Spirakis, "A graph-theoretic network security game," *Intl J. Auton. Adapt. Commun. Syst.*, **1**(4), 390–410, 2008.

[109] D. Meier, Y. A. Oswald, S. Schmid, and R. Wattenhofer, "On the windfall of friendship: Inoculation strategies on social networks," in *Proc. 9th ACM Conf. on Electronic Commerce (EC)*. New York, NY, USA: ACM, 2008, pp. 294–301.

[110] A. J. Menezes, P. C. van Oorschot, and S. A. Vanstone, *Handbook of Applied Cryptography*, ser. Discrete Mathematics and Its Applications. Boca Raton, FL: CRC Press, 1996. [Online]. Available: http://www.cacr.math.uwaterloo.ca/hac

[111] R. A. Miura-Ko and N. Bambos, "Dynamic risk mitigation in computing infrastructures," in *Proc. 3rd Intl Symp. on Information Assurance and Security*. IEEE, 2007, pp. 325–328.

[112] ——, "Securerank: A risk-based vulnerability management scheme for computing infrastructures." in *Proc. IEEE Conf. on Communication (ICC)*. IEEE, 2007, pp. 1455–1460.

[113] R. A. Miura-Ko, B. Yolken, N. Bambos, and J. Mitchell, "Security investment games of interdependent organizations," in *46th Annual Allerton Conf.*, Sept. 2008.

[114] R. A. Miura-Ko, B. Yolken, J. Mitchell, and N. Bambos, "Security decision-making among interdependent organizations," in *Proc. 21st IEEE Computer Security Foundations Symp. (CSF)*, June 2008, pp. 66–80.

[115] D. Moore, C. Shannon, and K. Claffy, "Code-Red: A case study on the spread and victims of an Internet worm," in *Proc. ACM SIGCOMM Workshop on Internet Measurement*, Marseille, France, 2002, pp. 273–284.

[116] D. Moore, C. Shannon, G. Voelker, and S. Savage, "Internet quarantine: Requirements for containing self-propagating code," in *Proc. 22nd IEEE Conf. on Computer Communications (Infocom)*, April 2003, vol. 3, pp. 1901–1910.

[117] T. Moore, J. Clulow, S. Nagaraja, and R. Anderson, "New strategies for revocation in ad-hoc networks," in *Proc. 4th Euro. Workshop on Security and Privacy in Ad Hoc and Sensor Networks (ESAS)*, Cambridge, England, July 2007, pp. 232–246.

[118] R. Morselli, J. Katz, and B. Bhattacharjee, "A game-theoretic framework for analyzing trust-inference protocols," in *Proc. 2nd Workshop on the Economics of Peer-to-Peer Systems*, Cambridge, MA, USA, June 2004.

[119] T. Moscibroda, S. Schmid, and R. Wattenhofer, "When selfish meets evil: Byzantine players in a virus inoculation game," in *Proc. 25th ACM Symp. on Principles of Distributed Computing (PODC)*. New York, NY, USA: ACM, Aug. 2006, pp. 35–44.

[120] P. Moulin, *ECE561 – Signal Detection and Estimation Lecture Notes*. Urbana-Champaign, IL, USA: University of Illinois at Urbana-Champaign, 2008.

[121] J. Mounzer, T. Alpcan, and N. Bambos, "Dynamic control and mitigation of interdependent IT security risks," in *Proc. IEEE Conf. on Communication (ICC)*. IEEE Communications Society, May 2010.

[122] D. R. Musicant and A. Feinberg, "Active set support vector regression," *IEEE Trans. Neural Networks*, **15**(2), 268–275, March 2004.

[123] R. B. Myerson, *Game Theory, Analysis of Conflict*. Cambridge, MA, USA: Harvard University Press, Sept. 1991.

[124] J. F. Nash, "Equilibrium points in n-person games," *Proc. Nat. Acad. Sci. USA*, **36**(1), 48–49, Jan. 1950. [Online]. Available: http://www.jstor.org/stable/88031

[125] ——, "Noncooperative games," *Ann. Math.*, **54**(2), 286–295, Sept. 1951. [Online]. Available: http://www.jstor.org/stable/1969529

[126] K. C. Nguyen, T. Alpcan, and T. Başar, "A decentralized Bayesian attack detection algorithm for network security," in *Proc. 23rd Intl Information Security Conf. (SEC 2008)*, Milan, Italy, Sept. 2008.

[127] ——, "Distributed hypothesis testing with a fusion center: The conditionally dependent case," in *Proc. 47th IEEE Conf. on Decision and Control (CDC)*, Cancun, Mexico, Dec. 2008.

[128] ——, "Security games with incomplete information," in *Proc. IEEE Conf. Commun. (ICC)*, Dresden, Germany, June 2009.

[129] ——, "Stochastic games for security in networks with interdependent nodes," in *Proc. Intl Conf. on Game Theory for Networks (GameNets)*, Istanbul, Turkey, May 2009.

[130] ——, "Fictitious play with time-invariant frequency update for network security," in *Proc. IEEE Multi-Conference on Systems and Control (MSC) and IEEE CCA*, Yokohama, Japan, September 2010.

[131] J. Nielsen, *Usability Engineering*. San Francisco, CA, USA: Morgan Kaufmann, 1994.

[132] N. Nisan, T. Roughgarden, E. Tardos, and V. V. Vazirani, *Algorithmic Game Theory*. New York, NY, USA: Cambridge University Press, 2007.

[133] P. Nurmi, "A Bayesian framework for online reputation systems," in *Proc. Advd Intl Conf. on Telecommunications and Intl Conf. on Internet and Web Applications and Services (AICT-ICIW)*. Washington, DC, USA: IEEE Computer Society, 2006, p. 121.

[134] K. Ogata, *Modern Control Engineering*, 4th edn. Upper Saddle River, NJ: Prentice Hall, 2002.

[135] M. J. Osborne and A. Rubinstein, *A Course in Game Theory*, ser. MIT Press Books. Cambridge, MA, USA: MIT Press, Dec. 1994, vol. 1, no. 0262650401.

[136] G. Owen, *Game Theory*, 3rd edn. New York, NY: Academic Press, 2001.

[137] L. Pontryagin, V. Boltyanskii, R. Gamkrelidze, and E. Mishchenko, *The Mathematical Theory of Optimal Processes*. New York, NY, USA: Interscience Publishers, 1962.

[138] V. H. Poor, *An Introduction to Signal Detection and Estimation*, 2nd edn. New York, NY, USA: Springer-Verlag, 1994.

[139] S. Radosavac, J. S. Baras, and I. Koutsopoulos, "A framework for MAC protocol misbehavior detection in wireless networks," in *Proc. 4th ACM Workshop on Wireless Security (WiSe '05)*. New York, NY, USA: ACM, Nov. 2005, pp. 33–42.

[140] T. Rappaport, *Wireless Communications: Principles and Practice*, 2nd edn. Upper Saddle River, NJ: Prentice Hall, 2002.

[141] M. Raya, M. H. Manshaei, M. Felegyhazi, and J.-P. Hubaux, "Revocation games in ephemeral networks," in *Proc. 15th ACM Conf. on Computer Security (CCS)*, Alexandria, Virginia, USA, Oct. 2008, pp. 199–210.

[142] M. Raya, R. Shokri, and J.-P. Hubaux, "On the tradeoff between trust and privacy in wireless ad hoc networks," in *Proc. 3rd ACM Conf. on Wireless Network Security (WiSec)*, March 2010.

[143] P. Resnick and R. Zeckhauser, "Trust among strangers in Internet transactions: Empirical analysis of eBay's reputation system," in *The Economics of the Internet and E-Commerce*, ser. Advances in Applied Microeconomics, M. R. Baye, ed. Amsterdam: Elsevier Science, 2002, vol. 11, pp. 127–157.

[144] P. Resnick, R. J. Zeckhauser, J. Swanson, and K. Lockwood, "The value of reputation on eBay: A controlled experiment," *SSRN eLibrary*, 2002.

[145] J. Robinson, "The mathematical theory of optimal processes," *Ann Math.*, **54**(2), 296–301, September 1951. [Online]. Available: http://www.jstor.org/stable/1969530

[146] K. Rohloff and T. Başar, "The detection of RCS worm epidemics," in *Proc. ACM Workshop on Rapid Malcode*, Fairfax, VA, 2005, pp. 81–86.

[147] J. B. Rosen, "Existence and uniqueness of equilibrium points for concave *n*-person games," *Econometrica*, **33**(3), 520–534, July 1965.

[148] S. M. Ross, *Applied Probability with Optimization Applications*. San Francisco: Holden-Day, Inc., 1970.

[149] K. Rozinov, "Are usability and security two opposite directions in computer systems?" Nov. 2004. [Online]. Available: http://rozinov.sfs.poly.edu/papers/security_vs_usability.pdf

[150] W. Saad, T. Alpcan, T. Başar, and A. Hjørungnes, "Coalitional game theory for security risk management," in *Proc. 5th Intl Conf. on Internet Monitoring and Protection (ICIMP)*, Barcelona, Spain, May 2010.

[151] W. Saad, Z. Han, M. Debbah, A. Hjørungnes, and T. Başar, "Coalitional game theory for communication networks: A tutorial," *IEEE Signal Process. Mag., Special issue on Game Theory in Signal Processing and Communications*, **26**(5), 77–97, Sept. 2009.

[152] F. A. Sadjadi, "Hypothesis testing in a distributed environment," *IEEE Trans. Aerosp. Electron. Syst.*, **AES-22**(2), 134–137, March 1986.

[153] Y. E. Sagduyu, R. Berry, and A. Ephremides, "MAC games for distributed wireless network security with incomplete information of selfish and malicious user types," in *Proc. IEEE Intl Conf. on Game Theory for Networks (GameNets)*, 2009.

[154] Y. E. Sagduyu and A. Ephremides, "A game-theoretic analysis of denial of service attacks in wireless random access," *J. Wireless Networks*, **15**, 651–666, July 2009.

[155] K. Sallhammar, "Stochastic models for combined security and dependability evaluation," PhD dissertation, Norwegian University of Science and Technology, Faculty of Information Technology, Mathematics and Electrical Engineering, Department of Telematics, 2007.

[156] K. Sampigethaya, L. Huang, M. Li, R. Poovendran, K. Matsuura, and K. Sezaki, "CARAVAN: Providing location privacy for VANET," in *Proc. Embedded Security in Cars Conf. (ESCAR)*, 2005.

[157] C. U. Saraydar, N. Mandayam, and D. Goodman, "Pricing and power control in a multicell wireless data network," *IEEE J. Sel. Areas Commun.*, pp. 1883–1892, Oct. 2001.

[158] K. Scheinberg, "An efficient implementation of an active set method for SVMs," *J. Mach. Learning Res.*, **7**, 2237–2257, Dec. 2006.

[159]   S. Schmidt, T. Alpcan, S. Albayrak, T. Başar, and A. Muller, "A malware detector placement game for intrusion detection," in *Proc. 2nd Intl Workshop on Critical Information Infrastructures Security (CRITIS)*, Malaga, Spain, Oct. 2007.

[160]   B. Schneier, "Schneier on security: A blog covering security and security technology," 1999–2009. [Online]. Available: http://www.schneier.com/blog/

[161]   ——, *Schneier on Security*. Indianapolis, IN, USA: Wiley Publishing, 2008.

[162]   B. Scholkopf and A. J. Smola, *Learning with Kernels: Support Vector Machines, Regularization, Optimization, and Beyond*. Cambridge, MA, USA: MIT Press, 2002.

[163]   J. S. Shamma and G. Arslan, "Unified convergence proofs of continuous-time fictitious play," *IEEE Trans. Autom. Control*, **49**(7), 1137–1141, July 2004.

[164]   ——, "Dynamic fictitious play, dynamic gradient play, and distributed convergence to nash equilibria," *IEEE Trans. Autom. Control*, **50**(3), 312–327, March 2005.

[165]   A. Simmonds, P. Sandilands, and L. van Ekert, "An ontology for network security attacks," in *Proc. 2nd Asian Applied Computing Conf. (AACC)*, ser. Lecture Notes in Computer Science. Kathmandu, Nepal: Springer, Oct. 2004, vol. 3285/2004, pp. 317–323.

[166]   S. Singh, *The Code Book: The Science of Secrecy from Ancient Egypt to Quantum Cryptography*. New York: Anchor Books, 1999.

[167]   P. Sommer, "Design and analysis of realistic mobility model for wireless mesh networks," Master's thesis, ETH Zürich, 2007.

[168]   M. Strasser, S. Capkun, C. Popper, and M. Cagalj, "Jamming-resistant key establishment using uncoordinated frequency hopping," in *Proc. IEEE Symp. on Security and Privacy (SP 2008)*, May 2008, pp. 64–78.

[169]   K. Suh, Y. Guo, J. Kurose, and D. Towsley, "Locating network monitors: Complexity, heuristics, and coverage," in *Proc. 24th IEEE Conf. on Computer Communications (Infocom)*, Miami, FL, March 2005, vol. 1, pp. 351–361.

[170]   N. N. Taleb, *The Black Swan: The Impact of the Highly Improbable*. New York, NY, USA: Random House Publishing, 2007.

[171]   A. S. Tanenbaum, *Computer Networks*, 2nd edn. Upper Saddle River, NJ, USA: Prentice-Hall, Inc., 1988.

[172]   R. Tenney and J. N. R. Sandell, "Detection with distributed sensors," *IEEE Trans. Aerosp. Electron. Syst.*, **AES-17**, 501–510, July 1981.

[173]   The-Gambit-Project, "Gambit game theory analysis software and tools," http://gambit.sourceforge.net/, 2002. [Online]. Available: http://gambit.sourceforge.net/

[174]   G. Theodorakopoulos and J. S. Baras, "Malicious users in unstructured networks," in *Proc. 22nd IEEE Conf. on Computer Communications (Infocom)*, May 2007, pp. 884–891.

[175]   J. N. Tsitsiklis, "Decentralized detection," in *Advances in Signal Processing*, H. V. Poor and J. B. Thomas, eds. Grenuich CT: JAI Press, 1993, vol. 2, pp. 297–344.

[176]   ——, "Extremal properties of likelihood-ratio quantizers," *IEEE Trans. Commun.*, **41**(4), 550–558, 1993.

[177]   J. N. Tsitsiklis and M. Athans, "On the complexity of decentralized decision making and detection problems," *IEEE Trans. Autom. Control*, **30**(5), 440–446, 1985. [Online]. Available: http://ieeexplore.ieee.org/xpls/abs_all.jsp?arnumber=1103988

[178]   J. Unnikrishnan and V. V. Veeravalli, "Decentralized detection with correlated observations," in *Proc. Asilomar Conf. on Signals, Systems, and Computers*, Pacific Grove, CA, USA, Nov. 2007.

[179]   J. von Neumann, "Zur Theorie der Gesellschaftsspiele," *Mathematische Annalen*, **100**(1), 295–320, Dec. 1928.

[180]   G. van Rossum, *An Introduction to Python*. Bristol: Network Theory Ltd, 2003.

[181] H. L. van Trees, *Detection, Estimation and Modulation Theory*. New York, NY, USA: John Wiley & Sons, 1968, vol. I.

[182] S. Vasudevan, J. Kurose, and D. Towsley, "On neighbor discovery in wireless networks with directional antennas," in *Proc. 24th IEEE Conf. on Computer Communications (Infocom)*, 2005.

[183] D. Vaughan, *The Challenger Launch Decision: Risky Technology, Culture, and Deviance at NASA*. Chicago, IL, USA: University of Chicago Press.

[184] V. V. Veeravalli, T. Başar, and H. V. Poor, "Decentralized sequential detection with a fusion center performing the sequential test," *IEEE Trans. on Inf. Theor.*, **39**(2), 433–442, 1993.

[185] ——, "Decentralized sequential detection with sensors performing sequential tests," *Math. Signals, Systs, and Control*, **7**(4), 292–306, 1994.

[186] ——, "Minimax robust decentralized detection," *IEEE Trans. Inf. Theor.*, **40**(1), 35–40, 1994.

[187] M. Vogt and V. Kecman, "Active-set methods for support vector machines," in *Support Vector Machines: Theory and Applications*, L. Wang, ed. Berlin, Heidelberg: Springer-Verlag, Aug. 2005, pp. 133–158.

[188] N. N. Vorobev, *Foundations of Game Theory: Noncooperative Games*. Boston: Birkhäuser, 1994.

[189] A. Wald, *Sequential Analysis*. New York, NY, USA: Dover Phoenix Editions, 2004.

[190] P. Willett, P. F. Swaszek, and R. S. Blum, "The good, bad, and ugly: Distributed detection of a known signal in dependent gaussian noise," *IEEE Trans. Signal Process.*, **48**, 3266–3279, Dec. 2000.

[191] D. Zamboni, "Using internal sensors for computer intrusion detection," PhD dissertation, Purdue University, August 2001.

[192] J. Zander, "Jamming games in slotted ALOHA packet radio networks," in *IEEE Military Communications Conference (MILCOM)*, Morleley, CA, 1990, vol. 2, Sept. 30–Oct. 3, pp. 830–834.

[193] D. Zhou, J. Weston, A. Gretton, O. Bousquet, and B. Schölkopf, "Ranking on data manifolds," in *Proc. NIPS*, 2004, pp. 169–176.

[194] F. Zhu and W. Zhu, "Rational exposure: A game theoretic approach to optimize identity exposure in pervasive computing environments," in *Proc. IEEE Intl Conf. on Pervasive Computing and Communications (PERCOM)*. Washington, DC, USA: IEEE Computer Society, 2009, pp. 1–8.

[195] M. Zinkevich, A. Greenwald, and M. L. Littman, "Cyclic equilibria in markov games," in *Proc. Neural Information Processing Systems (NIPS)*, Vancouver, BC, Canada, December 2005.

[196] C. Zou, W. Gong, and D. Towsley, "Worm propogation modeling and analysis under dynamic quarantine defense," in *Proc. ACM Workshop on Rapid Malcode*, Washington, DC, 2003, pp. 51–60.

Index